INTRODUCTION TO ORDINARY DIFFERENTIAL EQUATIONS

INTRODUCTION TO ORDINARY DIFFERENTIAL EQUATIONS

Rodney D. Driver

University of Rhode Island

HARPER & ROW, PUBLISHERS

New York   Hagerstown   San Francisco   London

To my mother, Marjorie E. Driver
and to the memory of my father, William T. Driver

Sponsoring Editor: Charlie Dresser
Project Editor: Pamela Landau
Designer: T. R. Funderburk
Production Supervisor: Marion Palen
Compositor: Santype Limited
Printer and Binder: The Maple Press Company
Art Studio: Danmark & Michaels Inc.

INTRODUCTION TO ORDINARY DIFFERENTIAL EQUATIONS
Copyright © 1978 by Rodney D. Driver

Library of Congress Cataloging in Publication Data

Driver, Rodney David, Date-
    Introduction to ordinary differential equations.

    Bibliography: p.
    Includes index.
    1. Differential equations.   2. Differential equa-
tions—Delay equations.   I. Title.
QA372.D73   515'.352   77-16789
ISBN 0-06-041738-2

CONTENTS

CHAPTER FOUR
LINEAR ORDINARY DIFFERENTIAL SYSTEMS 128

*CHAPTER FIVE
EXISTENCE AND COMPUTATIONS 180

CHAPTER SIX
DELAY DIFFERENTIAL EQUATIONS 206

CHAPTER SEVEN
POWER SERIES SOLUTION
OF LINEAR ORDINARY EQUATIONS 238

CHAPTER EIGHT
THE LAPLACE TRANSFORM METHOD
FOR LINEAR ORDINARY EQUATIONS 271

APPENDIXES

PREFACE

This text is designed for a first course in ordinary differential equations. The usual topics of such a course are included, and the only prerequisite is the standard freshman-sophomore calculus sequence.

What makes the book unique is its inclusion of a chapter on delay differential equations. Such equations, which model processes with time lags, are encountered more and more frequently in engineering, physics, and biology. But no other book (to my knowledge) has attempted to discuss such equations at an introductory level.

Real-world applications provide the central motivation for practically everything in the book. Emphasis is on the solution and the interpretation of solutions of applied problems.

But this does not imply a sacrifice of rigor. In fact, I hope that this book will convince its readers that the mathematical study of real-world problems *requires* a certain amount of "theory."

For example, in physics and engineering one gets accustomed to working with second-order differential equations. And one may begin to believe that *physics* dictates the specification of positions and velocities at $t = 0$ as the appropriate initial conditions for the universe. But, on the contrary, it is the theory of differential equations which asserts that these initial conditions are appropriate *for our models*. If the models were refined to account for the finite propagation speed of electromagnetic and gravitational interactions, then perhaps positions and velocities at $t = 0$ would no longer be appropriate initial conditions.

While, on the one hand, applied science requires rigorous mathematics (particularly the theory of differential equations), one can also argue conversely that the study of differential equations would be on weak ground without attention to real-world applications. For example, considerable mathematical effort has been devoted to studying the asymptotic behavior of the solutions of certain differential equations whose solutions are, in fact, "unstable." But an applied scientist might argue that such studies are a waste of time; for unstable solutions are not encountered in the real world, except perhaps approximately and briefly.

A course in differential equations can provide a natural transition from introductory calculus to the methods of advanced calculus. Using only the background of freshman-sophomore calculus, one can illustrate the need for existence and uniqueness theorems and then actually prove such theorems.

This leads quite naturally to some of the basic ideas of advanced calculus.

In this book the concepts of existence and uniqueness are introduced in Chapter One in conjunction with elementary examples. So readers should appreciate the significance of these matters when they get to Chapter Two where uniqueness theorems are presented together with fairly elementary proofs. The results are sufficient to establish the global existence and uniqueness and the basic properties of solutions of the many linear equations treated in Chapters Three and Four. The more difficult (and less important) general theorems on existence of solutions are discussed in an optional Chapter Five.

Chapter Six—the most unusual chapter in the book—introduces the subject of delay differential equations. Interest in delay differential equations has surged in recent years because of their occurrence in engineering (especially the study of feedback controls), physics, biology, physiology, and economics. But mathematicians have generally assumed that this was not appropriate material for an undergraduate course—least of all one which did not even assume advanced calculus. In this first effort to present delay differential equations at the most elementary level, emphasis is placed on elementary methods for extracting useful information in some practical examples.

Chapters Seven and Eight discuss power series and Laplace transform methods for solving linear ordinary differential equations. Their placement at the end of the book reflects a personal judgment of their importance relative to other topics treated. The instructor with a different opinion will find that these chapters (except for Section 42) can actually be presented anytime after Chapter Three.

The notation and theorems from calculus used in this book are listed for ready reference in the appendixes; and two theorems which are not covered in introductory calculus are proved in full there. In fact, proofs are provided throughout the book for the results used—although I omit many of these proofs when presenting material in class.

The material offered here is too much for a one-semester course meeting only three hours per week. Stars indicate sections, results, or proofs which may be naturally omitted from such a course. Most starred items are so designated because of their difficulty or the amount of time they might require. (But Section 5 is starred simply because I consider it unimportant.) One should probably omit starred Section 12 (except in an honors class or an extended course), and this omission will then dictate the omission also of the proofs of some theorems in Chapter Five and in Sections 40 and 41.

The following conventions are used for cross referencing. Within any section, Eq. (1), Problem 1, Figure 1, and Theorem A, refer to items in that same section. When it is necessary to refer to such items appearing in another section, say Section 8, they are called Eq. 8-(1), Problem 8-1, Figure 8-1, and Theorem 8-A, respectively.

This introductory text was written simultaneously with an intermediate book [4] (see references, page 312) on the same subject. There is necessarily a considerable overlap between the two books; and I am grateful to the two publishers, Harper & Row and Springer-Verlag, for accepting this arrangement. In particular, because book [4] was completed first, I appreciate the permission from Springer-Verlag New York Inc. to reprint with only minor changes the material that appears here in Sections 1-3, 8-12, 14-17, 19, 23-25, 30, and 31.

I am grateful to many people who contributed to this book. The students who used preliminary mimeographed versions of the manuscript as it developed in courses taught at the University of Rhode Island helped smooth out many rough spots. Further valuable suggestions were made by several colleagues and reviewers including Professors Kenneth L. Cooke, Michael Greenberg, Jack Hale, Kenneth M. Hoffman, Gerasimos Ladas, George R. Sell, and some reviewers who are anonymous to me. A particularly extensive and helpful list of corrections and suggestions for the entire manuscript was prepared by Professor William A. Smith of Georgia State University. The answers to most of the problems were carefully checked by Stephen F. Roehrig. The book was greatly improved by the contributions of all these people, and any obscurities which remain are probably due to my failure to use some of the reviewers' suggestions.

R. D. Driver

Chapter One

ELEMENTARY METHODS FOR ORDINARY DIFFERENTIAL EQUATIONS

Suppose that an object is moving along the x-axis and is located at the (unknown) position $x(t)$ at time t. If the velocity $f(t)$ at each instant is known, then

$$x'(t) = f(t),$$

where x' is the derivative of x. In this example we immediately conclude that x must be some indefinite integral (or antiderivative) of f.

More generally, in modeling real-world problems one may encounter equations of the form

$$x'(t) = f(t, x(t)),$$

where f is a given function. The objective is to find the unknown function x—a task which may be much more difficult than in the preceding paragraph.

Equations involving derivatives of unknown functions are called *differential equations*.

This book is concerned with techniques for finding solutions of certain types of differential equations, and for deciding whether a solution is unique and whether it makes sense. We shall motivate and illustrate several differential equations by describing their sources in physics, engineering, biology, or ecology.

1. EXAMPLES AND CLASSIFICATION

One of the simplest examples of a differential equation is the standard model for the decay of a radioactive material. If $x(t)$ is the quantity of some radioactive substance present at the instant of time t, then it is generally assumed that

$$x'(t) = -ax(t),$$

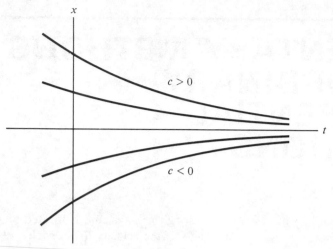

FIGURE 1

where a is a positive constant. This equation expresses the assertion that the rate of decay of the material at any instant is proportional to the quantity of material remaining at that instant.

It is easily verified that for any constant c the function defined by

$$x(t) = ce^{-at}$$

is a "solution" of this differential equation, that is,

$$x'(t) = -ace^{-at} = -ax(t).$$

Some members of this infinite family of solutions are indicated in Figure 1. Note that negative values of c are perfectly acceptable as far as the differential equation is concerned, although they would not be meaningful for our physical problem.

We apparently pulled the solutions ce^{-at} out of thin air. So it is natural to ask whether any other type of function, besides these exponentials, could also satisfy this same differential equation. The answer to this question is "No," and we shall prove it in Section 2. Thus the function defined by $x(t) = ce^{-at}$ with unspecified constant c is called the "general solution." Every possible solution can be obtained from this by an appropriate choice of c.

To uniquely determine a particular solution, some additional data is needed. This often takes the form of an "initial condition," say $x(t_0) = x_0$, the amount of material present at the "initial instant" $t = t_0$. Substituting $t = t_0$ in the general solution gives $x_0 = ce^{-at_0}$ or $c = x_0 e^{at_0}$. Thus we must have

$$x(t) = x_0 e^{-a(t-t_0)}.$$

This represents an exponential time decay from the initial amount.

Real-world problems are often represented by simple differential equations which can be solved exactly, as in the above example. But it should always be recognized that the differential equation itself is at best only an approximate representation of the problem being considered. Approximations might enter because of our imperfect knowledge of the laws governing the behavior of the real world or because of imperfect measurement of input data (such as initial conditions) for the problem. And in many cases we *deliberately* make approximations in order to reduce the mathematical equation to a tractable form.

For example, the representation of the decay of a radioactive material considered above is certainly not perfect. If we believe in atoms, we must admit that the function x will not even be continuous. And it certainly cannot have a derivative at points of discontinuity. If we overlook this difficulty, we should still admit that, for a random collection of atoms, the assertion that the rate of decay is some constant multiple of the number of atoms present is at best a statistical average. Such an approximation would become quite unacceptable if only a few atoms were involved.

In other situations more serious approximations enter.

One of the most pervasive approximating assumptions, often made unconsciously, is that of "simultaneity." One usually assumes that the values of the unknown functions and their derivatives, which occur in differential equations, should all be evaluated at the same instant. For example, in modeling radioactive decay we wrote $x'(t) = -ax(t)$ and not $x'(t) = -ax(t-1)$ or $x'(t) = -ax(t/2)$.

Thus most books on differential equations treat equations such as

$$x'(t) = -ax(t),$$

where a is a given constant, or

$$x''(t) = -\frac{K}{x^2(t)},$$

where K is a given constant, or

$$mx''(t) + bx'(t) + kx(t) = h(t),$$

where m, b, and k are given constants and h is a given function, or

$$\frac{\partial^2 u}{\partial x^2}(x, y) + \frac{\partial^2 u}{\partial y^2}(x, y) = 0,$$

where u is an unknown function of x and y.

Of these four examples, the first three, involving unknown functions of only a single variable, are called *ordinary differential equations*. The second of these equations describes the free fall of a mass under the sole influence of gravity. The third equation covers, for example, the oscillations of a mass on a simple spring.

Differential equations, involving partial derivatives of unknown functions of two or more variables are called *partial differential equations*. The fourth example given above is a famous partial differential equation of physics—Laplace's equation. This text does not treat partial differential equations.

It has been so universally assumed that all functions should be evaluated at the same point that one often omits the "independent variables" (arguments of the unknown functions) altogether. Thus the above equations are usually written simply as

$$x' = -ax,$$

$$x'' = -\frac{K}{x^2},$$

$$mx'' + bx' + kx = h(t),$$

and

$$\frac{\partial^2 u}{\partial x^2} + \frac{\partial^2 u}{\partial y^2} = 0.$$

However, since the middle of the eighteenth century, mathematicians have from time to time considered problems involving relations between unknown functions and their derivatives evaluated at different arguments. Thus one might encounter an equation such as

$$mx''(t) + bx'(t) + qx'(t - 1) + kx(t) = h(t).$$

Since the 1930s and 1940s the number of such "delay differential equations" arising in practical applications has escalated rapidly. Chapter Six introduces several examples in which delay differential equations arise and begins the study of their solutions.

The first five chapters are devoted to certain ordinary differential equations. In particular, the present chapter gives some elementary but useful methods for solving certain special types of ordinary differential equations. We shall use the accompanying examples as vehicles for introducing the concepts of "uniqueness" and "continuability" of solutions.

Some important terminology is as follows:

The *order* of a differential equation is the order of the highest derivative involved in that equation. We begin by considering ordinary differential equations of first order. Such equations have the general form $F(t, x(t), x'(t)) = 0$, where F is a given function. But we shall restrict ourselves to those which can be written in the more special form

$$x'(t) = f(t, x(t)),$$

or briefly,

$$x' = f(t, x),$$

where f is a given function. A *solution* is any differentiable function, on some open interval, which satisfies the given differential equation.

The differential equation is often considered together with an appropriate initial condition, say

$$x(t_0) = x_0,$$

to form an *initial-value problem*. In this case a *solution* is a differentiable function, on some open interval containing t_0, which satisfies both the differential equation and the initial condition. We shall say a solution x is *unique* if, for every other solution $\tilde{x}$ (pronounced "x tilde"), $\tilde{x}(t) = x(t)$ as far as both solutions are defined.

Throughout this chapter solutions will be real-valued functions.

Certain standard mathematical terms and symbols are described in Appendix 1 for reference as needed.

2. LINEAR EQUATIONS OF FIRST ORDER

The general *linear* ordinary differential equation of first order which we will consider is

$$x' + a(t)x = h(t). \tag{1}$$

Here a and h are given functions continuous on some open interval $J = (\alpha, \beta)$ where $-\infty \leq \alpha < \beta \leq \infty$. We shall often seek a solution of (1) which also satisfies the initial condition

$$x(t_0) = x_0 \quad \text{where} \quad t_0 \in J. \tag{2}$$

An equation of the form

$$b(t)x' + c(t)x = g(t),$$

where b, c, and g are given continuous functions, can be put in the form (1) on an interval J provided $b(t) \neq 0$ for all t in J. On such an interval one merely divides through by $b(t)$.

Equation (1) becomes particularly simple in the special case when $a(t) \equiv 0$. Indeed the resulting differential equation,

$$x' = h(t),$$

with initial condition (2) is solved at once using the fundamental theorem of calculus. One has

$$x(t) = x_0 + \int_{t_0}^{t} h(s)\, ds.$$

A more interesting example of a linear first-order equation was encountered in Section 1. In that example we had $a(t) = a$, a constant, and $h(t) = 0$. We shall now complete the analysis of that equation.

Example 1. Find all solutions (on an interval J) of the differential equation

$$x' + ax = 0$$

(without regard to any initial condition).

We proceed as follows. *Assume* that some function x is a solution. Then multiply both sides of the differential equation by the exponential e^{at}. (The reason for this will be seen shortly.) It follows that x must satisfy

$$x'e^{at} + axe^{at} = 0.$$

Now observe that the left-hand side of this last equation is just the derivative of the product $x(t)e^{at}$. Thus we have

$$\frac{d}{dt}[xe^{at}] = 0,$$

and the only functions with zero derivative (on an interval) are the constant functions. Hence it must be that

$$x(t)e^{at} = c$$

for some constant c. Multiply both sides by e^{-at} to find

$$x(t) = ce^{-at}.$$

In other words, the only functions that *can possibly* be solutions of our differential equation are of the form $x(t) = ce^{-at}$, where c is some (real) constant.

But we already agreed, in Section 1, that such exponentials *do* satisfy the differential equation. Thus the problem is completely solved. For each c, $x(t) = ce^{-at}$ is a solution, and no other type of function can be a solution. Hence we refer to ce^{-at} as the " general solution " of $x' + ax = 0$.

The key trick in the solution of Example 1 was the multiplication by e^{at}. This multiplicative factor yielded the combination

$$x'e^{at} + axe^{at} = \frac{d}{dt}[xe^{at}],$$

which is easily integrated. The multiplier e^{at} is an example of what we shall call an "integrating factor."

Students usually ask at this point, "How did you know in advance that multiplication by e^{at} would be helpful?"

We will not give a very satisfactory answer to this question. Let us just say that we have used a special case of a standard procedure for the solution of equations of this type. More generally, we shall always begin the analysis of Eq. (1) by multiplying both sides of the equation by the "integrating factor"

$$e^{v(t)} \quad \text{where} \quad v(t) = \int a(t) \, dt \tag{3}$$

is any indefinite integral of $a(t)$.

More than 200 years ago, mathematicians discovered that multiplication of Eq. (1) by $e^{v(t)}$ gives an integrable combination on the left-hand side, as in Example 1. Without worrying about how this trick was first discovered, let us accept it as a method worth using. We now show that it always works.

Justification of the Method

Assume, at first, that (1) has a solution x on some open interval $J_* \subset J$. (We refrain from assuming that a solution exists on the entire given interval J.) Let $v(t) = \int a(t) \, dt$, as in (3), and multiply Eq. (1) by $e^{v(t)}$ to obtain

$$x'e^{v(t)} + a(t)xe^{v(t)} = h(t)e^{v(t)},$$

or

$$\frac{d}{dt}[x(t)e^{v(t)}] = h(t)e^{v(t)}. \tag{4}$$

Verify this. Now integrate (4) to find

$$x(t)e^{v(t)} = \int h(t)e^{v(t)} \, dt + c,$$

or

$$x(t) = ce^{-v(t)} + e^{-v(t)} \int h(t)e^{v(t)} \, dt, \tag{5}$$

where c is an arbitrary constant.

If, in addition to satisfying (1), x satisfies $x(t_0) = x_0$ for some t_0 in J_*, we obtain a more explicit form for x as follows. Replace t by s in Eq. (4) and then integrate from t_0 to t to find

$$x(t)e^{v(t)} - x_0 e^{v(t_0)} = \int_{t_0}^{t} h(s)e^{v(s)} \, ds.$$

Thus

$$x(t) = x_0 e^{v(t_0) - v(t)} + e^{-v(t)} \int_{t_0}^{t} h(s)e^{v(s)} \, ds \tag{6}$$

on J_*.

It is easy to verify that (6) does reduce to $x(t_0) = x_0$ when $t = t_0$. However, it should be emphasized that we have *not* yet shown that either (5) or

(6) gives a solution of (1). What we have shown is that *if* Eq. (1) has a solution, *then* it must have the form (5) or (6).

Note that, for any given value of c, (5) can always be considered as a special case of (6) obtained by an appropriate choice of x_0. So, to complete our analysis, the reader is now asked to show that the function x defined by (6)—the sole *candidate* for a solution of (1) and (2)— actually *is a solution* of (1) and (2). You can do this either by justifying the reversal of each step in the derivation of (6), or by directly substituting (6) into Eq. (1). Furthermore, this calculation actually shows that Eq. (6) defines a solution on the entire interval J.

Thus we conclude that *Eqs.* (1) *and* (2) *have a solution on the entire interval J. This solution is unique and is given by* (6).

It should be noted that, in particular examples, the integrals occurring in Eqs. (5) and (6) may be difficult to evaluate. They may even be impossible to evaluate in terms of elementary functions. Thus, even though (6) exactly represents the unique solution of Eqs. (1) and (2), it may not be easy to interpret the result.

As already observed, (5) is implied by (6) if x_0 is considered arbitrary. Nevertheless, Eq. (5), which emphasizes the arbitrariness of the constant c in the absence of any specified initial condition, is the form often referred to as the " general solution " of (1).

When we described the integrating factor $e^{v(t)}$ in (3), $v(t)$ was "any" indefinite integral of $a(t)$. Suppose we had chosen a different $v(t)$. Why would this not affect the outcome? (Problem 1.)

In order to solve a specific linear, first-order equation one could simply substitute the appropriate functions a and h into (5) or (6). However, it is strongly recommended that, instead, the reader make a practice of using the *method* of the general case—and not the "formula" for the solution. Then all one must remember is that $\exp \int a(t)\, dt$ is an integrating factor for (1).

Example 2. Solve the differential equation

$$t^{-1}x' = 3 - t^{-2}x$$

with the initial condition

$$x(2) = 3.$$

We must first put the differential equation into the form of (1) by multiplying through by t and moving all terms involving x to the left-hand side. This yields

$$x' + \frac{1}{t}\, x = 3t.$$

Since $a(t) = t^{-1}$, $h(t) = 3t$, and $t_0 = 2$, we should take $J = (0, \infty)$. Why?

Taking $v(t) = \int t^{-1}\,dt = \ln|t| = \ln t$, we multiply the differential equation by the integrating factor $e^{v(t)} = e^{\ln t} = t$ to obtain

$$x't + x = 3t^2.$$

Now, as they should, the two terms on the left-hand side represent exactly the derivative of xt. (You should *always verify this step carefully*. For if there has been any mistake in the calculation of $e^{v(t)}$ this is your chance to catch it.) Thus

$$\frac{d}{dt}[xt] = 3t^2.$$

Change t to s, then integrate both sides from $s = 2$ to $s = t$ to get

$$tx(t) - 2x(2) = \int_2^t 3s^2\,ds = t^3 - 8,$$

or, since $x(2) = 3$,

$$x(t) = t^2 - \frac{2}{t} \quad \text{for} \quad t > 0.$$

It is *no longer essential* to verify that this x is a solution of our original problem, because our general analysis says it must be (if we have made no mistake). However, a direct substitution of x into the original equations does provide the ideal check on our work.

Example 3 (A Mixing Problem). A tank contains 500 liters of a salt water brine. An inlet pipe discharges into this tank, at the rate of 25 liters/min, brine containing 100 grams of salt per liter. At the same time fluid leaves the tank through an outlet hole at the same rate. The mixture in the tank is continually stirred so that the concentration can always be considered uniform throughout (Figure 1). If initially the mixture contains 50 g of salt per liter, determine the future concentration as a function of time.

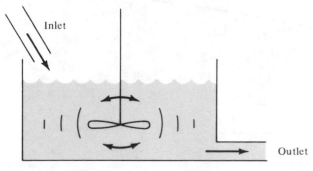

Inlet

Outlet

FIGURE 1

Let $x(t)$ be the concentration of salt in the tank in grams per liter at time t min after the initial instant $t = 0$. Then the total amount of salt in the tank at time t is $500x(t)$ g. Salt is being added to the tank at the rate of $25 \times 100 = 2500$ g/min, and is leaving at the rate of $25x(t)$ g/min. Thus

$$500x' = 2500 - 25x \quad \text{and} \quad x(0) = 50.$$

To put this differential equation into the form of Eq. (1), we divide through by 500 and get

$$x' + 0.05x = 5.$$

Then multiplying by the integrating factor $e^{0.05t}$, we find

$$\frac{d}{dt}[e^{0.05t}x] = 5e^{0.05t}$$

or, upon integrating,

$$e^{0.05t}x(t) - x(0) = 100(e^{0.05t} - 1).$$

Since $x(0) = 50$, this gives

$$x(t) = 100 - 50e^{-0.05t}.$$

This function describes, as would be expected, an increasing concentration starting with 50 g/liter at $t = 0$ and approaching 100 g/liter as $t \to \infty$. (See Figure 2.)

One assumption of the mathematical model in Example 3 appears particularly questionable. How can the brine always be so perfectly mixed that the liquid leaving the tank has the same concentration as does the tank as a whole? This assumption will be relaxed in a similar example in Chapter Six.

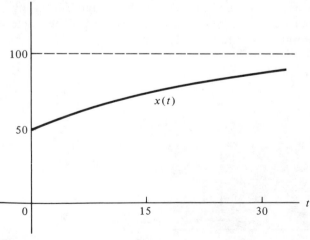

FIGURE 2

PROBLEMS

1. If in (3) we chose, as the indefinite integral of $a(t)$, some $v_1(t) \neq v(t)$, then $v_1(t) = v(t) + c_1$ for some constant c_1. Show that the use of $v_1(t)$ instead of $v(t)$ always leads to the same form for $x(t)$, namely (5) or (6).

2. Show that (6) is equivalent to

$$x(t) = x_0 e^{-\int_{t_0}^t a(s)ds} + \int_{t_0}^t h(s) e^{-\int_s^t a(u) du} ds.$$

Consider the equations in Problems 3 through 11 on $(-\infty, \infty)$. Find a solution for each problem and determine some interval J on which your solution is valid. Decide whether the solution is unique there.

3. $x' - x = 1$ with $x(2) = 3$.
4. $x' = (x + 1)\sin t$ with $x(t_0) = x_0$, where t_0 and x_0 are given constants.
5. $x' + 2x = t$.
6. $Li' + Ri = E$ with $i(0) = i_0$, where $L > 0$, $R \geq 0$, E and i_0 are given constants. This is the equation for the electrical current $i(t)$ (in amps) flowing in a series circuit having inductance L (in henries), resistance R (in ohms), and source E (in volts) (Figure 3).

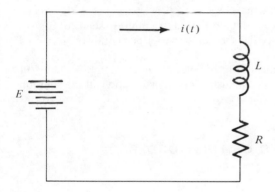

$i(t)$

L

E

R

FIGURE 3

7. (a) $tx' + x = t + 1$ with $x(1) = 0$.
   (b) $tx' + x = t + 1$ with $x(-1) = 2$.
   (c) Sketch graphs of the solutions of parts (a) and (b) and *describe the difference* between them.
8. $x' + 2tx = 1$ with $x(0) = 5$.
9. $tx' - x = t$.
10. $(t - 1)x' - x = t$ with $x(-1) = -2$.
11. $(\cos t)x' + (\sin t)x = 1$ with $x(0) = 1$.
*12. Show that Problem 11 has a solution on $(-\pi/2, 3\pi/2)$, but that the solution is *not unique* there. Why does this not contradict the assertions in the text?
13. (a) Find the *general solution* of

$$x' + x = \sin t \quad \text{on} \quad J = (-\infty, \infty).$$

That is, find a function involving an unspecified constant c which satisfies the differential equation and which gives as special cases all solutions of that equation. [*Hint:* Proceed as in the derivation of (5) and assume the validity of the assertion of part (c) below.]

(b) Verify your solution by substitution. Why can you be sure that an arbitrary initial condition of the form $x(t_0) = x_0$ will always uniquely determine c in your general solution?

(c) Using two integrations by parts, obtain

$$\int e^t \sin t \, dt = \frac{1}{2} e^t(\sin t - \cos t) + c.$$

14. Assume that the radioactive isotope strontium-90 decays exponentially according to the law $x' = -ax$ considered in Section 1. Use the fact that strontium-90 has a half-life of 28.1 years (that is, half of the original amount decays in 28.1 years) to evaluate a. How much time must elapse after an atomic explosion before the resulting strontium-90 is reduced to 10 percent of its initial amount?

15. An automobile cooling system contains 16 liters of solution, half of which is dirty 3-year-old antifreeze. In hopes of saving time flushing out the mess, the owner of the automobile decides to pour fresh water in while the old solution is running out. (Assume that the old solution runs out at a rate of 8 liters/min and the fresh water runs in at the same rate.) The owner keeps the engine running so that the solution is continually stirred to a uniform shade of brown. How long does it take to reduce the solution to only 1 percent dirty 3-year-old antifreeze?

*16. The solution of Problem 6 depends on the values of L, R, E, and i_0. Think of L, E, and i_0 as fixed while R is an adjustable parameter, and call the solution $i(t; R)$. Prove that $\lim_{R \to 0} i(t; R) = i(t; 0) = Et/L + i_0$ for each t.

3. SEPARABLE EQUATIONS OF FIRST ORDER

An equation of the form

$$M(t) + N(x)x' = 0, \tag{1}$$

where M and N are given functions, is said to have its variables (x and t) separated. And any equation which can be rewritten in this form is said to be *separable*.

For example, the equation

$$x' = a(t)h(x), \tag{1'}$$

where the functions a and h are given, is separable wherever $h \neq 0$; for it can be rewritten as

$$a(t) - \frac{1}{h(x)} x' = 0.$$

We shall generally assume M and N continuous with $N > 0$ or $N < 0$.

The method of solution called *separation of variables* is illustrated in the following example.

Example 1. Solve

$$x' = (at + b)x \quad \text{with} \quad x(0) = 1,$$

where a and b are constants. (Actually this equation is linear and hence could be solved by the method of Section 2.) Assume, at first, that a solution x exists on some interval. Since x must be a continuous function with $x(0) = 1$, it follows that $x(t) > 0$ for all t *sufficiently close to* 0. Thus we can divide the differential equation by x, to "separate" the variables, as long as x remains positive:

$$(at + b) - \frac{1}{x} x' = 0.$$

Now replace t by a dummy variable s and integrate from $t_0 = 0$ to t, obtaining

$$\int_0^t (as + b) \, ds - \int_0^t \frac{x'(s)}{x(s)} \, ds = 0.$$

If we make the change of variable $\xi = x(s)$ in the second integral we get, using $x(0) = 1$,

$$\int_0^t (as + b) \, ds - \int_1^x \xi^{-1} \, d\xi = 0,$$

or, as long as $x = x(t)$ remains positive,

$$\tfrac{1}{2}at^2 + bt - \ln x = 0.$$

Since the logarithm function has an inverse—the exponential function—we immediately find

$$x(t) = e^{\frac{1}{2}at^2 + bt}.$$

We have shown that the only candidate for a solution *near* $t_0 = 0$ is the function x defined above. It is left as an easy exercise for the reader to verify that x *is* a solution and, in fact, is valid for *all* t. (Problem 3(a).) Thanks to the positivity of x, it can also be shown that the solution is unique for all t. (Problem 3(b).)

*Justification of the Method

Let us now apply to the general case, Eq. (1), an argument similar to that used in the example. Let the given functions M and N be continuous with $N \neq 0$, and let

$$x(t_0) = x_0. \tag{2}$$

Assume at first that there is a solution x of (1) and (2) on some open interval J containing t_0. Then x is continuous and, since $x' = -M(t)/N(x)$, x' is continuous on J. Thus we obtain from (1)

$$\int_{t_0}^{t} M(s)\, ds + \int_{t_0}^{t} N(x(s))x'(s)\, ds = 0 \quad \text{for all} \quad t \quad \text{in} \quad J.$$

With the change of variable $\xi = x(s)$, this becomes

$$\int_{t_0}^{t} M(s)\, ds + \int_{x_0}^{x} N(\xi)\, d\xi = 0.$$

Now define $g(u) \equiv -\int_{x_0}^{u} N(\xi)\, d\xi$. Then

$$g(x) = \int_{t_0}^{t} M(s)\, ds, \tag{3}$$

where g is a continuously differentiable function with $g'(u) = -N(u)$. Hence, as long as $N(u) \neq 0$, it follows from the "inverse function theorem" that g^{-1} exists (and is differentiable) so that

$$x(t) = g^{-1}\left[\int_{t_0}^{t} M(s)\, ds \right] \quad \text{for} \quad t \quad \text{in} \quad J \tag{4}$$

is uniquely determined. The function x defined by (4) is the only possible *candidate* for a solution of (1) and (2) on J.

The reader should verify by implicit differentiation of (3) that if x is defined by (4) on some interval, then x is a solution of (1) and (2). (Problem 5.)

Even though Eq. (4) gives the exact solution of (1) and (2), we must recognize that only in very simple examples will $\int M(s)\, ds$ and $\int N(\xi)\, d\xi$ be elementary integrals and the inverse function g^{-1} be expressible in simple form.

Once again it is recommended that, in solving problems, the reader use the *method* described in this section and not the "formula" for the solution given by (4).

Example 2. Solve

$$x' = 1 + x^2 \quad \text{with} \quad x(0) = 1.$$

This equation is decidedly not linear, but it does satisfy all the requirements for the method of separation of variables (everywhere). From

$$\int_{1}^{x} \frac{d\xi}{1 + \xi^2} = \int_{0}^{t} ds,$$

we obtain[†]

$$\text{Arctan } x - \frac{\pi}{4} = t,$$

or

$$x(t) = \tan\left(\frac{\pi}{4} + t\right).$$

Therefore, that is the only candidate for a solution. Since the above steps are all reversible as long as $-3\pi/4 < t < \pi/4$, it follows that the function x which we have found is a solution. Thus we have a unique solution on $J = (-3\pi/4, \pi/4)$, but no further. Why? The fact that the solution cannot be continued as far back as $-3\pi/4$ or as far forward as $\pi/4$ was completely unpredictable from the form of the differential equation. We have seen that such an *unpredictable* ending of the solution cannot occur for the linear equations of Section 2.

Example 3. Solve

$$x' + x^2 \sin t = 3(tx)^2 \quad \text{with} \quad x(0) = x_0 > 0.$$

As long as $x(t) \neq 0$, the differential equation is separable since it can be rewritten as

$$x' = (3t^2 - \sin t)x^2.$$

Using the initial condition we find, for as long as $x(t)$ remains positive,

$$\int_{x_0}^{x} \xi^{-2}\, d\xi = \int_0^t (3s^2 - \sin s)\, ds,$$

or

$$-\frac{1}{x} + \frac{1}{x_0} = t^3 + \cos t - 1.$$

So

$$x(t) = \frac{1}{x_0^{-1} + 1 - t^3 - \cos t}.$$

The reader should verify directly that this is a solution of the given differential equation and the initial condition (Problem 7). Our derivation shows that it is unique, that is, no other function can be a solution as long as $x_0^{-1} + 1 - t^3 - \cos t > 0$. Moreover the solution is valid at least for all $t < x_0^{-1/3}$, and generally on a somewhat larger interval. Why?

[†]We use the notation arctan and Arctan for the inverse tangent and the principle value of the inverse tangent, respectively. Some books denote these by $\tan^{-1}$ and Tan^{-1}. Analogous notation is used for other inverse trigonometric functions.

Sometimes it is possible to "formally" apply a method even when the conditions which would justify that method are not fulfilled, as in the next examples.

Example 4. Solve

$$x' = x^2 \quad \text{with} \quad x(0) = x_0 = 0.$$

Since $h(0) = 0$, the condition $h \neq 0$ in (1') is *not* fulfilled. Nevertheless, formal solution with the initial condition $x(0) = x_0$ leads to

$$x(t) = \frac{x_0}{1 - x_0 t},$$

or $x(t) \equiv 0$ when $x_0 = 0$. It is easily verified by substitution that $x = 0$ *is* a solution. However, we cannot conclude, on the basis of what has been done here, that there is no other solution. (Problem 10.)

Example 5. Solve

$$x' = x^{2/3} \quad \text{with} \quad x(0) = 0.$$

Again $h(0) = 0$. However, formal application of the method of separation of variables (Problem 8) yields $x(t) = t^3/27$, which is easily verified to be a solution. In this case, not only have we *not proved* the solution to be unique, indeed the solution is *not* unique. For example, two more solutions, among infinitely many possibilities, are the functions defined by

$$x(t) = 0 \quad \text{for all} \quad t$$

and

$$x(t) = \begin{cases} 0 & \text{for } t \le 1, \\ \dfrac{(t-1)^3}{27} & \text{for } t > 1. \end{cases}$$

PROBLEMS

1. In each of the following use the method of separation of variables to find a solution on some interval J. Determine (or conjecture) whether the solution is unique there.

   (a) $x' - x = 1$ with $x(2) = 3$ (Problem 2–3 again).
   (b) $x' = (\sin t)(x + 1)$ with $x(t_0) = x_0$ (Problem 2–4 again).
   (c) $x' = x^2$ with $x(7) = 1$.
   (d) $x^2 x' + tx = x$ with $x(-1) = 2$.
   (e) $x' = \sqrt{|x|}$ with $x(0) = 0$.
   (f) $x' = -t/x$.
   (g) $x' = (1 + x^2)/(1 + t^2)$ with $x(1) = -1$.
   (h) $t^2 x' = x - tx$.

2. Solve Problem 1(h) by another method.
3. (a) Verify by substitution that the candidate for a solution x obtained in Example 1 really is a solution and is valid for all t.
   (b) Noting that $x(t) > 0$ for all t prove, via the same method already used in Example 1, that the solution is unique for *all* t. [*Hint:* Suppose (for contradiction) that there were another solution $\tilde{x}$. Then there would have to be some point $t_1 \neq 0$ at which $\tilde{x}(t_1) = x(t_1) > 0$, and yet $\tilde{x}(t) \neq x(t)$ for certain values of t arbitrarily close to t_1. Show that this is impossible.]
4. Solve the problem in Example 1 by the method given for linear equations in Section 2.
5. Verify that (4) defines a solution of Eqs. (1) and (2) by differentiating Eq. (3) which is equivalent to (4).
6. Sketch a graph of the solution of Example 2 to see what happens as t approaches $-3\pi/4$ or $\pi/4$.
7. Verify the solution found in Example 3 on $J = (-\infty, x_0^{-1/3})$.
8. Carry out the formal separation of variables calculation for Example 5.
9. Sketch the three solutions given for Example 5 plus a fourth solution of your own.
*10. A Challenge: The only solution of $x' = x^2$ with $x(0) = 0$ is $x \equiv 0$. Try to prove this without reading Chapter Two.

4. EXACT EQUATIONS OF FIRST ORDER

Consider a differential equation of the form

$$M(t, x) + N(t, x)x' = 0, \tag{1}$$

where M and N are given continuous functions on some "open rectangle" $D = (\alpha, \beta) \times (\gamma, \delta)$. Equation (1) is said to be *exact* if the left-hand side is *identically* $(d/dt)g(t, x(t))$ for some function g. Then

$$\frac{d}{dt} g(t, x) = 0, \tag{2}$$

which means that

$$g(t, x) = c \tag{3}$$

for some constant c. If an initial condition of the form $x(t_0) = x_0$ where $(t_0, x_0) \in D$ is also specified, then c will be determined. It remains to be decided in specific cases whether the "implicit equation" (3) determines a unique function x and whether such a function is a solution. These questions are sometimes nontrivial.

Example 1. The differential equation

$$x^3 \cos t + (3x^2 \sin t)x' = 0$$

is exact because it is equivalent to

$$\frac{d}{dt}(x^3 \sin t) = 0$$

or

$$x^3 \sin t = c.$$

In this case one finds $x(t) = (c/\sin t)^{1/3}$ as long as $\sin t \neq 0$, that is, on any interval $(n\pi, (n + 1)\pi)$ where n is an integer.

In more complicated cases it may not be obvious that an equation in the form (1) is really equivalent to Eq. (2) for some function g. However, there is a simple test:

Let g have continuous first partial derivatives

$$D_1 g(t, \xi) = \frac{\partial}{\partial t} g(t, \xi) \quad \text{and} \quad D_2 g(t, \xi) = \frac{\partial}{\partial \xi} g(t, \xi).$$

[Here and on future occasions we use a " dummy variable " such as ξ in place of x. This enables us to discuss the properties of a given function f without reference or restriction to some unknown function x.] Then by the chain rule, Theorem A2-C (i.e., Appendix 2, Theorem C),

$$\frac{d}{dt} g(t, x) = D_1 g(t, x) + D_2 g(t, x)x'.$$

So in order for Eq. (1) to be exact one must have

$$M = D_1 g = \frac{\partial g}{\partial t} \quad \text{and} \quad N = D_2 g = \frac{\partial g}{\partial \xi} \tag{4}$$

for some function g. But how can we determine, from the given functions M and N, whether such a g exists?

Let us assume, as is usually the case, that M and N themselves are continuously differentiable on D. Then *if* Eq. (1) is exact, so that Eqs. (4) hold for some g,

$$D_2 M = D_2 D_1 g = D_1 D_2 g = D_1 N,$$

or

$$\frac{\partial M}{\partial \xi} = \frac{\partial N}{\partial t} \quad \text{on} \quad D. \tag{5}$$

In other words, condition (5) is a *necessary* condition for exactness. It turns out happily that if M and N are continuously differentiable, then (5) is also a *sufficient* condition for exactness. The proof of this assertion is left as Problem 12. However, the following example shows a method for constructing the desired function g.

Example 2. Solve

$$2t \ln |x| - 1 + \frac{t^2 + 1}{x} x' = 0 \quad \text{with} \quad x(2) = -e.$$

[Note that the differential equation makes no sense if ever $x(t) = 0$.] This complicated looking equation is not linear and it is not separable. However, it turns out to be quite easy to handle. For it is in the form of (1) with

$$M(t, \xi) = 2t \ln |\xi| - 1 \quad \text{and} \quad N(t, \xi) = \frac{t^2 + 1}{\xi}.$$

And we find

$$\frac{\partial M}{\partial \xi} = \frac{2t}{\xi} = \frac{\partial N}{\partial t} \quad \text{for} \quad \xi \neq 0,$$

showing that the equation is exact for $\xi > 0$ or $\xi < 0$. We now seek to construct a function g such that

$$\frac{\partial g}{\partial t} = M \quad \text{and} \quad \frac{\partial g}{\partial \xi} = N.$$

We begin by considering the implications of either one of these last two equations, say the second. The relation $\partial g(t, \xi)/\partial \xi = N(t, \xi)$ holds if and only if

$$g(t, \xi) = \int N(t, \xi) \, d\xi = \int \frac{t^2 + 1}{\xi} \, d\xi,$$

an indefinite integral with respect to ξ for each *fixed* t. Thus we must have

$$g(t, \xi) = (t^2 + 1) \ln |\xi| + h(t),$$

where h is some as yet undetermined function. But g must also satisfy $\partial g(t, \xi)/\partial t = M(t, \xi)$. This requires

$$2t \ln |\xi| + h'(t) = 2t \ln |\xi| - 1$$

or

$$h'(t) = -1.$$

Thus $h(t) = -t + c_1$ for some constant c_1. This constant can be combined with the constant in (3) so that $g(t, x) = c$ becomes

$$(t^2 + 1) \ln |x| - t = c.$$

If we now invoke the initial condition, $x = -e$ when $t = 2$, we find $c = 3$. Therefore, if a solution exists it must satisfy the implicit equation

$$(t^2 + 1) \ln |x(t)| - t - 3 = 0.$$

In this example it is easy to solve for $x(t)$. We find

$$x(t) = \pm e^{(t+3)/(t^2+1)}.$$

But, since $x(2) = -e$ and x is to be continuous, we must discard the plus sign and conclude that the only possible candidate for a solution is

$$x(t) = -e^{(t+3)/(t^2+1)}.$$

This function is defined and nonzero for all t, and is easily verified by sub-stitution (Problem 1) to be a solution of the original equations. Thus x is the unique solution. Actually the easiest way to make this verification is to ob-serve that, since x as defined above is everywhere negative and differentiable, one can use "implicit differentiation" of the equation

$$(t^2+1)\ln(-x) - t - 3 = 0.$$

Example 3. Solve

$$x' = -\frac{2x+3}{2x+2t} \quad \text{with} \quad x(0) = \sqrt{2},$$

requiring $x(t) \neq -t$. Let us rewrite the equation in the form

$$2x + 3 + (2x + 2t)x' = 0.$$

This is a special case of Eq. (1) with

$$M(t, \xi) = 2\xi + 3, \qquad N(t, \xi) = 2\xi + 2t;$$

and we find that the differential equation is exact because

$$\frac{\partial M}{\partial \xi} = 2 = \frac{\partial N}{\partial t} \quad \text{for all} \quad (t, \xi).$$

In seeking a function g such that $\partial g/\partial t = M$ and $\partial g/\partial \xi = N$, let us this time begin with the relation $\partial g/\partial t = M$. This yields

$$g(t, \xi) = \int M(t, \xi)\, dt = \int (2\xi + 3)\, dt = 2\xi t + 3t + h(\xi),$$

where h is undetermined. But g must also satisfy $\partial g/\partial \xi = N$. So

$$2t + h'(\xi) = 2\xi + 2t.$$

This gives $h'(\xi) = 2\xi$, or

$$h(\xi) = \xi^2 + c_1$$

for some constant c_1. Thus (3) becomes $2xt + 3t + x^2 = c$ or since $x(0) = \sqrt{2}$ implies $c = 2$,

$$x^2(t) + 2tx(t) + 3t - 2 = 0.$$

It is easy to solve this quadratic equation for $x(t)$. We find

$$x(t) = -t \pm \sqrt{t^2 - 3t + 2};$$

and the initial condition $x(0) = \sqrt{2}$ forces us to discard the minus sign. Thus the *sole candidate* for a solution is defined by

$$x(t) = -t + \sqrt{t^2 - 3t + 2}.$$

The reader should verify that this x is a continuous real-valued function on $(-\infty, 1)$ which satisfies the differential equation and the initial condition. (Problem 6.) For $1 < t < 2$ the expression for $x(t)$ would become complex-valued. Note that 1, the right-hand end point of the interval $(-\infty, 1)$, happens to be exactly the point at which $x(t) = -t$ and the original differential equation becomes undefined.

Sometimes it is easy and convenient to bypass some of the formality, and find the function g essentially by inspection.

Example 4. Suppose in Example 3 that we have already verified that $\partial M/\partial \xi = \partial N/\partial t$, so that we know a function g exists. Then we consider the equation

$$2x + 3 + (2x + 2t)x' = 0.$$

Any terms which are functions of t alone or are of the form x' times a function of x alone can be integrated immediately. Thus, in this example, the terms 3 and $2xx'$ cause no trouble. They are the derivatives of $3t$ and x^2, respectively. Group the remaining terms and rewrite the differential equation as

$$3 + 2xx' + (2x + 2tx') = 0.$$

We can now observe that the terms in parentheses form exactly the derivative of $2tx$. Thus,

$$g(t, x) = 3t + x^2 + 2tx = c,$$

as before.

Example 5. Solve

$$t + xx' = 0 \quad \text{with} \quad x(0) = 0.$$

Since $M(t, \xi) = t$ and $N(t, \xi) = \xi$ it follows at once that $\partial M/\partial \xi = 0 = \partial N/\partial t$, so that the differential equation is exact. It is also easy to integrate and obtain

$$g(t, x) = \frac{t^2}{2} + \frac{x^2}{2} = c.$$

But now the initial condition gives $c = 0$ so that our only solution candidates must satisfy

$$x^2(t) = -t^2.$$

Since this equation has no (real) solutions anywhere except at $t = 0$, we conclude that our original problem has no solution.

By way of further examples we should note that every example and most of the problems in Sections 2 and 3 used the notion of exactness:

The linear equation 2-(1) is not exact, in general. But after multiplication by the appropriate integrating factor, $e^{\int a(t) dt}$, we get Eq. (1) of this section with

$$M(t, \xi) = e^{\int a(t) dt}[a(t)\xi - h(t)]$$

$$N(t, \xi) = e^{\int a(t) dt}.$$

The resulting equation *is* exact since $\partial M/\partial \xi = a(t)e^{\int a(t) dt} = \partial N/\partial t$.

Equation 3-(1), with variables separated, is exact since $\partial M/\partial \xi = 0 = \partial N/\partial t$.

PROBLEMS

1. Verify that $x(t) = -\exp[(t + 3)/(t^2 + 1)]$ does define a solution of Example 2.
2. Repeat Problem 1 using another method.
3. Construct the function g in Example 2 by starting with the relation $\partial g/\partial t = M$.
4. Construct the function g in Example 2 by grouping and then integrating by inspection as illustrated in Example 4.
5. Construct the function g in Example 3 by starting with the relation $\partial g/\partial \xi = N$.
6. Verify that the function x obtained as the candidate for a solution in Example 3 is continuous and real-valued on $(-\infty, 1]$, and that on $(-\infty, 1)$ it is a solution.

In Problems 7 and 9 show that the differential equation is exact (on some suitable rectangle). Then find the implicit equation, $g(t, x) = c$, which every solution must satisfy.

7. $e^x - \sin t + te^x x' = 0$.
8. Solve Problem 7 by another method.
9. $2te^x + x \cos t - 6t^2 + (t^2 e^x - e^x + \sin t)x' = 0$, using the method illustrated in Example 2 or 3.
10. Solve Problem 9 by grouping and then integrating by inspection as illustrated in Example 4.
11. In each of the following test the differential equation for exactness. For each exact equation find the implicit equation for solutions.
    (a) $xx' - t + x^2 = 0$.
    (b) $\sin t \sin x - x' \cos t \cos x + \tan t = 0$.
    (c) $(\cos t + 2tx + 1)x' + 1 + x^2 - x \sin t = 0$.
    (d) $x^2 + 1 + (2tx + 2)x' = 0$.
    (e) $x + (t + t^2 x)x' = 0$.
    (f) $(3t + 6)x^{1/2}x' + 2x^{3/2} - 2t = 0$.
    (g) $(t + x)x' + x - t = 0$.
12. Prove that if M and N are continuously differentiable, then condition (5) is *sufficient* to assure exactness of Eq. (1) on the *rectangle* $D = (\alpha, \beta) \times (\gamma, \delta)$.
13. Find two mistakes in the following argument. The equation $x' = 3x^{2/3}$ can be rewritten as $1 - \frac{1}{3}x^{-2/3}x' = 0$ which is an exact equation. Thus every solution

must satisfy $t - x^{1/3} = c$. In particular, if $x(0) = 0$, then we must have $t - x^{1/3} = 0$ or $x(t) = t^3$. [Clearly *something* is wrong since $x(t) \equiv 0$ is also a solution of $x' = 3x^{2/3}$ with $x(0) = 0$.]

When, in the following problems, we say "solve" a certain differential equation with specified initial condition this means find a solution (if one exists) on some interval J and determine whether or not it is unique.

14. Solve the differential equation of Problem 7 with $x(\pi) = 0$.
15. Solve the differential equation of Problem 11(f) with $x(-1) = 1$.
16. Solve the differential equation of Problem 11(f) with $x(2) = -1$.
*17. Solve the differential equation of Problem 11(d) with $x(1) = 1$. Be careful at $t = 0$.
*18. Solve the differential equation of Problem 9 with $x(0) = 0$. In this case you will find that the implicit equation for $x(t)$ is not solvable in terms of elementary functions. One then turns to the "implicit function theorem," which is proved in texts on advanced calculus. A statement of this theorem is given in Appendix 2, Theorem E.

*5. INTEGRATING FACTORS

Sometimes a differential equation in the form

$$M(t, x) + N(t, x)x' = 0 \tag{1}$$

which is not exact can be transformed into an exact equation through multiplication by an appropriate "integrating factor," say $\mu(t, x)$. An *integrating factor* for Eq. (1) is a (nonzero) function defined on some rectangle such that the new equation

$$\mu(t, x)M(t, x) + \mu(t, x)N(t, x)x' = 0 \tag{2}$$

is exact. Thus we want to find some continuously differentiable function μ such that

$$D_2(\mu M) = D_1(\mu N). \tag{3}$$

For example, we use the integrating factor $\mu(t) = \exp \int a(t)\, dt$ to transform the nonexact Eq. 2-(1) into the exact Eq. 2-(4). And we multiply Eq. 3-(1') by the integrating factor $\mu(x) = 1/h(x)$ to get the exact Eq. 3-(1). In each of these special cases the integrating factor is a function of only one variable. However, these are very special cases of Eq. (1) and we should not expect, in general, the existence of such a special type of integrating factor.

Let us try to determine the circumstances under which Eq. (1) does have an integrating factor depending on only one variable:

Suppose $\mu(t)$ satisfies condition (3). This becomes

$$\mu(t)\frac{\partial}{\partial \xi}M(t, \xi) = \mu'(t)N(t, \xi) + \mu(t)\frac{\partial}{\partial t}N(t, \xi)$$

or, assuming $\mu(t) \neq 0$ and $N(t, \xi) \neq 0$,

$$\frac{\mu'(t)}{\mu(t)} = \frac{1}{N(t, \xi)} \left[\frac{\partial}{\partial \xi} M(t, \xi) - \frac{\partial}{\partial t} N(t, \xi) \right]. \tag{4}$$

This is a differential equation for the unknown integrating factor, μ. And it can be solved *if* the right-hand side, $(D_2 M - D_1 N)/N$, turns out to be a continuous function of t alone. It cannot be solved if the right-hand side depends on ξ.

Similarly one finds (Problem 1) that Eq. (1) can be made exact by use of a continuously differentiable integrating factor $\mu(x)$ if and only if

$$\frac{\mu'(\xi)}{\mu(\xi)} = \frac{1}{M(t, \xi)} \left[\frac{\partial}{\partial t} N(t, \xi) - \frac{\partial}{\partial \xi} M(t, \xi) \right], \tag{5}$$

that is, if and only if $(D_1 N - D_2 M)/M$ is a continuous function of ξ alone.

The reader is not advised to memorize conditions (4) and (5). It is a simple matter to rederive them in any specific case.

Example 1. Solve

$$e^x + te^x - 1 + te^x x' = 0.$$

If we seek an integrating factor $\mu(t)$, then

$$\mu(t)M(t, \xi) = \mu(t)(e^\xi + te^\xi - 1) \quad \text{and} \quad \mu(t)N(t, \xi) = \mu(t)te^\xi$$

and condition (3) for exactness of Eq. (2) becomes

$$\mu(t)(e^\xi + te^\xi) = \mu'(t)te^\xi + \mu(t)e^\xi$$

or

$$\frac{\mu'(t)}{\mu(t)} = 1.$$

We can therefore take $\mu(t) = e^t$ and be assured that the equation

$$e^t e^x + e^t te^x - e^t + e^t te^x x' = 0$$

is exact. Using any of the methods in Section 4 this yields (Problem 2)

$$te^t e^x - e^t = c,$$

an equation which must be satisfied by every possible solution.

If, on the other hand, we had tried to find an integrating factor of the form $\mu(x)$ condition (3) would have become

$$\mu'(\xi)(e^\xi + te^\xi - 1) + \mu(\xi)(e^\xi + te^\xi) = \mu(\xi)e^\xi$$

or

$$\frac{\mu'(\xi)}{\mu(\xi)} = \frac{-te^\xi}{e^\xi + te^\xi - 1}.$$

Since the right-hand side cannot be expressed as a function of ξ alone this equation cannot be solved for $\mu(\xi)$. We conclude that, for this example, there can be no integrating factor which depends on x alone.

If it turns out that a particular differential equation has no integrating factor of the form $\mu(t)$ and no integrating factor of the form $\mu(x)$, this does not necessarily mean that it has no integrating factor at all. In certain other cases it will be possible to find, by judicious inspection, an integrating factor $\mu(t, x)$.

Example 2. Consider the equation of Problem 4-11(e),

$$x + (t + t^2x)x' = 0$$

This equation is not exact since Eq. 4-(5) does not hold. Indeed

$$\frac{\partial}{\partial \xi} M(t, \xi) = 1 \neq 1 + 2t\xi = \frac{\partial}{\partial t} N(t, \xi).$$

This last calculation shows that the differential equation would be exact if the term t^2xx' were not present. In fact, the equation can be rewritten as

$$\frac{d}{dt}(tx) + t^2xx' = 0.$$

But in this form we can easily recognize that division by $(tx)^2$ yields an exact equation, that is, $(tx)^{-2}$ is an integrating factor. Indeed, assuming $tx \neq 0$, we have

$$\frac{1}{(tx)^2}\frac{d}{dt}(tx) + \frac{1}{x}x' = 0$$

so that

$$\frac{d}{dt}\left(-\frac{1}{tx} + \ln|x|\right) = 0.$$

Thus, x can be a solution with $tx \neq 0$ only if

$$-\frac{1}{tx(t)} + \ln|x(t)| = c.$$

The specification of an initial condition $x(t_0) = x_0$ where $t_0 \neq 0$ and $x_0 \neq 0$ would determine c. In this case it is not possible to solve explicitly for x in terms of elementary functions. The interested reader who applies the implicit function theorem to complete the solution may find the result somewhat surprising. (Problem 9.)

Sometimes it is worthwhile to seek an integrating factor of the form $\mu(t, x) = t^mx^n$, where m and n are constants.

Example 3. Consider again the equation

$$x + (t + t^2 x)x' = 0.$$

We shall *try* to find an integrating factor of the form $\mu(t, x) = t^m x^n$, so that

$$\mu(t, \xi)M(t, \xi) = t^m \xi^{n+1} \quad \text{and} \quad \mu(t, \xi)N(t, \xi) = t^{m+1}\xi^n + t^{m+2}\xi^{n+1}.$$

Thus by condition (3), we must have

$$(n + 1)t^m \xi^n = (m + 1)t^m \xi^n + (m + 2)t^{m+1}\xi^{n+1}.$$

Now let us *try* to satisfy this equation by requiring that all terms involving any given powers of t and ξ should cancel. Thus, considering the terms in $t^m \xi^n$ we ask that

$$(n + 1)t^m \xi^n = (m + 1)t^m \xi^n$$

and considering the terms in $t^{m+1}\xi^{n+1}$ we try also to satisfy

$$0 = (m + 2)t^{m+1}\xi^{n+1}.$$

These two conditions are met if

$$n + 1 = m + 1 \quad \text{and} \quad 0 = m + 2,$$

in other words, $m = -2$ and $n = -2$. This leads, once again, to $\mu(t, x) = (tx)^{-2}$ as an integrating factor.

Note that we have not answered the two natural questions: When does an integrating factor exist for Eq. (1)? How can an integrating factor be found in general? We have only given a few tricks which might work in special cases. In other cases it may be possible to find integrating factors by perseverance, but this is *not* one of the major approaches to the study of differential equations.

PROBLEMS

1. Derive Eq. (5) for an integrating factor which is a function of x alone.
2. Verify the implicit equation $te^t e^x - e^t = c$ obtained in Example 1; then solve for $x(t)$.
3. What can you say about existence and uniqueness of the solution x in Example 1 if there is also specified the initial condition (a) $x(0) = 1$? (b) $x(-1) = \ln \dfrac{e - 2}{2}$?
4. If in Example 2 we impose the initial condition $x(1) = 0$, it is clear by inspection of the differential equation, $x + (t + t^2 x)x' = 0$, that a solution is $x(t) \equiv 0$ for all t. But this solution does not satisfy the implicit equation $-1/(tx) + \ln |x| = c$. Why is there no contradiction?
5. Find an integrating factor μ for each of the following differential equations. Then find all solutions of the differential equation either explicitly or implicitly.

(a) $t^2x^2 - tx + (t^2 + t^3x)x' = 0$. [*Hint:* Try $\mu(t, x) = t^m x^n$.]
(b) $e^t + x - 2 + x' = 0$.
(c) $x + (t - tx^2)x' = 0$.

(d) $x' + \dfrac{x}{t} = \sec(tx)$. [*Hint:* Since the combination tx occurs as the argument of

sec and since $x' + \dfrac{x}{t} = \dfrac{1}{t}\dfrac{d}{dt}(tx)$, try to find as an integrating factor a func-

tion of tx.]
(e) $x^2 - tx^2 - 1 + 2txx' = 0$. [*Hint:* Try the procedure used in the beginning
of Example 2.]
(f) $\cos t + (2x \sin t + 2x)x' = 0$.
(g) $tx' - x = x^2$.

(h) $tx' = x + t^3 + tx^2$. $\left[Hint:$ Note that $\dfrac{d}{dt}$ Arctan $\dfrac{x}{t} = \dfrac{tx' - x}{t^2 + x^2}.\right]$

*6. Solve the equation of Problem 5(d) with $x(1) = 7\pi/6$.
7. Convince yourself that it is relatively easy to pose and then solve difficult equations by finding an equation which can be solved using the integrating factor
$\mu(t, x) = e^{x \sin t}$. (The key is to start with the answer.)
8. Show that an equation can have many integrating factors by considering
Example 2. Let v be a given continuously differentiable function defined on **R**.

Then verify that $U(t, x) = v\left(-\dfrac{1}{tx} + \ln|x|\right)(tx)^{-2}$ is another integrating factor

for Example 2.
*9. Apply the implicit function theorem to complete the solution of Example 2
when $x(t_0) = x_0$ is also specified, with $t_0 \neq 0$ and $x_0 \neq 0$. What happens if
$x(\tfrac{1}{2}) = -2$?

6. SPECIAL SUBSTITUTIONS FOR FIRST-ORDER EQUATIONS

Many differential equations can be solved, or at least simplified, by the use
of some substitution or change of variables selected for the particular equation. Two such substitutions which have become recognized as "methods"
are described below.

Bernoulli's Equation

The equation

$$x' + a(t)x = h(t)x^n, \tag{1}$$

where $n \neq 0$ or 1, is nonlinear and is called Bernoulli's equation. (The constant n need not be an integer.) Let us assume that a and h are continuous on
some interval J. By means of a special substitution, we can transform Eq. (1)
into a linear equation. Assuming $x(t) > 0$, multiply through by x^{-n} to obtain

$$x^{-n}x' + a(t)x^{-n+1} = h(t).$$

Then, defining a new unknown function $y = x^{-n+1}$, we recognize the differential equation to be linear in y:

$$\frac{1}{1-n} y' + a(t)y = h(t). \tag{2}$$

If an initial condition $x(t_0) = x_0 > 0$ is given for some t_0 in J, we take $y(t_0) = y_0 = x_0^{-n+1}$. The new Eq. (2) can be solved for y by the method given for linear equations in Section 2. After finding y it is a simple matter to recover x. But some care is needed in determining how far the solution is valid:

Note that if $x_0 > 0$, then y_0 is well defined and it follows from Section 2 that the solution y of Eq. (2) is continuous and well defined on J. However, it is entirely possible that $y(t_1) = 0$ at some t_1. If that happens *and* if $n > 1$, then $x(t)$ must $\to \infty$ as $t \to t_1$, so that x cannot be continued to the point t_1. Thus even though y exists on all of J, x does not. See Problem 5.

We remark that when n is an integer, and in certain other cases, we can also allow $x_0 < 0$. The question of what to do in case $x_0 = 0$ will be treated in a problem at the end of Section 8.

Example 1 (A Model for Population Growth). The simplest natural mathematical model for the growth of a population, say of an isolated colony of insects or other creatures, assumes a rate of growth proportional to the existing population. Thus, if $N(t)$ is the population at time t, we might assume

$$N'(t) = kN(t),$$

where the constant $k > 0$ is a coefficient of growth (incorporating both birth and death rates). The solution of this simple linear equation, with the initial condition

$$N(0) = N_0 > 0$$

is found as in Section 1 or 2 to be

$$N(t) = N_0 e^{kt}.$$

This represents unbounded exponential growth as t increases.

Actually we know that such an isolated colony does not continue to increase exponentially, perhaps because of shortage of food or because of overcrowding. So let us try to find a more realistic model. Let us assume that the coefficient of growth is not a constant but rather a function that decreases to zero when $N(t)$ reaches a certain critical value P, and even becomes negative if $N(t)$ exceeds P. To make this more definite, but not too difficult, let us assume a coefficient of growth $k(1 - N(t)/P)$. Then the differential equation becomes

$$N' = kN - \frac{k}{P} N^2.$$

This is a Bernoulli equation with $n = 2$. We subtract kN from both sides and then divide through by N^2 to get

$$\frac{1}{N^2} N' - \frac{k}{N} = -\frac{k}{P}.$$

Now introduce $y = 1/N$ to find

$$y' + ky = \frac{k}{P}.$$

The method of Section 2 gives

$$\frac{d}{dt}(ye^{kt}) = \frac{k}{P} e^{kt},$$

from which we find

$$ye^{kt} - y_0 = \frac{1}{P}(e^{kt} - 1),$$

where $y_0 = y(0) = 1/N_0$. This gives

$$\frac{1}{N(t)} = y(t) = e^{-kt}\left(\frac{1}{N_0} + \frac{1}{P} e^{kt} - \frac{1}{P}\right).$$

Thus

$$N(t) = \frac{N_0 e^{kt}}{1 + (N_0/P)(e^{kt} - 1)}.$$

The reversibility of all steps guarantees that this candidate for a solution really *is* the solution.

If N_0/P is small compared to 1 and t is close to 0, the solution is like $N_0 e^{kt}$, as one might expect. But as $t \to \infty$, $N(t) \to P$ regardless of the size of N_0. Thus, in all cases, $N(t)$ approaches the "equilibrium value" of P.

The solution curves are sketched in Figure 1 for $N_0 = P/10$, $P/2$, P, and $3P/2$.

In a case like this, where all solutions are asymptotic as $t \to \infty$ to a particular solution $N(t) \equiv P$, it is customary to say the solution $N(t) \equiv P$ is "asymptotically stable."

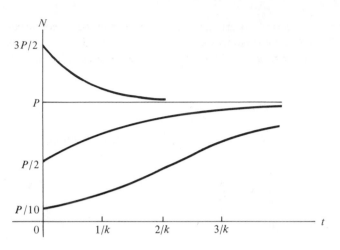

Homogeneous Coefficients

Consider an equation of the form

$$x' = h\left(\frac{x}{t}\right),\tag{3}$$

where h is a given continuous function.

This is an equation of the form $x' = f(t, x)$ in which the right-hand side depends on t and x only in the combination x/t (or t/x). Examples would include

$$x' = \frac{x}{t} - 2\frac{x^3}{t^3}\quad\text{and}\quad x' = \tan\frac{x}{t}.$$

To solve Eq. (3) it is useful to introduce a new unknown function defined by $y = x/t$. Then we find that y must satisfy the differential equation

$$y' = \frac{x'}{t} - \frac{x}{t^2} = \frac{1}{t}[h(y) - y]\quad\text{for}\quad t \neq 0.\tag{4}$$

But Eq. (4) is separable, provided $h(y(t)) - y(t) \neq 0$, and so we can apply the method of Section 3 to find y. (Of course, if $h(\xi) \equiv \xi$, Eq. (4) is especially simple.) Then $x(t) = ty(t)$.

Now consider a differential equation of the form

$$M(t, x) + N(t, x)x' = 0,\tag{5}$$

and assume that for some real number k

$$M(pt, p\xi) = p^k M(t, \xi)\quad\text{and}\quad N(pt, p\xi) = p^k N(t, \xi)\tag{6}$$

for all $p \geq 0$. We then say that Eq. (5) has *homogeneous coefficients of the same degree*, k. In this case, and assuming $t > 0$ and $N(t, x) \neq 0$, we can rewrite Eq. (5) in the form

$$x' = -\frac{M(t, x)}{N(t, x)} = -\frac{t^k M(1, x/t)}{t^k N(1, x/t)} = -\frac{M(1, x/t)}{N(1, x/t)} = h\left(\frac{x}{t}\right),$$

Eq. (3) again. Thus conditions (6), if satisfied, permit us to rewrite Eq. (5) as Eq. (3).

One can often treat Eq. (5) similarly for $t < 0$.

Example 2. The equation

$$t^2 x - 2x^3 - t^3 x' = 0$$

satisfies conditions (6), with $k = 3$. Indeed we can rewrite this equation as

$$x' = \frac{x}{t} - 2\left(\frac{x}{t}\right)^3.$$

Defining $y - x/t$ one then finds

$$y' = -\frac{2}{t} y^3$$

which has solution candidates

$$y(t) = 0 \quad \text{and} \quad y(t) = \pm(c + 4 \ln |t|)^{-1/2} \quad \text{for} \quad t \neq 0.$$

Hence potential solutions of the original equation include

$$x(t) = 0 \quad \text{and} \quad x(t) = \pm t(c + 4 \ln |t|)^{-1/2}.$$

These should be tested by substitution into the original differential equation.

PROBLEMS

For Problems 1 through 6, find either explicit or implicit equations for solution candidates. Use the methods of this section unless otherwise specified.

1. $xx' - t + x^2 = 0$. Same as Problem 4-11(a).
2. Solve Problem 1 directly using an integrating factor.
3. $(t + x)x' + x - t = 0$. Same as Problem 4-11(g).
4. $(3t + 6)x^{1/2}x' + 2x^{3/2} - 2t = 0$. Same as Problem 4-11(f).
5. $x' = x/t + tx^3$ with $x(\frac{1}{2}) = \sqrt{\frac{8}{3}}$. Where is the solution valid?
6. $tx' = x + (t^2 - x^2)^{1/2}$.
7. Solve $N' = kN - (k/P)N^2$, the equation of Example 1, as a separable equation.
8. Prove that the solutions of $N' = kN - (k/P)N^2$ can never actually equal P unless $N_0 = P$.
9. By computing N'', verify the general shape of the solution curves in Figure 1.

10. Consider the equation

$$x' = f\left(\frac{a_1 t + b_1 x + c_1}{a_2 t + b_2 x + c_2}\right),$$

where a_2, b_2, and c_2 are not all zero.

(a) Show that if $a_1 b_2 = a_2 b_1$, then the substitution $v \equiv a_2 t + b_2 x + c_2$ (in case either a_2 or $b_2 \neq 0$) or $v \equiv a_1 t + b_1 x + c_1$ (in case a_1 or $b_1 \neq 0$) leads to a separable equation for v. Why do we not worry about the case $a_1 = a_2 = b_1 = b_2 = 0$?

(b) If $a_1 b_2 \neq a_2 b_1$ we can find constants h and k such that $a_1 h + b_1 k + c_1 = 0$ and $a_2 h + b_2 k + c_2 = 0$. Show that by introducing $s = t - h$ and $z(s) = x(s + h) - k$ we obtain an equation with homogeneous coefficients.

11. Find substitutions which will reduce each of the following equations to one of the cases already discussed.

(a) $x' = -\left(\dfrac{t + 5x + 1}{t + 9x + 3}\right)^2$.

(b) $x' = \dfrac{t - 5x + 1}{20x - 4t + 4}$.

12. Solve

$$x' = \frac{1}{t + x} \quad \text{with} \quad x(1) = 0.$$

13. Solve

$$x' = (t + x - 1)^2 \quad \text{with} \quad x(0) = 1.$$

14. Solve $x' - 2x = 2x^{1/2}$, where $x(t) > 0$.

15. Using the general solution obtained in Problem 14, solve $x' - 2x = 2x^{1/2}$ with $x(0) = 4$. Be careful. You may at first find *two* solution candidates.

16. If we modify the Bernoulli equation with $n = 2$ by adding a function of t, we get the *Ricatti equation*

$$x' + f(t)x = h(t)x^2 + g(t).$$

If some particular solution x_1 is known and if x is an arbitrary solution, define $y = x - x_1$ and show that y satisfies the Bernoulli equation

$$y' + [f(t) - 2x_1(t)h(t)]y = h(t)y^2.$$

17. Use Problem 16 to solve $x' + x^2 = -1/4t^2$ with $x(1) = 1$. [*Hint:* $x_1(t) = 1/2t$ is a particular solution of the differential equation.]

7. FIRST-ORDER METHODS FOR CERTAIN HIGHER-ORDER EQUATIONS

Some special cases of ordinary differential equations of second order or higher can be solved, or at least simplified, by the methods which we have already considered for first-order equations.

Dependent Variable Missing

Consider an nth-order equation of the form

$$f(t, x', x'', \ldots, x^{(n)}) = 0 \tag{1}$$

which does not involve the unknown function or "dependent variable" x but only its derivatives. Then the substitution $v = x'$ reduces (1) to an equation of order $n - 1$, $f(t, v, v', \ldots, v^{(n-1)}) = 0$.

Example 1. The second-order equation

$$x'' = g - \frac{b}{m}(x')^2$$

is one possible model for the free fall of an object through a resisting medium. Here $x(t)$ is the distance fallen at time t, g is the acceleration of gravity, m is the mass of the object, and b is a coefficient of resistance—say due to air friction. The three constants g, m, and b are all positive. Defining $v = x'$ we obtain the separable first-order equation

$$v' = g - \frac{b}{m}v^2.$$

With $a = \sqrt{mg/b}$, we obtain

$$\frac{v'}{a^2 - v^2} = \frac{g}{a^2},$$

as long as $|v(t)| < a$. Rewriting the left-hand side in partial fractions gives

$$\frac{v'}{a + v} + \frac{v'}{a - v} = \frac{2g}{a}.$$

Now letting the initial velocity be $v(t_0) = v_0$, where $|v_0| < a$, one obtains

$$\ln \frac{a + v}{a + v_0} - \ln \frac{a - v}{a - v_0} = \frac{2g(t - t_0)}{a},$$

or

$$\frac{a + v}{a + v_0} = \frac{a - v}{a - v_0} e^{2g(t - t_0)/a}.$$

This yields

$$v(t) = a \frac{(a + v_0)e^{2g(t - t_0)/a} - a + v_0}{(a + v_0)e^{2g(t - t_0)/a} + a - v_0}.$$

In order to compute $x(t)$, one integrates $x'(t) = v(t)$ and uses another initial condition, say $x(t_0) = x_0$.

In this example of a second-order equation we have specified two initial conditions, namely $x(t_0) = x_0$ and $x'(t_0) = v_0$, in order to determine the solution. It is typical, as we shall see later, that an nth-order equation should be considered together with n initial conditions if the solution is to be uniquely determined.

The Equation $x'' = f(x)$

A second-order equation of the form

$$x'' = f(x) \tag{2}$$

can often be solved by the following trick. Multiply both sides of (2) by x' to get

$$x'x'' - f(x)x' = 0. \tag{3}$$

This is now an "exact" equation since it is equivalent to

$$\frac{d}{dt}\left[\frac{x'^2}{2} - g(x)\right] = 0,$$

where g is any indefinite integral of f. Thus $x'^2/2 - g(x) = c$, or

$$x' = \pm[2g(x) + 2c]^{1/2}. \tag{4}$$

If one can solve the separable Eq. (4) (with an appropriate choice for the $+$ or $-$ sign), then perhaps x will also be a solution of Eq. (2).

Note that Eqs. (2) and (3) are equivalent only so long as $x'(t) \neq 0$. In this case x' can be considered as an integrating factor for Eq. (2).

Example 2 (Escape Velocity). Let us apply this trick to the simplified equation for a "free-falling" object or projectile traveling radially toward or away from the center of the earth—for example, a burned-out rocket which had been traveling radially outward before burnout. If we ignore air resistance and the gravitational attraction of all celestial bodies other than the earth, we adopt the differential equation

$$x'' = -\frac{kM}{x^2}.$$

Here $x(t)$ is the distance from the center of the earth at time t, M is the mass of the earth, and k is the constant of gravitation. If the radius of the earth is R and the acceleration of gravity at the earth's surface is g, then $kM = gR^2$.

Let us take as initial conditions the position and velocity at $t = t_0$, say $x(t_0) = x_0$ and $x'(t_0) = v_0$. If $x_0 > R$, $v_0 \neq 0$, and if a solution exists, we can be sure that $x(t) > R$ and $x'(t) \neq 0$ in some open interval J containing t_0.

Why? As long as $x'(t) \neq 0$ we can multiply the differential equation by x' to find

$$x'x'' = -\frac{gR^2}{x^2} x'$$

or

$$\frac{1}{2} x'^2(t) - \frac{1}{2} v_0^2 = \frac{gR^2}{x(t)} - \frac{gR^2}{x_0}. \tag{5}$$

Thus,

$$x'(t) = (\text{sgn } v_0)\left[v_0^2 + \frac{2gR^2}{x(t)} - \frac{2gR^2}{x_0} \right]^{1/2},$$

where sgn v_0 (read "signum v_0" or "sign of v_0") is $v_0/|v_0|$.

Without any further integration, we can use the above to compute "escape velocities." For example, we can determine the value of $v_0 > 0$ required in order that a projectile fired, say, radially outward from the surface of the earth ($x_0 = R$) should continue traveling outward indefinitely. We require that $x'(t)$ remain positive forever, which will be the case if $v_0^2 - 2gR \geq 0$. Taking $g - 9.81$ m/sec^2 (or 0.00981 km/sec^2) and $R = 6370$ km we find the minimum escape velocity from the earth's surface, $v_0 = \sqrt{2gR}$, to be 11.2 km/sec (or about 7 miles/sec).

In order to continue on to determine $x(t)$ we should solve the differential equation (5), which becomes for $v_0 > 0$

$$x'(t) = \left[v_0^2 + \frac{2gR^2}{x(t)} - \frac{2gR^2}{x_0} \right]^{1/2} \quad \text{with} \quad x(t_0) = x_0 > R.$$

This differential equation is immediately seen to be separable. It has a unique solution, although x is not expressible as a simple function. (Problem 4.)

PROBLEMS

Find the general solutions of the equations in Problems 1, 2, and 3.
1. $tx'' + x' = 0$.
2. $x''' + x'' = 0$.
3. $x'' + tx' = t$.
4. Complete the arguments to show that the differential equation of Example 2, $x'' = -gR^2/x^2$, with initial conditions $x(t_0) = x_0 \neq 0$ and $x'(t_0) = v_0$ has a unique solution.
5. Using a method of this section, find a solution of $x'' + \omega^2 x - 0$ with $x(0) = 0$ and $x'(0) = 2$.
6. An electron (assumed to be a point charge) is fired from a point $x_0 > 0$ (at $t = 0$) with velocity $v_0 < 0$ along the x-axis toward another charge of like sign fixed at the origin. If x_0 is large enough and $|v_0|$ is small compared to the speed of light, the equation of motion can be considered to be $x'' = K/x^2$ for some

constant $K > 0$. Prove that the moving electron comes to rest when $x(t) = Kx_0/(K + \frac{1}{2}x_0 v_0^2)$; and thereafter accelerates away from the origin.

7. The position x of an electron moving in a uniform electric field along the x-axis may be assumed to satisfy the equation

$$\frac{x''}{(1 - x'^2/c^2)^{3/2}} = K,$$

where $K \neq 0$ is a constant and $c > 0$ is the speed of light. (We do not assume "small" velocity in this problem, but only $|x'(t)| < c$.) Determine the behavior of $x'(t)$ as $t \to \infty$. Your result should be independent of the initial conditions.

Chapter Two

UNIQUENESS FOR ORDINARY DIFFERENTIAL EQUATIONS

We have considered a variety of special cases in which one can explicitly determine a solution of a differential equation with initial condition(s) *and* conclude that the solution is unique. Many other special cases can also be handled exactly (see Kamke [8]).

But, in general, differential equations do *not* have simple solutions. That is, in general, one cannot hope to deduce that some particular simple function *is* a solution and *the only* solution of a given differential equation with initial condition(s). Then we seek instead to obtain information indirectly about the problem.

The two most basic questions to ask about a given problem are: "Does a solution exist?" and "Can there be more than one solution?"

The reader should note that, in most of the examples considered thus far, we have actually considered the second question first. In the present chapter we shall pursue this uniqueness question further, and shall find rather simple conditions which suffice to assure that a given problem has at most one solution. Later (in Section 28) we shall discuss existence of solutions of differential equations.

8. FIRST-ORDER SCALAR EQUATIONS

For most differential equations with initial conditions, including the examples which we have considered, it is not at all obvious in advance that one and only one solution will exist. It is certainly not obvious (nor even true) for the general first-order problem of the form

$$x' = f(t, x) \tag{1}$$

with initial condition

$$x(t_0) = x_0. \tag{2}$$

We shall sometimes refer to Eq. (1) as a "scalar" equation since it involves only one scalar-valued unknown function, x.

Equation (1) asserts that the rate of change of the unknown function x at the instant t will be determined if the *value* of x at t is known. Thus the problem of finding the unknown function would sound, to the uninitiated, like a hopeless chicken-and-egg problem: We would like to find x by integrating x' as computed from Eq. (1). But in order to compute x' from Eq. (1) we must first know x.

Nevertheless, we have already encountered many examples in which Eqs. (1) and (2) do determine a unique solution. Thus we forge on to seek some simple conditions on the equation itself which will assure that a unique solution is determined.

The information contained in Eq. (1) can be displayed graphically as a *direction field*: To each point (t, ξ) in the plane $\mathbf{R}^2$, or some appropriate subset of $\mathbf{R}^2$, we assign a number $f(t, \xi)$ which is thought of as the value of the slope at that point. For graphing purposes we draw a short line segment having slope $f(t, \xi)$ at the point (t, ξ). Then any solution (or "integral curve") x of Eq. (1) must be tangent to these segments at all points $(t, x(t))$. Figures 1, 2, and 3 indicate the direction fields and one solution for the equations $x' = tx$, $x' = x^{2/3}$, and $x' = x^{4/3}$, respectively. These types of equations are all familiar. Their solutions subject to the initial condition $x(0) = x_0$ are briefly reviewed below.

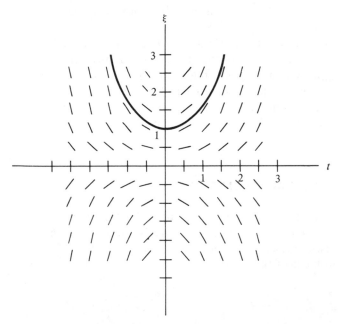

FIGURE 1 $f(t, \xi) = t\xi$

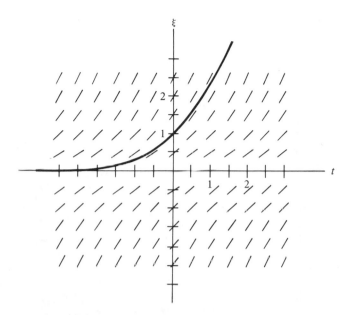

FIGURE 2 $f(t, \xi) = \xi^{2/3}$

Example 1 (Figure 1). The equation $x' = tx$ with $x(0) = x_0$ has a unique solution, defined by $x(t) = x_0 e^{t^2/2}$, for all t. The reader should verify this (using the method of Section 2), and verify that the curve shown in Figure 1 is a graph of the solution when $x_0 = 1$. (Problem 1.)

Example 2 (Figure 2). The equation $x' = x^{2/3}$ with $x(0) = x_0 > 0$ has a unique solution defined by

$$x(t) = \left(x_0^{1/3} + \frac{t}{3}\right)^3 \quad \text{as long as} \quad t > -3x_0^{1/3}.$$

The curve in Figure 2 is the graph of the solution when $x_0 = 1$. However, if $x(0) = 0$, the equation will have solutions defined by $x(t) = t^3/27$, $x(t) = 0$, *and* infinitely many others (valid for all t). These assertions should be proved, as Problem 2, using the method of Section 3 (see Example 3-5).

Example 3 (Figure 3). The equation $x' = x^{4/3}$ with $x(0) = x_0 > 0$ has a unique solution defined by

$$x(t) = \frac{1}{(x_0^{-1/3} - t/3)^3} \quad \text{as long as} \quad t < 3x_0^{-1/3}.$$

Using separation of variables the reader should derive this solution and then verify the fact that the curve shown in Figure 3 is the graph of the solution when $x_0 = 1$. (Problem 3.) In the case when $x(0) = 0$, a solution is clearly

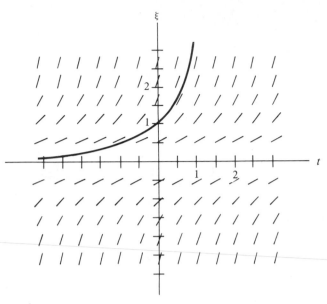

FIGURE 3 $f(t, \xi) = \xi^{4/3}$

seen to be given by $x(t) = 0$ for all t. However, it is not clear whether this solution is unique.

We call attention especially to Examples 2 and 3 in the case when $x(0) = 0$. We shall soon discover that the solution $x = 0$ *is* unique in the case of Example 3. Yet we know that the solution $x = 0$ in Example 2 is *not* unique. How could we possibly answer the uniqueness question by looking at either the differential equations involved or their direction fields (Figures 2 and 3)? Why are the results of Examples 2 and 3 so different while the differential equations involved are so similar?

The analogous general questions regarding existence of a solution of (1) need not bother us yet. For in the examples and problems presented in the first four chapters of this book we will be able to explicitly find solutions. In particular, we have explicitly found the zero solutions in Examples 2 and 3 above.

In the present section we introduce methods and notation which will be used again later, and prove a simple but useful "uniqueness theorem" for Eq. (1) with initial condition (2). More precisely, we shall show that a certain problem can have *at most one solution*.

Our proof will not exhibit a candidate for a solution nor will it even guarantee that a solution exists. However, having proved uniqueness for a particular problem, one then has the right to seek a solution by *any* means whatsoever—even a random guess or a dubious calculation. If we find *a* solution, then we have found *the* solution.

We consider Eq. (1) with initial condition (2). Let f be a given real-valued function defined on some "open set" $D \subset \mathbf{R}^2$, and assume that $(t_0, x_0) \in D$.

The term *open set* is defined in Appendix 1. However, the reader who is unfamiliar with this concept can think of D as an *open rectangle*, say $D = (\alpha, \beta) \times (\gamma, \delta)$. An open rectangle will generally be quite adequate, and the concept is simpler than that of a more general open set. In Examples 1, 2, and 3 one can take $D = \mathbf{R}^2$.

We can now state precisely what we mean by a solution, and by a unique solution.

DEFINITION

A *solution* of Eqs. (1) and (2) is a differentiable real-valued function x on an interval $J = (\alpha_1, \beta_1)$ such that

(i) $t_0 \in J$ and $x(t_0) = x_0$, and
(ii) for all t in J, $(t, x(t)) \in D$ and $x'(t) = f(t, x(t))$.

Note that if f is continuous (as will usually be the case), then it follows that x' must also be continuous. Why?

We shall frequently use the following fact: If f is continuous and x is a continuous function such that $(t, x(t)) \in D$ for all t in J, then x is a solution of Eqs. (1) and (2) on J if and only if

$$x(t) = x_0 + \int_{t_0}^{t} f(s, x(s))\, ds \quad \text{for all} \quad t \text{ in } J. \tag{3}$$

Verify this important assertion using the fundamental theorems of calculus. (Problem 6.)

A solution of Eqs. (1) and (2) has been defined as a function x on a certain interval J. Thus, if we change J, the domain of x, then strictly speaking, we are no longer talking about the same function. For instance, in Example 1, one solution is defined by

$$x(t) = x_0 e^{t^2/2} \quad \text{for} \quad -2 < t < 2,$$

and another is defined by

$$x(t) = x_0 e^{t^2/2} \quad \text{for} \quad -1 < t < \pi.$$

And yet we want to say the solution is "unique."

Actually there is no problem if we recall the definition mentioned in Section 1:

DEFINITION

A solution x of Eqs. (1) and (2) is said to be *unique* if, for every other solution $\tilde{x}$, $\tilde{x}(t) = x(t)$ wherever both solutions are defined.

If we do not know that a solution exists, we will say Eqs. (1) and (2) have *at most one solution* on an interval J if it can be shown that any two solutions (which might exist) would have to agree wherever both are defined on J.

Now, with an example for motivation, we shall lead up to the simple test for uniqueness which is the main goal of this section. This example looks more complicated, but it will actually turn out to be a little simpler than Example 3.

Example 4. Consider the first-order equation

$$x' = 2 \sin t \sin x \quad \text{with} \quad x(t_0) = x_0, \tag{4}$$

where t_0 and x_0 are given real numbers. For this special case of Eqs. (1) and (2) one can take $D = \mathbf{R}^2$.

The differential equation can be solved by separation of variables *so long as* sin $x(t)$ *does not become zero*. In case $x_0 = 0$, or $\pm \pi$, or $\pm 2\pi, \ldots$, although separation of variables is not applicable, we can see by inspection that $x(t) \equiv$ constant is a solution. But how can we decide in these cases whether or not there is also some other solution?

The calculations which follow will show that, regardless of the values of t_0 and x_0, the given problem (4), has at most one solution on *any* open interval J.

As a matter of fact the proof which we shall give works for more general equations. Let D be any open *rectangle*—not necessarily $\mathbf{R}^2$. Then we can treat Eq. (1) provided f is continuous and $D_2 f$ exists and is bounded on D. Thus we shall proceed to consider Eq. (1) assuming

$$|D_2 f(t, \xi)| = \left| \frac{\partial f}{\partial \xi} (t, \xi) \right| \le K \quad \text{on} \quad D. \tag{5}$$

Note that these conditions are satisfied for our particular example. Indeed (4) satisfies condition (5) with $K = 2$.

Suppose Eqs. (1) and (2) have two solutions x and $\tilde{x}$ on some interval $J = (\alpha_1, \beta_1)$. It follows from (3) that for all t in J,

$$x(t) = x_0 + \int_{t_0}^{t} f(s, x(s)) \, ds$$

and

$$\tilde{x}(t) = x_0 + \int_{t_0}^{t} f(s, \tilde{x}(s)) \, ds.$$

Subtracting these two equations we have

$$x(t) - \tilde{x}(t) = \int_{t_0}^{t} [f(s, x(s)) - f(s, \tilde{x}(s))] \, ds$$

for all t in J. From this it follows that

$$|x(t) - \tilde{x}(t)| \le \left| \int_{t_0}^{t} |f(s, x(s)) - f(s, \tilde{x}(s))| \, ds \right|.$$

Why do we keep the absolute value signs outside this last integral?

Now apply the mean value theorem (Theorem A2-A) to conclude that, for each fixed s, there is some number θ between $x(s)$ and $\tilde{x}(s)$ such that

$$f(s, x(s)) - f(s, \tilde{x}(s)) = D_2 f(s, \theta)[x(s) - \tilde{x}(s)].$$

Then, since $|D_2 f(s, \theta)| \leq K$ throughout D, it follows that

$$|f(s, x(s)) - f(s, \tilde{x}(s))| \leq K|x(s) - \tilde{x}(s)|.$$

Hence, for all t in J,

$$|x(t) - \tilde{x}(t)| \leq K \left| \int_{t_0}^{t} |x(s) - \tilde{x}(s)|\, ds \right|. \tag{6}$$

We shall complete the proof by showing, as a consequence of inequality (6), that $|x(t) - \tilde{x}(t)| = 0$ for all t in J. Then it will follow that $x = \tilde{x}$, that is, there cannot be two different solutions on J. The cases $t_0 \leq t < \beta_1$ and $\alpha_1 < t \leq t_0$ are treated separately:

For $t_0 \leq t < \beta_1$, define $Q(t) = K \int_{t_0}^{t} |x(s) - \tilde{x}(s)|\, ds$. Thus (6) is transformed into the linear "differential inequality"

$$Q'(t) = K|x(t) - \tilde{x}(t)| \leq K Q(t)$$

or

$$Q'(t) - K Q(t) \leq 0,$$

with $Q(t_0) = 0$. This inequality will be "solved" by the same method used for linear differential *equations* in Section 2. Multiply both sides by the (positive) integrating factor e^{-Kt} to obtain

$$\frac{d}{dt}[Q(t)e^{-Kt}] \leq 0.$$

Integration, or application of the mean value theorem, between t_0 and t then gives

$$Q(t)e^{-Kt} - Q(t_0)e^{-Kt_0} \leq 0.$$

Since $Q(t_0) = 0$ and $Q(t) \geq 0$, it follows that $Q(t) = 0$. Thus from (6),

$$|x(t) - \tilde{x}(t)| \leq Q(t) = 0 \quad \text{for} \quad t_0 \leq t < \beta_1.$$

This proves that

$$x(t) = \tilde{x}(t) \quad \text{for} \quad t_0 \leq t < \beta_1.$$

In case $\alpha_1 < t \leq t_0$, with $Q(t) \equiv K \int_{t_0}^{t} |x(s) - \tilde{x}(s)|\, ds$ as before, (6) gives

$$Q'(t) \leq -K Q(t),$$

or

$$Q'(t) + K Q(t) \leq 0,$$

with $Q(t_0) = 0$. Now multiply by e^{Kt} to obtain

$$\frac{d}{dt}[Q(t)e^{Kt}] \leq 0,$$

with $Q(t_0) = 0$. Thus, integrating from t to t_0 (the positive direction),

$$Q(t_0)e^{Kt_0} - Q(t)e^{Kt} \leq 0.$$

Then, since $Q(t_0) = 0$ and since in this case $Q(t) \leq 0$, it again follows that $Q(t) \equiv 0$, which proves that $x(t) = \tilde{x}(t)$ for $\alpha_1 < t \leq t_0$.

If we try to apply the above argument to prove uniqueness for Example 3 (or even Example 1), we run into difficulty. The trouble is that condition (5), the boundedness of $D_2 f$, fails. In Examples 1 and 3 we have, respectively,

$$D_2 f(t, \xi) = t \quad \text{and} \quad D_2 f(t, \xi) = \tfrac{4}{3}\xi^{1/3}.$$

In one $D_2 f$ is unbounded as $|t|$ becomes large, and in the other $D_2 f$ is unbounded as $|\xi|$ becomes large.

Actually this difficulty is not serious. We can overcome it by considering appropriate smaller open rectangles instead of the entire set $D = \mathbf{R}^2$. The following theorem uses this idea to remove condition (5) from the stated requirements.

THEOREM A

If f and $D_2 f$ are continuous on D and if $(t_0, x_0) \in D$, then Eqs. (1) and (2) have at most one solution on any open interval J containing t_0.

Remarks. This is probably the best known and most easily applied of all uniqueness theorems. It certainly establishes uniqueness of the solutions in Examples 1 and 3 and in Problem 3-10.

But why does Theorem A not also assert uniqueness for Example 2, which would of course be false? In Example 2, $f(t, \xi) = \xi^{2/3}$ which is continuous on $\mathbf{R}^2$. But $D_2 f$ is undefined at points $(t, 0)$, so it is certainly not continuous there.

*Proof of Theorem A. Suppose (for contradiction) that there are two solutions x and $\tilde{x}$ on (α_1, β_1) with $x \not\equiv \tilde{x}$. We must, of course, have $x(t_0) = x_0 = \tilde{x}(t_0)$. Let us assume $x(t) \neq \tilde{x}(t)$ for some t in (t_0, β_1). [The case of $x(t) \neq \tilde{x}(t)$ for some t in (α_1, t_0) is handled similarly. (Problem 13.)]

Let

$$t_1 = \inf \{t \in (t_0, \beta_1): x(t) \neq \tilde{x}(t)\},$$

where "inf" stands for infimum or greatest lower bound. Then, since x and $\tilde{x}$ are both continuous, it follows that $x(t_1) = \tilde{x}(t_1) \equiv x_1$. Also $t_1 < \beta_1$, since we have assumed $x(t) \neq \tilde{x}(t)$ for some t in (t_0, β_1).

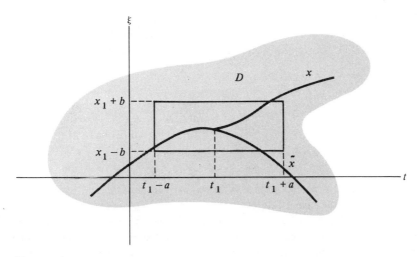

FIGURE 4

Now choose positive numbers a and b such that

$$A = [t_1 - a, t_1 + a] \times [x_1 - b, x_1 + b] \subset D.$$

Since $D_2 f$ is continuous on the closed bounded rectangle A, it follows from Theorem A2-G that

$$|D_2 f(t, \zeta)| \leq K \quad \text{for all} \quad (t, \zeta) \quad \text{in} \quad A,$$

for some K. Thus condition (5) is fulfilled on the new open set $D_* = (t_1 - a, t_1 + a) \times (x_1 - b, x_1 + b)$. (Figure 4.)

But, by continuity, the points $(t, x(t))$ and $(t, \tilde{x}(t))$ must remain in D_* on *some* open interval J_* about t_1. Thus we can apply the computation of Example 4 with (t_1, x_1) playing the role of (t_0, x_0). Specifically, we obtain condition (6) with t_1 in place of t_0. It follows that $x(t) = \tilde{x}(t)$ on J_*. But this contradicts the definition of t_1. ∎

In the proof of uniqueness we did not actually use the partial derivative $D_2 f$ itself. What we used was inequality (6) which is a consequence of the boundedness of $D_2 f$. It will sometimes be useful to have isolated out the actual condition on f which is used. This is called a " Lipschitz condition."

DEFINITION

Let $A \subset \mathbf{R}^2$. Then a function f is said to satisfy a *Lipschitz condition on A* if there exists a constant K (called a *Lipschitz constant*) such that

$$|f(t, \xi) - f(t, \tilde{\xi})| \leq K|\xi - \tilde{\xi}| \tag{7}$$

whenever (t, ξ) and $(t, \tilde{\xi}) \in A$. We shall also say f is *Lipschitzian on A*.

If condition (5) holds on A, then we certainly get the Lipschitz condition (7). On the other hand, a Lipschitz condition is slightly weaker than condition (5). (Problem 16.) Hence the following theorem is slightly stronger than Theorem A.

THEOREM B

Let f be continuous on D and Lipschitzian on each closed bounded rectangle $A \subset D$, and let $(t_0, x_0) \in D$. Then Eqs. (1) and (2) have at most one solution on any open interval J containing t_0.

Remarks. The Lipschitz constant K for f will, in general, be different for different closed bounded rectangles A.

The proof of Theorem B is virtually the same as that of Theorem A. Actually it is slightly simpler since in the proof of Theorem A we first had to essentially produce the Lipschitz condition from the properties of $D_2 f$ in order to get inequality (6).

PROBLEMS

1. Verify the assertions in Example 1.
2. Verify the assertions in Example 2.
3. Verify the assertions in Example 3.
4. Show that the equation $x' = x^{2/3}$, of Example 2, with initial condition $x(0) = 1$ has a solution for all t in $(-\infty, \infty)$, but the solution is not unique on this interval. On what interval (if any) is it unique?
5. Sketch the direction field for the equation $N' = kN - kN^2/P$ and compare with Figure 6-1.
6. Verify the assertion that a continuous function x on J such that $(t, x(t))$ remains in D satisfies Eqs. (1) and (2) if and only if it satisfies Eq. (3).
7. For each of the following, find an appropriate set D, and prove, *using the methods or theorems of this section*, that the problem has at most one solution in D.

    (a) $x' + \dfrac{1}{t} x = 3t$ with $x(2) = 3$ (Example 2-2).

    (b) $x' = (x + 1)\sin t$ with $x(t_0) = x_0$ (Problem 2-4).

    (c) $(\cos t)x' + (\sin t)x = 1$ with $x(0) = 1$ (Problem 2-11).

    (d) $x' = 1 + x^2$ with $x(0) = 1$ (Example 3-2).

    (e) $x^2x' + tx = x$ with $x(-1) = 2$ [Problem 3-1(d)].

    (f) $2t \ln|x| - 1 + \dfrac{t^2 + 1}{x} x' = 0$ with $x(2) = -e$ (Example 4-2).

    (g) $N' = kN - kN^2/P$, where k and P are psotive constants, with $N(0) = N_0$ arbitrary (Example 6-1).

    (h) $x' = (1 + t^3)x^{4/3}$ with $x(0) = 0$.

8. Show that if $x_0 \neq 0$ in Example 4, then $x(t)$ can never become zero.
*9. As noted in Problem 5-4, the equation $x + (t + t^2x)x' = 0$ with $x(1) = 0$ has the solution $x = 0$ for all t. Is that solution unique?

10. Show in detail why Theorem B is not contradicted by the nonuniqueness encountered in Example 2 with $x(0) = 0$.

11. By considering the equation $x' = 1 + x^{2/3}$ with $x(0) = 0$, show that a Lipschitz condition is not a necessary condition for uniqueness. [*Hint:* Apply the method of solution of Section 3.]

12. Consider Bernoulli's equation, $x' + a(t)x = h(t)x^n$, with $x(t_0) = 0$, and a and h continuous. Show that for $n = 2, 3, \ldots$ the only solution is $x = 0$. Also show that for $n = \frac{1}{3}$ the problem makes sense, but the solution may not be unique. (These are cases that were left unresolved in Section 6.)

*13. Complete the proof of Theorem A by treating the case $x(t) \neq \tilde{x}(t)$ for some t in (α_1, t_0). [*Hint:* Define $t_1 = \sup \{t \in (\alpha_1, t_0): x(t) \neq \tilde{x}(t)\}$, where "sup" stands for supremum or least upper bound.]

*14. Consider the differential equation $x' = x^{p/q}$ where p and q are positive integers with q odd. (This includes the equations of Examples 2 and 3 as special cases.)
  (a) If $x(0) = 0$, for what p/q is the solution $x = 0$ unique?
  (b) If $x(0) = x_0 > 0$, solve and determine where the solution is valid for various values of p/q.

*15. Complete the following alternate proof for Example 4. If x and $\tilde{x}$ are two solutions of Eqs. (1) and (2), let $v(t) = [x(t) - \tilde{x}(t)]^2$. Then

$$v'(t) = 2[x(t) - \tilde{x}(t)][f(t, x(t)) - f(t, \tilde{x}(t))] \leq 2Kv(t).$$

16. Show that the condition "$D_2 f$ exists and is continuous on A" is *not a necessary condition* for a Lipschitz condition. This will show that Theorem B is stronger than Theorem A. [*Hint:* Consider $f(t, \xi) = |\xi|$ on $\mathbf{R}^2$.]

*17. Let $f: D \to \mathbf{R}$ for some open set (or open rectangle) $D \subset \mathbf{R}^2$. Assume $D_2 f$ is unbounded somewhere on D. Then prove that f cannot be Lipschitzian on D.

9. SYSTEMS OF EQUATIONS (VECTOR SPACE NOTATION)

Frequently in applications one encounters problems involving more than one unknown function and more than one differential equation.

Example 1. The two electric currents, $x_1(t)$ and $x_2(t)$, in the circuit of Figure 1 satisfy a pair of simultaneous differential equations

$$2x_1' + 4(x_1 - x_2) = 10 \sin 2t,$$

$$2x_2' + 6x_2 - 4(x_1 - x_2) = 0.$$

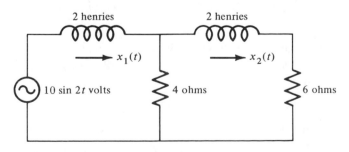

FIGURE 1

These differential equations can be rewritten in the form

$$x_1' = -2x_1 + 2x_2 + 5 \sin 2t,$$
$$x_2' = 2x_1 - 5x_2.$$

More generally, let us consider a system of n first-order equations involving n unknown functions, $x_1, \ldots, x_n$,

$$x_1' = f_1(t, x_1, \ldots, x_n),$$
$$\vdots$$
$$x_n' = f_n(t, x_1, \ldots, x_n),$$

where each f_i is a given real-valued function defined on some open set $D \subset \mathbf{R}^{1+n}$. Again you may assume for simplicity that D is an "open rectangle," say $D = (\alpha, \beta) \times (\gamma_1, \delta_1) \times \cdots \times (\gamma_n, \delta_n)$.

We can write the above system of n equations more briefly as

$$x_i' = f_i(t, x_1, \ldots, x_n) \quad \text{for} \quad i = 1, \ldots, n.$$

We shall consider this system together with n initial conditions

$$x_i(t_0) = x_{0i} \quad \text{for} \quad i = i, \ldots, n,$$

where $(t_0, x_{01}, \ldots, x_{0n})$ is a given point in D.

It is convenient to introduce vector notation and represent the n equations even more compactly as

$$\mathbf{x}' = \mathbf{f}(t, \mathbf{x}) \tag{1}$$

with initial conditions

$$\mathbf{x}(t_0) = \mathbf{x}_0. \tag{2}$$

The boldface symbols $\mathbf{x}_0$, $\mathbf{x}(t)$, $\mathbf{x}'(t)$, and $\mathbf{f}(t, \xi)$ now stand for vectors in $\mathbf{R}^n$ which we shall represent as columns

$$\begin{pmatrix} x_{01} \\ \cdot \\ \cdot \\ \cdot \\ x_{0n} \end{pmatrix}, \quad \begin{pmatrix} x_1(t) \\ \cdot \\ \cdot \\ \cdot \\ x_n(t) \end{pmatrix}, \quad \begin{pmatrix} x_1'(t) \\ \cdot \\ \cdot \\ \cdot \\ x_n'(t) \end{pmatrix}, \quad \text{and} \quad \begin{pmatrix} f_1(t, \xi) \\ \cdot \\ \cdot \\ \cdot \\ f_n(t, \xi) \end{pmatrix}, \quad \text{where} \quad \xi = \begin{pmatrix} \xi_1 \\ \cdot \\ \cdot \\ \cdot \\ \xi_n \end{pmatrix},$$

respectively.

To save space we will often write $\mathbf{x}_0 = \mathrm{col}\,(x_{01}, \ldots, x_{0n})$, $\mathbf{x}(t) = \mathrm{col}\,(x_1(t), \ldots, x_n(t))$, and so on. Thus, $\mathbf{f}(t, \xi) = \mathrm{col}\,(f_1(t, \xi), \ldots, f_n(t, \xi))$.

Given two vectors in $\mathbf{R}^n$,

$$\xi = \mathrm{col}\,(\xi_1, \ldots, \xi_n) \quad \text{and} \quad \eta = \mathrm{col}\,(\eta_1, \ldots, \eta_n),$$

we say $\xi = \eta$ if and only if $\xi_i = \eta_i$ for $i = 1, \ldots, n$. We define

$$\xi + \eta = \mathrm{col}\,(\xi_1 + \eta_1, \ldots, \xi_n + \eta_n)$$

and

$$c\xi = \text{col}\,(c\xi_1, \ldots, c\xi_n),$$

where c is any scalar (real number).

It is easy to verify (Problem 1) that under the above definitions, addition of vectors and multiplication by scalars satisfy the usual rules of associativity, commutativity, and distributivity for a *linear vector space*. These are as follows.

LINEAR VECTOR SPACE PROPERTIES

Whenever ξ, η, and $\zeta \in \mathbf{R}^n$ and c, c_1, and $c_2 \in \mathbf{R}$, then

(i) $\xi + \eta = \eta + \xi$ (commutative law of addition),
(ii) $(\xi + \eta) + \zeta = \xi + (\eta + \zeta)$ (associative law of addition),
(iii) There exists a unique vector $\mathbf{0}$, called the zero vector, such that $\xi + \mathbf{0} = \xi$ for all ξ in $\mathbf{R}^n$,
(iv) For each ξ in $\mathbf{R}^n$ there exists a unique vector $-\xi$ in $\mathbf{R}^n$ such that $\xi + (-\xi) = \mathbf{0}$,
(v) $c_1(c_2\,\xi) = (c_1 c_2)\xi$ (associative law of scalar multiplication),
(vi) $(c_1 + c_2)\xi = c_1\xi + c_2\,\xi$ $\left.\begin{array}{l}\\\\\end{array}\right\}$ (distributive laws),
(vii) $c(\xi + \eta) = c\xi + c\eta$
(viii) $1\xi = \xi$.

In conditions (iii) and (iv) the symbols $\mathbf{0}$ and $-\xi$ stand for col $(0, \ldots, 0)$ and col $(-\xi_1, \ldots, -\xi_n)$, respectively.

Subtraction of vectors is defined by

$$\xi - \eta \equiv \xi + (-1)\eta,$$

so that $\xi - \eta = (\xi_1 - \eta_1, \ldots, \xi_n - \eta_n)$.

A solution of Eqs. (1) and (2) must be an n-vector-valued function. But, instead of saying "$\mathbf{x}$ is an n-vector-valued function on (α, β)," we shall say more briefly, $\mathbf{x}$ maps (α, β) into $\mathbf{R}^n$ or $\mathbf{x}: (\alpha, \beta) \to \mathbf{R}^n$. Similarly, instead of saying "$\mathbf{f}$ is an n-vector-valued function on D," we shall simply say $\mathbf{f}$ maps D into $\mathbf{R}^n$ or $\mathbf{f}: D \to \mathbf{R}^n$.

An n-vector-valued function $\mathbf{y}$ is said to be *continuous* (or *differentiable* or *integrable*) if each of its components, y_i, is continuous (or differentiable or integrable, respectively). If $\mathbf{y}$ is differentiable or integrable on $[a, b]$ we write, respectively, $\mathbf{y}' = \text{col}\,(y'_1, \ldots, y'_n)$ or

$$\int_a^b \mathbf{y}(s)\,ds = \text{col}\left(\int_a^b y_1(s)\,ds, \ldots, \int_a^b y_n(s)\,ds\right).$$

The reader should verify that differentiation and integration of vector-valued functions are *linear operations*. That is, if n-vector-valued functions $\mathbf{y}$ and $\mathbf{z}$ are differentiable or Riemann integrable, then so is $c_1\mathbf{y} + c_2\,\mathbf{z}$ and

$$\frac{d}{dt}(c_1\mathbf{y} + c_2\,\mathbf{z}) = c_1\mathbf{y}' + c_2\,\mathbf{z}' \tag{3}$$

or

$$\int_a^b (c_1\mathbf{y} + c_2\,\mathbf{z})(s)\ ds = c_1 \int_a^b \mathbf{y}(s)\ ds + c_2 \int_a^b \mathbf{z}(s)\ ds \qquad (4)$$

for arbitrary scalar constants c_1 and c_2. (Problem 2.)

As Problem 3, verify two other basic properties of vector integration and differentiation—the following analogs of the fundamental theorems of calculus:

If $\mathbf{y}$ is a continuous function mapping an interval $J \to \mathbf{R}^n$ and if t_0, $t \in J$, then

$$\frac{d}{dt} \int_{t_0}^t \mathbf{y}(s)\ ds = \mathbf{y}(t) \qquad (5)$$

(where a one-sided derivative is understood if t happens to be an endpoint of the interval J).

If $\mathbf{y}:[a, b] \to \mathbf{R}^n$ has a continuous (or merely integrable) derivative $\mathbf{y}'$ on $[a, b]$ (implying one-sided derivatives at a and b), then

$$\int_a^b \mathbf{y}'(s)\ ds = \mathbf{y}(b) - \mathbf{y}(a). \qquad (6)$$

With this notation available we are now ready to define a "solution" of (1) and (2) and then to present an easily applied condition for uniqueness.

DEFINITION

A *solution* of (1) and (2) is a differentiable function $\mathbf{x}$ mapping an open interval $J \to \mathbf{R}^n$ such that

(i) $t_0 \in J$ and $\mathbf{x}(t_0) = \mathbf{x}_0$, and
(ii) for all t in J, $(t, \mathbf{x}(t)) \in D$ and $\mathbf{x}'(t) = \mathbf{f}(t, \mathbf{x}(t))$.

This is a natural generalization of the definition given for scalar equations. In fact it reads almost word-for-word the same as the definition in Section 8. The difference lies in the meaning of the symbols $\mathbf{f}$, $\mathbf{x}$, and $\mathbf{x}_0$.

Uniqueness is defined exactly as in Section 8.

If $\mathbf{f}$ is continuous and $\mathbf{x}:J \to \mathbf{R}^n$ is a continuous function such that $(t, \mathbf{x}(t)) \in D$ for all t in J, then $\mathbf{x}$ is a solution of Eqs. (1) and (2) on J if and only if

$$\mathbf{x}(t) = \mathbf{x}_0 + \int_{t_0}^t \mathbf{f}(s, \mathbf{x}(s))\ ds \quad \text{for all } t \text{ in } J. \qquad (7)$$

This analog of 8-(3) is an easy consequence of Eqs. (5) and (6).

The following uniqueness theorem—completely analogous to Theorem 8-A—is stated here without proof. The proof can be found later (or now if desired) in Section 12.

THEOREM A

If $\mathbf{f}$ and its partial derivatives $D_{1+j} f_i$ for $i, j = 1, \ldots, n$ are continuous on D and if $(t_0, \mathbf{x}_0) \in D$, then Eqs. (1) and (2) have at most one solution on any open interval J. (We assume nothing about $D_1 f_i = \partial f_i / \partial t$.)

Note that in Example 1

$$\mathbf{f}(t, \xi) = \text{col} \, (-2\xi_1 + 2\xi_2 + 5 \sin 2t, \, 2\xi_1 - 5\xi_2),$$

and $\mathbf{f}$ can be considered to map $\mathbf{R}^3 \to \mathbf{R}^2$. The function $\mathbf{f}$ is continuous on $D = \mathbf{R}^3$; and the partial derivatives required in Theorem A are

$$D_2 f_1 = -2, \quad D_3 f_1 = 2, \quad D_2 f_2 = 2, \quad \text{and} \quad D_3 f_2 = -5,$$

which are also clearly continuous. Hence the system in Example 1 has at most one solution if $\mathbf{x}(t_0) = \mathbf{x}_0 = \text{col} \, (x_{01}, x_{02})$ is given.

The system of differential equations in Example 1 is called a "linear" system. In general, system (1) is said to be *linear* if it has the form

$$x_i' = \sum_{j=1}^{n} a_{ij}(t)x_j + h_i(t) \quad \text{for} \quad i = 1, \ldots, n, \tag{8}$$

where each a_{ij} and each h_i $(i, j = 1, \ldots, n)$ is a given function. For a linear system we will usually have $D = J \times \mathbf{R}^n$, for some open interval $J = (\alpha, \beta)$.

For such a linear system,

$$f_i(t, \xi) = \sum_{j=1}^{n} a_{ij}(t)\xi_j + h_i(t) \quad \text{for} \quad i = 1, \ldots, n.$$

So

$$D_{1+j} f_i(t, \xi) = a_{ij}(t).$$

Thus the conditions for uniqueness in Theorem A will be satisfied if each a_{ij} and each h_i is continuous on J, and we have the following.

COROLLARY B

Let each a_{ij} and each h_i be a continuous function mapping $J \to \mathbf{R}$. Then Eqs. (8) and (2) have at most one solution on any open subinterval of J.

PROBLEMS

1. Verify that vector addition and multiplication of vectors by scalars, as defined in $\mathbf{R}^n$, have the properties (i) through (viii) listed.
2. Let $\mathbf{y}$ and $\mathbf{z}$ be n-vector-valued functions on an interval J, and let c_1 and c_2 be real numbers.

    (a) Prove that differentiation is a linear operation, as asserted by Eq. (3).
    (b) Prove that integration is also a linear operation, as asserted by Eq. (4).

3. Prove the validity of Eqs. (5) and (6) by using the corresponding properties of scalar functions.

4. If $\mathbf{f}(t, \xi_1, \xi_2) \equiv \operatorname{col}(t\xi_1 + \xi_2^2, 1 + 2\xi_2 - 5e^{\xi_1})$ on $D = (-1, 2) \times (-2, -1) \times (0, 1)$, what can you say about uniqueness of solutions of the system $\mathbf{x}' = \mathbf{f}(t, \mathbf{x})$?

5. Find the largest open "rectangle" $D \subset \mathbf{R}^{1+n}$ you can for each of the following systems such that Theorem A applies.

   (a) $\mathbf{x}' = \operatorname{col}(t\sqrt{x_1}, x_1 x_2)$ with $\mathbf{x}(2) = \operatorname{col}(1, 0)$.
   (b) $\mathbf{x}' = \operatorname{col}(x_2, -(x_1^2 + x_2^2)^{-1})$ with $\mathbf{x}(0) = \operatorname{col}(-2, -2)$. (There is more than one correct answer.)
   (c) $\mathbf{x}' = \operatorname{col}(x_2, x_3, (2 - x_1^2)^{1/3})$ with $\mathbf{x}(0) = \operatorname{col}(1, 2, 3)$.
   (d) $\mathbf{x}' = \operatorname{col}(x_1 + x_2, t^{-1}x_1)$ with $\mathbf{x}(-1) = \operatorname{col}(0, 0)$.

10. HIGHER-ORDER EQUATIONS

The theorems of the previous section are easily applied to scalar equations of order $n \geq 2$. We illustrate the method by analyzing a certain second-order equation. We will not only prove uniqueness in this example, but will actually find the unique solution.

   Example 1. Consider an object of mass $m > 0$ bouncing up and down on the end of a spring hanging from the ceiling (Figure 1). Assume the restoring force of the spring is proportional to the amount it is stretched from its normal length l, with proportionality constant k (the "spring constant").

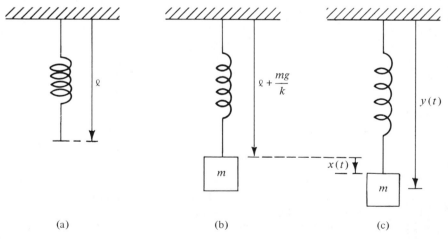

(a)   (b)   (c)

FIGURE 1

(a) Unstretched spring

(b) Equilibrium position

(c) In motion

Let $y(t)$ be the position of the mass below the fixed support at time t. Then the net downward force on m due to gravity pulling down and the spring pulling up is $mg - k[y(t) - l]$, where g is the acceleration of gravity. In addition, let us assume a frictional force $-by'(t)$ proportional to the speed of the mass and in the direction opposing the motion. Then the equation of motion is

$$my''(t) = mg - k[y(t) - l] - by'(t).$$

This equation will be simplified if we define

$$x = y - \left(l + \frac{mg}{k}\right).$$

The resulting simpler equation is

$$mx'' + bx' + kx = 0, \tag{1}$$

where m, b, and k are positive constants. Equation (1) is called a "second-order linear equation with constant coefficients." We shall show first that, regardless of the values of m, b, and k, if t_0, x_0, and v_0 are arbitrary real numbers, then Eq. (1) has at most one solution such that

$$x(t_0) = x_0 \quad \text{and} \quad x'(t_0) = v_0. \tag{2}$$

In other words, if the position and velocity of the object are specified at some instant, there can be at most one solution.

Defining $v = x'$ we convert the given second-order equation (1) into an equivalent system of two first-order equations:

$$x' = v,$$
$$v' = -\frac{k}{m}x - \frac{b}{m}v. \tag{3}$$

System (3) is said to be *equivalent* to Eq. (1) since any solution of (3) provides a solution of (1) and conversely. For if x and v satisfy (3), then $x'' = v' = -(k/m)x - (b/m)x'$, which is Eq. (1). While if x is any solution of (1) and if we define $v = x'$, then it follows that x and v satisfy (3).

Thus it suffices to show that system (3) with the initial conditions

$$x(t_0) = x_0, \qquad v(t_0) = v_0$$

has at most one solution. We can take $D = \mathbf{R}^3$ so that $(t_0, x_0, v_0) \in D$ is clear. Then we find that $\mathbf{f}$, defined by

$$\mathbf{f}(t, \xi_1, \xi_2) = \operatorname{col}\left(\xi_2, -\frac{k}{m}\xi_1 - \frac{b}{m}\xi_2\right)$$

satisfies all the conditions of Theorem 9-A. It follows that system (3) with the given initial conditions has at most one solution. Thus Eqs. (1) and (2) also have at most one solution.

The standard approach for actually finding solutions of homogeneous, linear differential equations with constant coefficients is to seek exponential solutions, that is, solutions of the form $e^{\lambda t}$, where λ is a constant. Substituting $x(t) = e^{\lambda t}$ into Eq. (1), we find

$$(m\lambda^2 + b\lambda + k)e^{\lambda t} = 0.$$

Since $e^{\lambda t}$ cannot vanish we require

$$m\lambda^2 + b\lambda + k = 0.$$

This quadratic equation, called the *characteristic equation* for (1), has solutions

$$\lambda_1 = \frac{-b + (b^2 - 4mk)^{1/2}}{2m} \quad \text{and} \quad \lambda_2 = \frac{-b - (b^2 - 4mk)^{1/2}}{2m}.$$

Let us assume that $b^2 > 4mk$ so that λ_1 and λ_2 are real and distinct. (We will treat the other cases in Section 14.) Then both $e^{\lambda_1 t}$ and $e^{\lambda_2 t}$ represent solutions of Eq. (1). In fact one can readily verify that a "linear combination" of $e^{\lambda_1 t}$ and $e^{\lambda_2 t}$,

$$x(t) = c_1 e^{\lambda_1 t} + c_2 e^{\lambda_2 t}, \tag{4}$$

defines a solution for any constants c_1 and c_2. (Problem 1.)

Now let arbitrary initial conditions of the form

$$x(t_0) = x_0, \qquad x'(t_0) = v_0$$

be specified. Assuming x is given by (4), these conditions become

$$c_1 e^{\lambda_1 t_0} + c_2 e^{\lambda_2 t_0} = x_0,$$

$$c_1 \lambda_1 e^{\lambda_1 t_0} + c_2 \lambda_2 e^{\lambda_2 t_0} = v_0.$$

We want to solve for c_1 and c_2. Since the determinant of the coefficients of the unknowns, c_1 and c_2, is

$$\begin{vmatrix} e^{\lambda_1 t_0} & e^{\lambda_2 t_0} \\ \lambda_1 e^{\lambda_1 t_0} & \lambda_2 e^{\lambda_2 t_0} \end{vmatrix} = (\lambda_2 - \lambda_1)e^{(\lambda_1 + \lambda_2)t_0} \neq 0,$$

it follows that one *can* always solve for c_1 and c_2. In other words, there does exist *a* solution of Eq. (1) satisfying the given initial conditions (2); and, from the uniqueness argument, we know that this is *the* solution.

The procedure used in Example 1 suggests a simple corollary of Theorem 9-A for the nth-order equation

$$x^{(n)} = f(t, x, x', \ldots, x^{(n-1)}) \tag{5}$$

with initial conditions

$$x(t_0) = b_0, \quad x'(t_0) = b_1, \quad \ldots, \quad x^{(n-1)}(t_0) = b_{n-1}. \tag{6}$$

Assuming that f is a given function mapping $D \to \mathbf{R}$ for some open set $D \subset \mathbf{R}^{1+n}$ and that $(t_0, b_0, b_1, \ldots, b_{n-1}) \in D$, we define a solution of Eqs. (5) and (6) as follows.

DEFINITION

A *solution* of (5) and (6) is a function x mapping an open interval $J \to \mathbf{R}$ such that

(i) $t_0 \in J$ and (6) is satisfied, and
(ii) for all t in J, $(t, x(t), x'(t), \ldots, x^{(n-1)}(t)) \in D$ and (5) is satisfied.

It is implicit in this definition that $x^{(n)}$ exists on J, and this in turn implies that $x, x', \ldots,$ and $x^{(n-1)}$ are all continuous. Thus, if the given function f is continuous, it follows from (5) that $x^{(n)}$ is also continuous on J.

THEOREM A

If f, $D_2 f$, $D_3 f$, $\ldots,$ and $D_{1+n} f$ are continuous on D and $(t_0, b_0, b_1, \ldots, b_{n-1}) \in D$, then Eqs. (5) and (6) have at most one solution on any open interval J.

Proof. Define n new unknown functions as follows:

$$x_1 = x, \quad \text{and} \quad x_i = x^{(i-1)} \quad \text{for} \quad i = 2, \ldots, n, \tag{7}$$

and consider the system

$$x_1' = x_2,$$
$$\vdots$$
$$x_{n-1}' = x_n, \tag{8}$$
$$x_n' = f(t, x_1, x_2, \ldots, x_n).$$

System (8) with initial conditions

$$x_i(t_0) = b_{i-1} \quad \text{for} \quad i = 1, \ldots, n \tag{9}$$

is equivalent to Eq. (5) with initial conditions (6) in the sense that any solution of one yields, through (7), a solution of the other. (Problem 6.) Since Eqs. (8) and (9) have at most one solution, by Theorem 9-A it follows that Eqs. (5) and (6) have at most one solution on J. ∎

We shall say Eq. (5) is a *linear* nth-order equation if it takes the form

$$a_n(t)x^{(n)} + a_{n-1}(t)x^{(n-1)} + \cdots + a_1(t)x' + a_0(t)x = h(t),$$

where $a_n, \ldots, a_1, a_0,$ and h are given functions, with $a_n(t) \neq 0$. In this case we can, without loss of generality, assume $a_n(t) \equiv 1$. For otherwise we could divide through by $a_n(t)$ and then relabel the coefficients. Thus we consider the equation

$$x^{(n)} + a_{n-1}(t)x^{(n-1)} + \cdots + a_1(t)x' + a_0(t)x = h(t). \tag{10}$$

For a linear equation we can take the set D in Theorem A to be $(\alpha, \beta) \times \mathbf{R}^n$. The following uniqueness result for Eqs. (10) and (6) is then a consequence of Theorem A. It would have nicely covered uniqueness for Eq. (1).

COROLLARY B

Let $a_0, a_1, \ldots, a_{n-1}$, and h be continuous real-valued functions on (α, β), and let $(t_0, b_0, b_1, \ldots, b_{n-1})$ be an arbitrary point in $(\alpha, \beta) \times \mathbf{R}^n$. Then there is at most one solution of Eqs. (10) and (6) on any subinterval of (α, β).

The proof is left as Problem 7.

Example 2. Solve the linear second-order equation

$$x'' + \omega^2 x = 0, \tag{11}$$

where ω is a positive constant. [This is a special case of Eq. (1), but we no longer have $b^2 > 4mk$ as we did in Example 1.]

Let us use the procedure suggested in Problem 7-5, multiplying through by x'. This gives

$$x'x'' + \omega^2 x x' = 0$$

or

$$(x')^2 + \omega^2 x^2 = a^2,$$

where a^2 is a constant of integration. Now our problem is reduced to a pair of separable first-order equations

$$x' = \pm \sqrt{a^2 - \omega^2 x^2}.$$

Proceeding formally without worrying about possible division by zero, we find

$$\frac{x'}{\sqrt{a^2 - \omega^2 x^2}} = \pm 1$$

or

$$\int \frac{(\omega x'/a)\, dt}{\sqrt{1 - \omega^2 x^2 / a^2}} = \pm \int \omega\, dt,$$

which says

$$\text{Arcsin } \frac{\omega x}{a} = \pm \omega t + b,$$

where b is another constant of integration. Solving for x, we find

$$x(t) = \frac{a}{\omega} \sin{(b \pm \omega t)}$$

or

$$x(t) = c_1 \cos \omega t + c_2 \sin \omega t, \tag{12}$$

where the constants c_1 and c_2 are related to a and b by

$$c_1 = \frac{a}{\omega} \sin b \quad \text{and} \quad c_2 = \pm \frac{a}{\omega} \cos b.$$

Our derivation of the family of solution candidates given by (12) was not at all rigorous. If you are particularly unhappy about the derivation, forget it. You can consider (12) to have been pulled out of a hat instead. In any case it is a simple exercise to verify by substitution that $x(t)$ defined by (12) *does indeed satisfy* Eq. (11) for all t, regardless of the values of c_1 and c_2. (Problem 8.)

But do we have the right to call (12) the general solution of Eq. (11)? That is, are all possible solutions of (11) contained in (12) by appropriate choice of c_1 and c_2? This question can be answered with the aid of our knowledge of uniqueness.

We know, from Corollary B, for example, that given any $(t_0, b_0, b_1) \in \mathbf{R}^3$, there is at most one solution of (11) on any interval containing t_0 subject to

$$x(t_0) = b_0 \quad \text{and} \quad x'(t_0) = b_1. \tag{13}$$

The question then is: Can we always choose c_1 and c_2 so that the solution given by (12) satisfies conditions (13)? The answer to this is also "yes" because conditions (13) applied to (12) become

$$c_1 \cos \omega t_0 + c_2 \sin \omega t_0 = b_0,$$

$$-\omega c_1 \sin \omega t_0 + \omega c_2 \cos \omega t_0 = b_1.$$

These two linear algebraic equations for the two unknowns c_1 and c_2 are always solvable (uniquely) since the determinant of the coefficients is

$$\begin{vmatrix} \cos \omega t_0 & \sin \omega t_0 \\ -\omega \sin \omega t_0 & \omega \cos \omega t_0 \end{vmatrix} = \omega \neq 0.$$

The situation is now similar to what it was in Example 1, and we can say that x defined by (12) is the "general solution" of Eq. (11).

The trick of transforming a single higher-order equation into a system of first-order equations is the foundation of the proof of Theorem A. The next example illustrates how this idea can also be applied to a *system* of higher-order equations.

Example 3. Consider the pair of differential equations

$$x_1'' = -\frac{kM}{(x_1^2 + x_2^2)^{3/2}} x_1 \quad \text{and} \quad x_2'' = -\frac{kM}{(x_1^2 + x_2^2)^{3/2}} x_2 \tag{14}$$

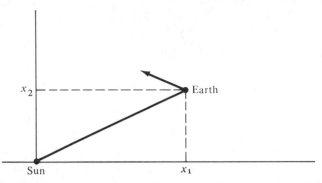

FIGURE 2

as representing the motion of the earth at (x_1, x_2) about the sun, assumed stationary at $(0, 0)$ and having mass M (Figure 2). Here k is the "constant of gravitation." (We assume there are no other celestial bodies.) We shall not undertake to solve these equations. We will not even prove that a solution exists. But we shall prove that there can be at most one solution having a specified position and velocity at t_0. More precisely, we shall show that if

$$x_1(t_0) = x_{10}, \qquad x_2(t_0) = x_{20},$$

$$x_1'(t_0) = v_{10}, \quad \text{and} \quad x_2'(t_0) = v_{20} \tag{15}$$

are given with $x_{10}^2 + x_{20}^2 \neq 0$, then, as long as the earth does not collide with the sun, it can have at most one trajectory.

Once again, we replace the given differential equations (14) by an equivalent system of first-order equations. This time we get the four first-order equations,

$$x_1' = v_1, \qquad x_2' = v_2,$$

$$v_1' = -\frac{kM}{(x_1^2 + x_2^2)^{3/2}} x_1, \quad \text{and} \quad v_2' = -\frac{kM}{(x_1^2 + x_2^2)^{3/2}} x_2. \tag{16}$$

To relate to the notation of Section 9, we define the vector-valued function $\mathbf{f}$ by

$$\mathbf{f}(t, \xi_1, \xi_2, \xi_3, \xi_4) = \begin{pmatrix} \xi_3 \\ \xi_4 \\ -kM\xi_1/(\xi_1^2 + \xi_2^2)^{3/2} \\ -kM\xi_2/(\xi_1^2 + \xi_2^2)^{3/2} \end{pmatrix}. \tag{17}$$

Then $\mathbf{f}$ is continuous and has continuous derivatives as long as $\xi_1^2 + \xi_2^2 > 0$. Now, with an appropriate choice of the open set D, one can apply Theorem 9-A. (Problem 9.) Thus there can be at most one solution, as long as the earth and sun do not collide.

Remarks. In connection with Example 3 it might be appropriate to insert some comments on the relationship between "real-world" phenomena and the associated mathematical models. In the above discussion is the statement: "We will not even prove that a solution exists." A natural reaction of the nonmathematician to this statement is: "Why bother? Everbody knows a solution exists for the motion of the earth about the sun. And, for that matter, everyone knows it is unique."

The flaw in such reasoning is simply this: While we may "know" that the earth travels around the sun, we also "know" just as surely that the *mathematical model* presented in Example 3 is *wrong*. Indeed, we can be sure that, no matter how much we improve it, it will still be wrong. For example, we have ignored the influence of all other celestial bodies; we have assumed that the earth and the sun are *homogeneous spherical solids**; we have ignored "relativity"; and we have assumed that the sun is not influenced by the motion of the earth. This last objection would not count (see Problem 10) if the gravitational interaction between the earth and the sun really propagated instantaneously as we have assumed. But it is no longer believed that this is the case. In other words, our equations of motion should also incorporate time delays.

Every mathematical model should be considered (at best) as an approximation to some real-world phenomenon. If we have confidence that the real phenomenon which we are trying to study is well behaved, then we should test our model to make certain *it* is also well behaved. The study of existence and uniqueness of solutions of differential equations can be regarded as one way of testing mathematical models. If no solution exists for a differential equation we are using as a model, or if uniqueness, or some other property which we demand, fails to hold, then we better change our model.

As a matter of fact there is an even more basic need for existence and uniqueness theorems. We need to discover what kind of questions one should ask for the given differential equations. For example, given the linear second-order Eq. (1), how do we know whether to specify as initial conditions

$$x(t_0) = x_0, \tag{18}$$

or the conditions $x(t_0) = x_0$ and $x'(t_0) = v_0$ of (2), or

$$x(t_0) = x_0, \quad x'(t_0) = v_0, \quad \text{and} \quad x''(t_0) = w_0, \tag{19}$$

or something else. Why are (18) and (19) inappropriate? (Problem 11.)

It is sometimes assumed that "physical intuition" suggests the appropriateness of the initial conditions (2). But, quite the contrary, it is actually the theory of differential equations which shows that these are appropriate "physical" initial conditions—assuming the physical problem can indeed be represented by a second-order ordinary differential equation such as Eq. (1).

---

* Actually we assumed them to be point particles, but this defect can be remedied by defining a collision as

$$x_1^2(t) + x_2^2(t) \leq R \quad \text{for some} \quad R > 0 \quad \text{instead of} \quad R = 0.$$

Section 11 will discuss still another type of question one might pose for an equation like (1).

PROBLEMS

1. Verify that (4) defines a solution of Eq. (1) regardless of the values of the constants c_1 and c_2.
2. (Review of solution of simultaneous linear algebraic equations.) Solve for c_1 and c_2:

$$c_1 + 2c_2 = 3,$$

$$2c_1 - 3c_2 = 5.$$

   (a) Use Cramer's rule.   (b) Use some other method.
3. Solve the equation

$$x'' + 3x' + 2x = 0$$

   with $x(0) = 1$ and $x'(0) = -2$. Graph the solution for $t \geq 0$.
4. Show that the general solution of $x'' - k^2x = 0$ (where k is a positive constant) is $x(t) = c_1 e^{kt} + c_2 e^{-kt}$.
5. State a uniqueness result for

   (a) $x'' + (\cot t)x' + (\cos t)x = 0$.
   (b) $x'' = g - (b/m)(x')^2$, where g, b, and m are positive constants (Example 7-1).
   (c) $x'' = -kM/x^2$, where k and M are positive constants (Example 7-2).
   (d) $x'' + x(x' - 1)^{2/3} = 0$.
   (e) $1 + (x')^2 + xx'' = 0$ with $x(2) = 2$ and $x'(2) = \frac{1}{2}$.

6. Show that system (8) is equivalent to Eq. (5).
7. Prove Corollary B.
8. Verify by substitution that (12) does define a solution of Eq. (11) regardless of the values of c_1 and c_2.
*9. Verify the applicability of Theorem 9-A to system (16) when $D \equiv \{(t, \xi_1, \xi_2, \xi_3, \xi_4) \in \mathbf{R}^5 : \xi_1^2 + \xi_2^2 > 0\}$. The ambitious student will want to verify that this set D is open as defined in Appendix 1.
*10. Actually the sun is not stationary, as assumed in Example 3. In the (idealized) two-body problem the sun moves because of the influence of the earth. Write the equations of motion without assuming the sun stationary. Let $(y_1(t), y_2(t))$ be the position of the earth and $(z_1(t), z_2(t))$ be the position of the sun. Show from your equations that $my_i''(t) + Mz_i''(t) = c_i$, a constant for $i = 1, 2$, where m is the mass of the earth. The constant vector (c_1, c_2) is the momentum of the "center of mass" of the system.
   Now define $x_i(t) = y_i(t) - z_i(t)$ for $i = 1, 2$ and show that your equations reduce to those in Example 3 except that M is replaced with $M + m$.
11. Discuss the solution of Eq. (1) subject to

   (a) condition (18),
   (b) conditions (19).

11. BOUNDARY VALUE PROBLEMS

Sections 8 through 10 dealt only with "initial-value problems." From these one might jump to the conclusion that the solution to an nth-order equation, or a system of n first-order equations, will be unique whenever n auxiliary conditions are specified.

To dispel this notion, and to re-emphasize that existence and uniqueness should not be considered obvious, we will briefly discuss another type of problem for differential equations. This important topic, "boundary value problems," will not be studied in this text beyond the brief introduction it gets in the present section.

Consider the second-order linear homogeneous differential equation

$$x'' + \omega^2 x = 0, \tag{1}$$

where ω is a given positive constant. The general solution of (1) was shown in Example 10-2 to be given by

$$x(t) = c_1 \cos \omega t + c_2 \sin \omega t, \tag{2}$$

where c_1 and c_2 are constants.

Let us seek a function $x\colon [\alpha, \beta] \to \mathbf{R}$, where $\alpha < \beta$, which satisfies Eq. (1) together with certain "boundary conditions." The latter will be restrictions on x or x' at α and β, the endpoints or boundaries of the interval. Thus, instead of imposing two conditions at one point t_0, we will now impose one condition at each of two points, α and β.

When we talk about a solution of a differential equation on a *closed* interval, such as $[\alpha, \beta]$, it should be understood that any derivatives are interpreted as one-sided derivatives at the endpoints of the interval.

For simplicity we shall take $\alpha = 0$ (and $\beta > 0$). It can be shown that this does not restrict the generality of our conclusions.

Keep in mind that every solution of Eq. (1) must be a special case of (2).

Example 1. Find a solution of Eq. (1) on $[0, \beta]$ which satisfies the boundary conditions

$$x(0) = 0 \quad \text{and} \quad x(\beta) = 0. \tag{3}$$

Putting $t = 0$ in Eq. (2), the first boundary condition, $x(0) = 0$, reduces to the requirement $c_1 = 0$. Thus any solution of (1) and (3) must be of the form

$$x(t) = c_2 \sin \omega t.$$

Now impose the second boundary condition $x(\beta) = 0$ to find

$$c_2 \sin \omega \beta = 0.$$

If $\omega \beta$ is not an integral multiple of π, then $\sin \omega \beta \neq 0$ and we must have

$c_2 = 0$. But if $\omega\beta = n\pi$ for some integer $n = 1, 2, \ldots$, then $\sin \omega\beta = 0$, and any value of c_2 will be acceptable. Thus:

(i) If $\omega\beta \neq n\pi$ for $n = 1, 2, \ldots$, the solution of Eqs. (1) and (3) is *unique*, $x(t) \equiv 0$.

(ii) If $\omega\beta = n\pi$ for some $n = 1, 2, \ldots$, then Eqs. (1) and (3) have *infinitely many solutions* of the form $x(t) = c_2 \sin \omega t$.

Physically, one can think of Eq. (1) as describing the motion of an object bouncing on the end of a spring with no damping—an idealized special case of Example 10-1. Then the boundary conditions (3) specify that the object must be at its equilibrium position at the instants $t = 0$ and $t = \beta$.

Example 2. Let the boundary conditions for Eq. (1) be

$$x(0) = 0 \quad \text{and} \quad x(\beta) = 1. \tag{4}$$

Then again the first of these conditions alone reduces the possible solutions to

$$x(t) = c_2 \sin \omega t.$$

The second boundary condition in (4) now becomes

$$c_2 \sin \omega\beta = 1. \tag{5}$$

If $\omega\beta \neq n\pi$ for $n = 1, 2, \ldots$, then this yields

$$c_2 = \frac{1}{\sin \omega\beta}.$$

But if $\omega\beta = n\pi$ for some $n = 1, 2, \ldots$, there is no way to satisfy Eq. (5). Thus:

(i) If $\omega\beta \neq n\pi$ for $n = 1, 2, \ldots$, then Eqs. (1) and (4) have a *unique* solution $x(t) = (\sin \omega t)/(\sin \omega\beta)$.

(ii) If $\omega\beta = n\pi$ for some $n = 1, 2, \ldots$, then Eqs. (1) and (4) have *no solution*.

Admittedly, in these two examples it is only in the exceptional cases

$$\beta = \frac{n\pi}{\omega} \quad \text{for} \quad n = 1, 2, \ldots$$

that one fails to get a unique solution. And, indeed, it is easy to find examples which always have unique solutions (Problem 1).

However, using a nonlinear equation, we can also exhibit examples in which the solution of a boundary value problem is never unique, regardless of the value of β.

* Example 3. The equation

$$x'' + x^2 = 0 \tag{6}$$

with boundary conditions

$$x(0) = 0, \qquad x(\beta) = 0 \tag{7}$$

clearly has the trivial solution $x(t) \equiv 0$ on $[0, \beta]$. Moreover, if we write (6) as $x'' = f(t, x) = -x^2$, f and $D_2 f$ are continuous everywhere.

Nevertheless, we will now show that, regardless of the values of $\beta > 0$, there is also a nontrivial solution of (6) and (7).

To find a nontrivial solution let us apply the method introduced in Section 7 and obtain $x'x'' + x^2 x' = 0$, or

$$\frac{(x')^2}{2} + \frac{x^3}{3} = \text{constant}.$$

For convenience let us write the constant as $c^3/3$. Then

$$x' = \pm(\tfrac{2}{3})^{1/2}(c^3 - x^3)^{1/2}.$$

To get a nontrivial solution of (6) and (7) we must have $x'(0) > 0$. Why? Thus, replace the $\pm$ sign by $+$ (for sufficiently small $t > 0$). Then, by separation of variables,

$$\int_0^{x(t)} (c^3 - \xi^3)^{-1/2} \, d\xi = (\tfrac{2}{3})^{1/2} t. \tag{8}$$

Let us try to construct a solution having the general form indicated in Figure 1. It should increase as $t \to \beta/2$ to the value $x(\beta/2) = c$ with $x'(\beta/2) = 0$. Then it should decrease symmetrically as t goes from $\beta/2$ to β. All this can be achieved if we can choose c in (8) such that $x(\beta/2) = c$, that is,

$$\int_0^c (c^3 - \xi^3)^{-1/2} \, d\xi = \frac{\beta}{6^{1/2}}.$$

Introducing $u = \xi/c$ this condition becomes

$$\int_0^1 (1 - u^3)^{-1/2} \, du = \beta \left(\frac{c}{6}\right)^{1/2}. \tag{9}$$

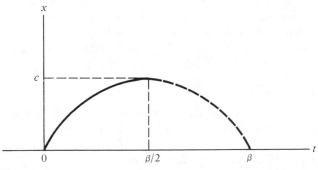

FIGURE 1

It is left as an exercise for the reader to show that the improper integral converges, and hence the appropriate value of c can be determined from (9) for each given $\beta > 0$. (Problem 7.) Then Eq. (8) determines the desired solution $x(t) \not\equiv 0$ implicitly on $[0, \beta/2]$.

PROBLEMS

1. The equation of motion for a free-falling body at a distance $x(t)$ above the earth's surface ignoring friction is

$$x'' = -g,$$

where g is a positive constant. Show that a unique solution is determined by any given boundary conditions of the form

$$x(0) = x_0 \quad \text{and} \quad x(\beta) = x_\beta \quad \text{where} \quad \beta > 0.$$

2. Consider Eq. (1), $x'' + \omega^2 x = 0$ (where $\omega > 0$) together with boundary conditions of the form $x(0) = x_0$ and $x'(\beta) = v_\beta$.

    (a) Under what conditions on ω and β is a unique solution determined?
    (b) What is the solution?
    (c) What happens if the condition you found in part (a) is not satisfied?

3. Answer the questions of Problem 2 for $x'' + \omega^2 x = 0$ (where $\omega > 0$) with $x(0) + x'(0) = 1$ and $x(\beta) = 0$.

4. Answer the questions of Problem 2 for the equation $x'' - k^2 x = 0$ (where $k > 0$) with $x(0) = x_0$ and $x(\beta) = x_\beta$. (You may use the result of Problem 10-4.)

5. Answer the questions of Problem 2 for $x'' - k^2 x = 0$ (where $k > 0$) with $x(0) = x_0$ and $x'(\beta) = v_\beta$.

6. Answer the questions of Problem 2 for $x'' - k^2 x = 0$ (where $k > 0$) with $x(0) + x'(0) = 1$ and $x(\beta) = 0$.

*7. Prove the convergence of the improper integral in (9).

*12. A VALUABLE LEMMA AND ITS APPLICATIONS

The type of trick used for analyzing inequality 8-(6) in the basic uniqueness proof for a scalar equation dates back at least to G. Peano in 1885. It was formalized as a lemma by T. H. Gronwall in 1918 and then generalized by W. T. Reid in 1930. This lemma, which we now present essentially in Reid's form, is used repeatedly in the study of differential equations.

LEMMA A (GRONWALL-REID).

Let C be a given constant and k a given nonnegative continuous function on an interval J. Let $t_0 \in J$. Then if $v : J \to [0, \infty)$ is continuous and

$$v(t) \leq C + \left| \int_{t_0}^{t} k(s)v(s)\, ds \right| \quad \text{for all} \quad t \quad \text{in} \quad J, \tag{1}$$

it follows that

$$v(t) \le C e^{\left| \int_{t_0}^t k(s)\, ds \right|} \quad \text{for all } t \text{ in } J. \tag{2}$$

Remarks. The hypothesis, (1), involves the unknown function v on both the left- and right-hand sides. However, the conclusion, (2), gives an unambiguous upper bound for $v(t)$.

Note that if $C = 0$ and $k(s) \equiv K$, then (1) is equivalent to inequality 8-(6) with $v(t) = |x(t) - \tilde{x}(t)|$. Then Lemma A gives $v(t) = 0$ for all t in J as desired.

Proof of Lemma A. The procedure is similar to that used in Example 8-4. If $t \ge t_0$ (with $t \in J$), inequality (1) can be rewritten as

$$k(t)v(t) - k(t)\Big[C + \int_{t_0}^t k(s)v(s)\, ds\Big] \le 0.$$

Introducing $Q(t) \equiv C + \int_{t_0}^t k(s)v(s)\, ds$, this becomes

$$Q'(t) - k(t)Q(t) \le 0.$$

Multiply through by the integrating factor $\exp\{-\int_{t_0}^t k(s)\, ds\} > 0$ to find

$$\frac{d}{dt}\Big[Q(t)e^{-\int_{t_0}^t k(s)\, ds}\Big] \le 0.$$

Integrating this last inequality from t_0 to t, and noting that $Q(t_0) = C$, we find

$$Q(t)e^{-\int_{t_0}^t k(s)\, ds} - C \le 0$$

or

$$Q(t) \le C e^{\int_{t_0}^t k(s)\, ds}$$

Now substitute this estimate for $Q(t)$ into (1) to obtain

$$v(t) \le Q(t) \le C e^{\int_{t_0}^t k(s)\, ds},$$

which is inequality (2).

If $t \le t_0$, the proof that (1) yields (2) is left to the reader as Problem 1. ∎

We are going to apply this lemma to prove Theorem 9-A, the uniqueness theorem for a system of differential equations

$$\mathbf{x}' = \mathbf{f}(t, \mathbf{x}) \tag{3}$$

with initial condition

$$\mathbf{x}(t_0) = \mathbf{x}_0. \tag{4}$$

Recall that if $\mathbf{f}$ and $\mathbf{x}$ are continuous functions, then Eqs. (3) and (4) are equivalent to

$$\mathbf{x}(t) = \mathbf{x}_0 + \int_{t_0}^{t} \mathbf{f}(s, \mathbf{x}(s)) \, ds. \tag{5}$$

To proceed we must now generalize the concept of "absolute value" to vectors.

If $\xi \in \mathbf{R}^n$, we shall define the *norm* of ξ to be

$$\|\xi\| = \sum_{i=1}^{n} |\xi_i| = |\xi_1| + \cdots + |\xi_n|. \tag{6}$$

The norm of a vector as defined in (6) can be considered as an extension of the concept of absolute value because of the following properties.

PROPERTIES OF THE NORM

Let the norm of a vector in $\mathbf{R}^n$ be defined by Eq. (6). Then

(i) $\|\xi\| \geq 0$ for all ξ in $\mathbf{R}^n$,
(ii) $\|\xi\| = 0$ if and only if $\xi = \mathbf{0}$ (the zero vector),
(iii) $\|c\xi\| = |c| \cdot \|\xi\|$ for all ξ in $\mathbf{R}^n$ and all c in $\mathbf{R}$, and
(iv) $\|\xi + \eta\| \leq \|\xi\| + \|\eta\|$ for all ξ, η in $\mathbf{R}^n$ (the "triangle inequality").

The verification of these properties is left as Problem 3.

Remark. Equation (6) is by no means the only suitable definition for extending the concept of absolute value to vectors in $\mathbf{R}^n$. Other possible definitions include

$$\|\xi\|_\infty = \max_{i=1,\ldots,n} |\xi_i| \tag{6a}$$

and

$$\|\xi\|_2 = \left[\sum_{i=1}^{n} \xi_i^2 \right]^{1/2}. \tag{6b}$$

Both of these "norms" can also be shown to have the properties (i) through (iv) listed above. (Problem 4.) We shall use only the norm defined in Eq. (6), however.

Regarding integration of vector-valued functions, the important property of the norm is this. If $\mathbf{y}$ is a continuous (or merely integrable) n-vector-valued function on $[a, b]$, then

$$\left\| \int_{a}^{b} \mathbf{y}(s) \, ds \right\| \leq \int_{a}^{b} \|\mathbf{y}(s)\| \, ds. \tag{7}$$

This is proved as follows (assuming the result for scalar-valued functions):

$$\left\| \int_a^b \mathbf{y}(s) \, ds \right\| = \sum_{i=1}^n \left| \int_a^b y_i(s) \, ds \right| \le \sum_{i=1}^n \int_a^b |y_i(s)| \, ds$$

$$= \int_a^b \sum_{i=1}^n |y_i(s)| \, ds = \int_a^b \|\mathbf{y}(s)\| \, ds.$$

Using the norm notation, let us now extend the meaning of "Lipschitz condition" (from Section 8) to vector-valued functions $\mathbf{f}$ mapping a subset of $\mathbf{R}^{1+n}$ into $\mathbf{R}^n$, as occurs in Eq. (3).

DEFINITION

Let $A \subset \mathbf{R}^{1+n}$. Then a function $\mathbf{f} : A \to \mathbf{R}^n$ is said to satisfy a *Lipschitz condition* on A if there exists a constant K (called a *Lipschitz constant*) such that

$$\|\mathbf{f}(t, \xi) - \mathbf{f}(t, \tilde{\xi})\| \le K\|\xi - \tilde{\xi}\| \tag{8}$$

for all (t, ξ) and $(t, \tilde{\xi})$ in A. We also say $\mathbf{f}$ is *Lipschitzian* on A.

Under the hypotheses of Theorem 9-A, that each $D_{1+j} f_i$ is continuous on an open set $D \subset \mathbf{R}^{1+n}$, it follows that $\mathbf{f}$ is Lipschitzian on each closed bounded $(n + 1)$-dimensional "rectangle"

$$A = [\bar{\alpha}, \bar{\beta}] \times [\bar{\gamma}_1, \bar{\delta}_1] \times \cdots \times [\bar{\gamma}_n, \bar{\delta}_n] \subset D.$$

We show this as follows.

Let A be a closed bounded $(n + 1)$-dimensional rectangle in D and let any (t, ξ) and $(t, \tilde{\xi})$ in A be given. Then it follows from the mean value theorem, Theorem A2-D, that, for each i,

$$f_i(t, \xi) - f_i(t, \tilde{\xi}) = \sum_{j=1}^n (D_{1+j} f_i)(t, \boldsymbol{\theta})(\xi_j - \tilde{\xi}_j),$$

where $(t, \boldsymbol{\theta}) \in A$. This means that $(t, \boldsymbol{\theta})$ is also in D, as it must be in order that $(D_{1+j} f_i)(t, \boldsymbol{\theta})$ be defined.

Since A is closed and bounded, Theorem A2-G asserts that each of the continuous functions $D_{1+j} f_i$ is bounded on A. Let B be the largest of the bounds for $|D_{1+j} f_i|$ $(i, j = 1, \ldots, n)$. Then we have for each i

$$|f_i(t, \xi) - f_i(t, \tilde{\xi})| \le \sum_{j=1}^n B|\xi_j - \tilde{\xi}_j| = B\|\xi - \tilde{\xi}\|.$$

Consequently,

$$\sum_{i=1}^n |f_i(t, \xi) - f_i(t, \tilde{\xi})| \le nB\|\xi - \tilde{\xi}\|,$$

which is Lipschitz condition (8) with $K = nB$.

Now, instead of proving Theorem 9-A, we shall prove the following slightly more general theorem—a vector analog of Theorem 8-B.

THEOREM B

Let $\mathbf{f}$ be continuous on D and Lipschitzian on each closed bounded $(n + 1)$-dimensional rectangle A in D, and let $(t_0, \mathbf{x}_0) \in D$. Then Eqs. (3) and (4) have at most one solution on any open interval J containing t_0.

Proof. (Compare with the proof of Theorem 8-A.) Suppose (for contradiction) that there are two solutions $\mathbf{x}$ and $\tilde{\mathbf{x}}$ on (α_1, β_1) with $\mathbf{x} \neq \tilde{\mathbf{x}}$. Let us assume $\mathbf{x}(t) \neq \tilde{\mathbf{x}}(t)$ for some t in (t_0, β_1). [The case of $\mathbf{x}(t) \neq \tilde{\mathbf{x}}(t)$ for some t in (α_1, t_0) is handled similarly.] Let

$$t_1 = \inf \{t \in (t_0, \beta_1): \mathbf{x}(t) \neq \tilde{\mathbf{x}}(t)\}.$$

Then, since $\mathbf{x}$ and $\tilde{\mathbf{x}}$ are both continuous, it follows that $t_1 < \beta_1$ and $\mathbf{x}(t_1) = \tilde{\mathbf{x}}(t_1) \equiv \mathbf{x}_1$.

Now choose positive numbers a and b such that

$$A = [t_1 - a, t_1 + a] \times \mathop{\times}_{i=1}^{n} [x_{1i} - b, x_{1i} + b] \subset D.$$

Compare Figure 8-4. Then it follows that $\mathbf{f}$ satisfies a Lipschitz condition (8) on A.

Now regard $\mathbf{x}$ and $\tilde{\mathbf{x}}$ as solutions of Eqs. (3) and (4) with the new open set

$$D_* = (t_1 - a, t_1 + a) \times \mathop{\times}_{i=1}^{n} (x_{1i} - b, x_{1i} + b)$$

playing the role of D and with the point $(t_1, \mathbf{x}_1)$ playing the role of $(t_0, \mathbf{x}_0)$.

By continuity, the points $(t, \mathbf{x}(t))$ and $(t, \tilde{\mathbf{x}}(t))$ must remain in D_* on some open interval J_* about t_1. Moreover, it follows from (5), with t_1 in place of t_0, that for all t in J_*,

$$\mathbf{x}(t) = \mathbf{x}_1 + \int_{t_1}^{t} \mathbf{f}(s, \mathbf{x}(s)) \, ds$$

and

$$\tilde{\mathbf{x}}(t) = \mathbf{x}_1 + \int_{t_1}^{t} \mathbf{f}(s, \tilde{\mathbf{x}}(s)) \, ds.$$

Subtraction of these two equations gives

$$\mathbf{x}(t) - \tilde{\mathbf{x}}(t) = \int_{t_1}^{t} [\mathbf{f}(s, \mathbf{x}(s)) - \mathbf{f}(s, \tilde{\mathbf{x}}(s))] \, ds.$$

Then, applying inequality (7) and the Lipschitz condition, we obtain for all t in J_*,

$$\|\mathbf{x}(t) - \tilde{\mathbf{x}}(t)\| = \left| \int_{t_1}^t \|\mathbf{f}(s, \mathbf{x}(s)) - \mathbf{f}(s, \tilde{\mathbf{x}}(s))\| \, ds \right|$$

$$\leq K \left| \int_{t_1}^t \|\mathbf{x}(s) - \tilde{\mathbf{x}}(s)\| \, ds \right|.$$

(9)

But (9) is the analog of inequality 8-(6). Proceeding as in that case, or simply applying Lemma A, one finds $\|\mathbf{x}(t) - \tilde{\mathbf{x}}(t)\| = 0$ or $\mathbf{x}(t) = \tilde{\mathbf{x}}(t)$ for all t in J_*. This contradicts the definition of t_1. ∎

As a matter of fact it is practically no more difficult to prove a useful theorem regarding the "growth of errors" in the solution of a system of differential equations.

If we think of system (3) as representing some physical process, then the n initial conditions, Eq. (4), become input data. If the conditions $\mathbf{x}(t_0) = \mathbf{x}_0$ have been measured somehow, then of course they have only been evaluated to the accuracy of the measuring equipment. Assuming we actually measured $\tilde{\mathbf{x}}_0$ instead of $\mathbf{x}_0$ the question is this:

If $\tilde{\mathbf{x}}_0$ is "close" to $\mathbf{x}_0$, will the corresponding solution $\tilde{\mathbf{x}}$ be close to the solution $\mathbf{x}$?

Theorem C asserts that the answer is "yes" if $\mathbf{f}$ satisfies a Lipschitz condition and if $|t - t_0|$ is not too large.

THEOREM C

Let $\mathbf{f}$ map $D \to \mathbf{R}^n$ for some open set $D \subset \mathbf{R}^{1+n}$ and let $\mathbf{f}$ be continuous and satisfy the Lipschitz condition (8) on D. Let $(t_0, \mathbf{x}_0) \in D$ and $(t_0, \tilde{\mathbf{x}}_0) \in D$ and let $\mathbf{x}$ and $\tilde{\mathbf{x}}$ be solutions of Eq. (3) on an interval J such that $\mathbf{x}(t_0) = \mathbf{x}_0$ and $\tilde{\mathbf{x}}(t_0) = \tilde{\mathbf{x}}_0$. Then

$$\|\mathbf{x}(t) - \tilde{\mathbf{x}}(t)\| \leq \|\mathbf{x}_0 - \tilde{\mathbf{x}}_0\| e^{K|t - t_0|} \quad \text{for all} \quad t \quad \text{in} \quad J. \tag{10}$$

Proof. For t in J

$$\mathbf{x}(t) = \mathbf{x}_0 + \int_{t_0}^t \mathbf{f}(s, \mathbf{x}(s)) \, ds$$

and

$$\tilde{\mathbf{x}}(t) = \tilde{\mathbf{x}}_0 + \int_{t_0}^t \mathbf{f}(s, \tilde{\mathbf{x}}(s)) \, ds.$$

Subtracting these two equations we obtain the estimate

$$\|\mathbf{x}(t) - \tilde{\mathbf{x}}(t)\| = \|\mathbf{x}_0 - \tilde{\mathbf{x}}_0 + \int_{t_0}^{t} [\mathbf{f}(s, \mathbf{x}(s)) - \mathbf{f}(s, \tilde{\mathbf{x}}(s))] \, ds\|$$

$$\leq \|\mathbf{x}_0 - \tilde{\mathbf{x}}_0\| + \left| \int_{t_0}^{t} K \|\mathbf{x}(s) - \tilde{\mathbf{x}}(s)\| \, ds \right|$$

for all t in J. But from this Lemma A gives inequality (10) at once. ∎

The following corollary applies this result to a system of n linear equations with variable coefficients

$$x_i' = \sum_{j=1}^{n} a_{ij}(t)x_j + h_i(t) \quad \text{for} \quad i = 1, \ldots, n. \tag{11}$$

COROLLARY D

Let each a_{ij} and each h_i be a continuous function mapping some interval $(\alpha, \beta) \to \mathbf{R}$ and let $\mathbf{x}$ and $\tilde{\mathbf{x}}$ be two solutions of (11) on (α_1, β_1) where $\alpha < \alpha_1 < \beta_1 < \beta$. Then if $\mathbf{x}(t_0) = \mathbf{x}_0$ and $\tilde{\mathbf{x}}(t_0) = \tilde{\mathbf{x}}_0$,

$$\|\mathbf{x}(t) - \tilde{\mathbf{x}}(t)\| \leq \|\mathbf{x}_0 - \tilde{\mathbf{x}}_0\| e^{K|t - t_0|} \quad \text{for} \quad \alpha_1 < t < \beta_1$$

for some constant $K > 0$.

Proof. Since each a_{ij} is continuous on the closed bounded interval $[\alpha_1, \beta_1]$, there must exist some $B > 0$ such that

$$|a_{ij}(t)| \leq B \quad \text{for} \quad \alpha_1 \leq t \leq \beta_1, \quad i, j = 1, \ldots, n.$$

But then system (11) is a special case of (3) with

$$|f_i(t, \xi) - f_i(t, \tilde{\xi})| = |\sum_{j=1}^{n} a_{ij}(t)[\xi_j - \tilde{\xi}_j]| \leq B\|\xi - \tilde{\xi}\|$$

for all (t, ξ) and $(t, \tilde{\xi})$ in $[\alpha_1, \beta_1] \times \mathbf{R}^n$ and $i = 1, \ldots, n$. Thus,

$$\|\mathbf{f}(t, \xi) - \mathbf{f}(t, \tilde{\xi})\| \leq nB\|\xi - \tilde{\xi}\|,$$

and the assertion of the corollary follows from Theorem C. ∎

Lemma A is one of the best-known examples of a theorem on "integral inequalities." This result and generalizations of it have been exploited widely in the study of ordinary differential equations, functional differential equations, partial differential equations, and integral equations.

A minor generalization of Lemma A which will be of use in Section 40 is presented as a corollary:

COROLLARY E

Let M and k be given nonnegative functions with k continuous on an interval J. Let $t_0 \in J$ and assume $M(t)$ is nondecreasing as $|t - t_0|$ increases.

Then if v is any nonnegative continuous function such that

$$v(t) \leq M(t) + \left| \int_{t_0}^{t} k(s)v(s) \, ds \right| \quad \text{for all} \quad t \quad \text{in} \quad J, \tag{12}$$

it follows that

$$v(t) \leq M(t)e^{\left| \int_{t_0}^{t} k(s) \, ds \right|} \quad \text{for all} \quad t \quad \text{in} \quad J. \tag{13}$$

Proof. Let $t_0 \leq t \leq t_1$ with $t_1 \in J$. Then from (12) we have

$$v(t) \leq M(t_1) + \int_{t_0}^{t} k(s)v(s) \, ds,$$

where t_1 is now regarded as a constant. Thus Lemma A gives

$$v(t) \leq M(t_1)e^{\left| \int_{t_0}^{t} k(s) \, ds \right|},$$

and (13) follows by putting $t = t_1$. The proof is similar when $t_1 \leq t \leq t_0$. ∎

PROBLEMS

1. Complete the proof of Lemma A by treating the case $t \leq t_0$. (Be careful of the algebraic signs at each step of your proof.)
2. If we add the hypothesis "v is continuously differentiable," what is wrong with the following simpler "proof" for Lemma A? For $t > t_0$ inequality (1) becomes

$$v(t) \leq C + \int_{t_0}^{t} k(s)v(s) \, ds.$$

   Differentiate this with respect to t and rearrange to find

$$v'(t) - k(t)v(t) \leq 0 \quad \text{with} \quad v(t_0) = v_0 \leq C.$$

   Now multiply by $\exp\{-\int_{t_0}^{t} k(s) \, ds\}$ and integrate from t_0 to t, getting

$$v(t)e^{-\int_{t_0}^{t} k(s) \, ds} - v(t_0) \leq 0.$$

   This immediately gives (2).
3. Prove that the "norm" defined by Eq. (6) has the properties (i) through (iv) listed.
*4. Prove that $\|\cdot\|_\infty$ and $\|\cdot\|_2$ as defined by (6a) and (6b) also have the properties (i) through (iv). The proof of property (iv) for $\|\cdot\|_2$ is nontrivial, but can be found in books on linear algebra or advanced calculus.
5. For any norm, having the properties (i) through (iv) listed, prove that

$$\|\xi - \eta\| \geq \big| \, \|\xi\| - \|\eta\| \, \big|.$$

6. (a) The system of equations for the electric currents in Example 9-1 was

$$x_1' = -2x_1 + 2x_2 + 5\sin 2t,$$

$$x_2' = 2x_1 - 5x_2.$$

Prove that the corresponding **f** is Lipschitzian on $\mathbf{R}^3$ with $K = 7$.

(b) Prove that **f** in Problem 9-4 is Lipschitzian on its given domain D.

7. Consider a scalar linear equation,

$$x' + a(t)x = h(t),$$

where a and h are continuous on $\mathbf{R}$ with $|a(t)| \leq B$ for all t. Compute the difference between two solutions, x and $\tilde{x}$, in terms of $x(t_0) = x_0$ and $\tilde{x}(t_0) = \tilde{x}_0$ proceeding directly from the general solution (Problem 2-2). How does your result compare with the estimate obtained using Theorem C?

8. Assume that the system in Problem 6(a) has two solutions on $J = \mathbf{R}$, $\mathbf{x}(t) = \mathrm{col}\,(x_1(t), x_2(t))$ satisfying initial conditions $\mathbf{x}(t_0) = \mathrm{col}\,(x_{01}, x_{02})$, and $\tilde{\mathbf{x}}(t) = \mathrm{col}\,(\tilde{x}_1(t), \tilde{x}_2(t))$ satisfying $\tilde{\mathbf{x}}(t_0) = \mathrm{col}\,(\tilde{x}_{01}, \tilde{x}_{02})$. Then show that for all t in $\mathbf{R}$,

$$|x_1(t) - \tilde{x}_1(t)| + |x_2(t) - \tilde{x}_2(t)| \leq (|x_{01} - \tilde{x}_{01}| + |x_{02} - \tilde{x}_{02}|)e^{7|t - t_0|}.$$

9. Obtain an estimate for the rate of growth of the error in the solution of $x''' - x'' = 0$ due to incorrect initial data. [*Hint:* Convert to an equivalent system of first-order equations.]

*10. For the equation $(1 - t)x' = x - 1$ with $x(0) = 1$, it is easily seen that a solution is defined by $x(t) = 1$ for all $t < 1$.

(a) Prove that this solution is unique.

(b) Nevertheless, if these equations represent any real-world physical system, then that system will surely explode as $t \to 1$. Show clearly that this is true. Or is it?

(c) Why does the assertion in (b) not contradict Theorem C?

*11. Discuss the assertions and questions of Problem 10 for the equation $(1 - t)^{1/3}x' = x - 1$ with $x(0) = 1$.

Chapter Three

LINEAR ORDINARY DIFFERENTIAL EQUATIONS OF ORDER n

The examples in Chapters One and Two have introduced special methods for finding exact solutions of some special equations. Generally speaking, the discovery of exact solutions of a differential equation (when this is possible) does depend on the use of ad hoc methods or tricks, often requiring considerable ingenuity.

However, in the case of linear equations and linear systems a well-organized general theory exists. And in the special case of linear equations with constant coefficients,

$$a_n x^{(n)} + a_{n-1} x^{(n-1)} + \cdots + a_1 x' + a_0 x = h(t),$$

where $h(t)$ is reasonably well behaved, the general theory provides exact solutions.

Linear systems and linear equations of order n are important in physics, engineering, and other applied sciences. In this chapter we shall concentrate on the scalar linear equation of order n. Systems of linear first-order equations will be discussed in Chapter Four.

13. COMPLEX FUNCTIONS

The reader is assumed to be familiar with the algebra of complex numbers of the form

$$z = a + ib,$$

where a and b are real and $i^2 = -1$, and with the definitions of the complex conjugate and absolute value:

$$\bar{z} = a - ib, \qquad |z| = \sqrt{a^2 + b^2}.$$

We call a and b the real and imaginary parts of z, respectively, and write

$$\operatorname{Re} z = a, \qquad \operatorname{Im} z = b.$$

Recall also that

$$z = 0 \quad \text{if and only if} \quad \text{Re } z = 0 \quad \text{and} \quad \text{Im } z = 0$$

Some review exercises are provided in Problems 1 and 2.

Those who are also acquainted with complex-valued functions of a real variable can omit this section. Otherwise read on.

You should verify (Problem 3) that if z_1 and z_2 are any two complex numbers, then

$$\left.\begin{array}{c} \overline{z_1 + z_2} = \overline{z_1} + \overline{z_2} \\[2mm] \overline{z_1 z_2} = \overline{z_1}\,\overline{z_2}, \text{ and} \\[2mm] \overline{z_1^n} = (\overline{z_1})^n, \end{array}\right\} \tag{1}$$

for every $n = 1, 2, \ldots$, and for $n = 0, -1, -2, \ldots$ if $z_1 \neq 0$. It is easy also to show (Problem 4) that

$$|z_1|^2 = z_1 \overline{z_1}, \quad |z_1 z_2| = |z_1|\,|z_2|, \quad \text{and} \quad |z_1 + z_2| \le |z_1| + |z_2|. \tag{2}$$

The third of these properties is called the *triangle inequality*.

The complex numbers with which we shall be concerned will usually be zeros of algebraic polynomials of the form

$$p(z) = a_n z^n + a_{n-1} z^{n-1} + \cdots + a_1 z + a_0. \tag{3}$$

Recall that λ is a zero of p, or a root (or solution) of

$$p(z) = 0, \tag{4}$$

if and only if $z - \lambda$ is a factor of $p(z)$.

If, for some integer $m \ge 1$, $(z - \lambda)^m$ divides $p(z)$ while $(z - \lambda)^{m+1}$ does not, we say λ is a *root of multiplicity m* of Eq. (4). In this case

$$p(z) = (z - \lambda)^m q(z),$$

where $z - \lambda$ is not a factor of the polynomial $q(z)$.

In most applications the polynomial p will have real coefficients. If the coefficients $a_n, a_{n-1}, \ldots, a_0$ in (3) are all real, then any complex roots of (4) must occur in complex conjugate pairs. More precisely, if λ is a complex root with multiplicity m of the polynomial equation (4), then $\bar{\lambda}$ is also a root with multiplicity m. This is shown as follows.

Given that λ is a solution, it follows from (1) that

$$a_n(\bar{\lambda})^n + a_{n-1}(\bar{\lambda})^{n-1} + \cdots + a_1 \bar{\lambda} + a_0$$

$$= \overline{a_n \lambda^n + a_{n-1}\lambda^{n-1} + \cdots + a_1 \lambda + a_0} = 0.$$

Thus $\bar{\lambda}$ is a solution. Hence $(z - \lambda)$ and $(z - \bar{\lambda})$ are both factors of $p(z)$. Since the product

$$(z - \lambda)(z - \bar{\lambda}) = z^2 - (\lambda + \bar{\lambda})z + \lambda\bar{\lambda}$$

is a polynomial in z with real coefficients, it follows that the quotient

$$b_{n-2}\, z^{n-2} + \cdots + b_1 z + b_0 \equiv \frac{p(z)}{z^2 - (\lambda + \bar{\lambda})z + \lambda\bar{\lambda}}$$

is also polynomial with real coefficients. We can now argue that if $m \geq 2$, then λ is also a root of

$$b_{n-2}\, z^{n-2} + \cdots + b_1 z + b_0 = 0,$$

and hence so also is $\bar{\lambda}$. The proof is completed by an induction argument.

The fact that complex roots occur in conjugate pairs is easily seen in the special case of a second-degree equation such as

$$m\lambda^2 + b\lambda + k = 0$$

encountered in Example 10-1. For if $b^2 < 4mk$ the two roots λ_1 and λ_2 found via the quadratic formula are clearly complex conjugates.

Example 1. (Some manipulations to motivate a definition.) Consider the equation

$$x'' + x = 0. \tag{5}$$

If we set out to find exponential solutions $e^{\lambda t}$, as in Example 10-1, we are led to the requirement $\lambda^2 + 1 = 0$, the solutions of which are $\lambda = \pm i$. But, unless we have already defined and studied e^{it} and e^{-it}, this makes no sense. Is there some way we could give meaning to such an undefined symbol as e^{it} so that it would represent a meaningful solution of (5)?

Let us proceed formally as though the "imaginary exponential" e^{it} could be differentiated and otherwise manipulated just like an ordinary real exponential. Then the "solution" $x(t) = e^{it}$ of (5) is seen to "satisfy" the initial conditions

$$x(0) = 1 \quad \text{and} \quad x'(0) = i.$$

But we know, from Example 10-2, that the general solution of (5) is given by

$$x(t) = c_1 \cos t + c_2 \sin t.$$

Let us try to choose c_1 and c_2 so that this solution "satisfies" the same initial conditions as does e^{it}. Setting $t = 0$ we get $c_1 = x(0) = 1$, while differentiating and then setting $t = 0$ gives $c_2 = x'(0) = i$, so that

$$x(t) = \cos t + i \sin t.$$

These formal manipulations, of course, do not prove that $e^{it} = \cos t + i \sin t$, but they do suggest that we try making this a definition.

Thus for all real θ we will *define*

$$e^{i\theta} \equiv \cos \theta + i \sin \theta. \tag{6}$$

This is called Euler's formula.

Using this definition, the reader is asked to verify (Problem 6) that

$$e^{i\theta_1}e^{i\theta_2} = e^{i(\theta_1+\theta_2)} \tag{7}$$

for all real values of θ_1 and θ_2. The validity of (7) gives us more confidence that (6) is a reasonable definition.

Attempting to preserve the familiar multiplicative properties of exponentials still further we are led to define

$$e^{\mu+i\theta} \equiv e^{\mu}e^{i\theta} = e^{\mu}(\cos\theta + i\sin\theta) \tag{8}$$

whenever μ and θ are real.

The reader should verify (Problem 7) that as a consequence of this definition we always have for arbitrary complex numbers λ, λ_1, and λ_2,

$$\left.\begin{aligned} &|e^{\lambda}| = e^{\operatorname{Re}\lambda} \neq 0, \\[4pt] &e^{\lambda_1}e^{\lambda_2} = e^{\lambda_1+\lambda_2}, \\[4pt] &\frac{e^{\lambda_1}}{e^{\lambda_2}} = e^{\lambda_1-\lambda_2}, \\[4pt] &(e^{\lambda})^n = e^{n\lambda}, \quad \text{for} \quad n = 0, \pm1, \pm2, \ldots \end{aligned}\right\} \tag{9}$$

We next show that any complex number $z = a + ib$, where a and b are real, can be rewritten as

$$z = re^{i\theta}, \tag{10}$$

where $r = |z| > 0$ and θ is real. First observe that, given any two real numbers, a and b, it is always possible to choose real numbers $r > 0$ and θ such that

$$a = r\cos\theta \quad \text{and} \quad b = r\sin\theta,$$

namely, $r = \sqrt{a^2 + b^2}$ and $\theta = \arctan^{-1} b/a$ with θ lying in the appropriate quadrant. (In case $a = b = 0$, $r = 0$ and θ is arbitrary.) Thus we can write

$$a + ib = r(\cos\theta + i\sin\theta) = re^{i\theta}.$$

A complex number written $re^{i\theta}$ is said to be in *polar form*.

The relationship between a, b, r, and θ is easily exhibited by plotting the point $a + ib$ in the "complex plane." (Figure 1.)

The value of θ in (10) is certainly not unique. In fact any acceptable θ can be changed into a different but equally acceptable value by the addition or subtraction of 2π. For convenience we usually choose θ such that $0 \leq \theta < 2\pi$. If we require that $a + ib = r(\cos\theta + i\sin\theta)$ with $r > 0$ and $0 \leq \theta < 2\pi$, then θ *is* uniquely determined.

Noting, from Eq. (6), that

$$e^{i\theta} = e^{i(\theta+2k\pi)} \quad \text{for each} \quad k = 0, \pm1, \pm2, \ldots$$

we can use the polar form (10) to show, in a straightforward manner, that

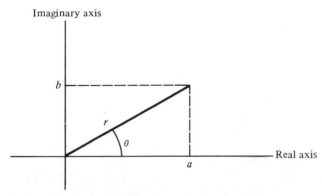

Imaginary axis

b

r

θ

a

Real axis

FIGURE 1

every real or complex number $z \neq 0$ has n distinct nth roots. One merely expresses z in the form $z = re^{i\theta}$. Then it follows that every number of the form

$$r^{1/n}e^{i(\theta + 2k\pi)/n} \quad \text{for} \quad k = 0, 1, \ldots, n - 1$$

(where $r^{1/n}$ denotes the positive nth root of r) is an nth root of z. These n numbers are all distinct; but the use of any values of the integer k besides 0, $1, \ldots, n - 1$ will only give a repetition of roots already listed.

Example 2. To find the fourth roots of $-2 + i2\sqrt{3}$ we first write

$$-2 + i2\sqrt{3} = 4e^{i2\pi/3}.$$

Then the fourth roots are

$$\sqrt{2}\, e^{i(\pi/6 + k\pi/2)} \quad \text{for} \quad k = 0, 1, 2, 3,$$

that is,

$$\pm\sqrt{2}\left(\frac{\sqrt{3}}{2} + i\frac{1}{2}\right) \quad \text{and} \quad \pm\sqrt{2}\left(-\frac{1}{2} + i\frac{\sqrt{3}}{2}\right).$$

If the set of all complex numbers is denoted by $\mathbf{C}$, then a *complex-valued function* on an interval J is any function f mapping $J \to \mathbf{C}$.

DEFINITIONS

Let J be an interval in $\mathbf{R}$

(i) If f_1 and f_2 are two functions mapping $J \to \mathbf{C}$ and if c is a complex constant, we define new functions $f_1 + f_2$, cf_1, and $f_1 f_2$ mapping $J \to \mathbf{C}$ in the natural manner:

$$(f_1 + f_2)(t) \equiv f_1(t) + f_2(t),$$

$$(cf_1)(t) \equiv cf_1(t),$$

and

$$(f_1 f_2)(t) \equiv f_1(t) f_2(t),$$

for t in J.

(ii) If $f: J \to \mathbb{C}$, we define two functions mapping $J \to \mathbb{R}$ denoted by $\operatorname{Re} f$ and $\operatorname{Im} f$, called the *real and imaginary parts of f*, by

$$(\operatorname{Re} f)(t) \equiv \operatorname{Re}[f(t)], \qquad (\operatorname{Im} f)(t) \equiv \operatorname{Im}[f(t)]$$

for t in J.

(iii) We say $f: J \to \mathbb{C}$ is *continuous* (or *differentiable* or *integrable*) if $\operatorname{Re} f$ and $\operatorname{Im} f$ are both continuous (or differentiable or integrable). If f is differentiable, we define $f': J \to \mathbb{C}$ (also denoted by Df or df/dt) by

$$f'(t) = (\operatorname{Re} f)' + i(\operatorname{Im} f)'(t) \quad \text{for} \quad t \quad \text{in} \quad J.$$

If f is integrable on $J = [a, b]$, we define

$$\int_a^b f(s)\, ds = \int_a^b (\operatorname{Re} f)(s)\, ds + i \int_a^b (\operatorname{Im} f)(s)\, ds.$$

Under definition (iii), the familiar rules of differential and integral calculus carry over to complex-valued functions of a real variable. In particular, let f_1 and f_2 be differentiable (or integrable) functions mapping $J \to \mathbb{C}$ for some interval J. Then, for any complex constants c_1 and c_2, the new function $c_1 f_1 + c_2 f_2$ is differentiable (or integrable) and $f_1 f_2$ is differentiable on J, and

$$(c_1 f_1 + c_2 f_2)' = c_1 f_1' + c_2 f_2' \tag{11}$$

and

$$(f_1 f_2)' = f_1' f_2 + f_1 f_2' \quad \text{on} \quad J \tag{12}$$

or, if $J = [a, b]$,

$$\int_a^b (c_1 f_1 + c_2 f_2)(s)\, ds = c_1 \int_a^b f_1(s)\, ds + c_2 \int_a^b f_2(s)\, ds. \tag{13}$$

Equations (11) and (13) were encountered earlier, as Eqs. 9-(3) and 9-(4), for vector-valued functions. To prove (11), (12), and (13), one considers the real and imaginary parts of each equation. (Problem 9.)

If f_1 and f_2 are two differentiable functions mapping $J \to \mathbb{C}$ for some interval $J \subset \mathbb{R}$ and if

$$f_1'(t) = f_2'(t) \quad \text{for all} \quad t \quad \text{in} \quad J, \tag{14}$$

then there exists some $c \in \mathbb{C}$ such that

$$f_1(t) = f_2(t) + c \quad \text{for all} \quad t \quad \text{in} \quad J. \tag{15}$$

The proof again depends on consideration of the real and imaginary parts. (Problem 10.)

The fundamental theorems of calculus in our complex setting are easily obtained using the corresponding theorems for the real and imaginary parts. These results are as follows, just as for vector-valued functions in Section 9.

If f is a continuous function mapping an interval $J \to \mathbf{C}$ and if $t_0,\, t \in J$, then

$$\frac{d}{dt} \int_{t_0}^{t} f(s)\, ds = f(t) \tag{16}$$

(understanding, as usual, a one-sided derivative if t is an endpoint of J).

If $f: [a, b] \to \mathbf{C}$ has a continuous (or merely integrable) derivative on $[a, b]$, then

$$\int_{a}^{b} f'(s)\, ds = f(b) - f(a). \tag{17}$$

With the aid of these results we can now extend the uniqueness theorems for linear differential equations and linear differential systems to the case of complex-valued functions.

THEOREM A

The uniqueness assertion of Corollary 9-B (or Corollary 10-B) remains valid if the given functions a_{ij} and h_i for $i, j = 1, \ldots, n$ and the initial values, $x_{01}, \ldots, x_{0n}$ (or a_i and h for $i = 1, \ldots, n$ and $b_0, \ldots, b_{n-1}$), are complex-valued. Now a solution x maps an interval into $\mathbf{C}^n$ (or into $\mathbf{C}$).

Proof. Admitting complex values as indicated, consider the linear system

$$x'_k = \sum_{j=1}^{n} a_{kj}(t)x_j + h_k(t) \quad \text{for} \quad k = 1, \ldots, n. \tag{18}$$

with initial conditions

$$x_k(t_0) = x_{0k} \quad \text{for} \quad k = 1, \ldots, n. \tag{19}$$

For any solution x mapping $(\alpha, \beta) \to \mathbf{C}^n$ let us introduce $u_k(t) = \operatorname{Re} x_k(t)$ and $v_k(t) = \operatorname{Im} x_k(t)$ so that

$$x_k(t) = u_k(t) + iv_k(t) \quad \text{for} \quad k = 1, \ldots, n.$$

Now system (18) will be satisfied if and only if the real and imaginary parts of both sides of (18) agree, that is,

$$\left.\begin{aligned}
u'_k &= \sum_{j=1}^{n} [\operatorname{Re} a_{kj}(t)]u_j - \sum_{j=1}^{n} [\operatorname{Im} a_{kj}(t)]v_j + \operatorname{Re} h_k(t) \\
v'_k &= \sum_{j=1}^{n} [\operatorname{Im} a_{kj}(t)]u_j + \sum_{j=1}^{n} [\operatorname{Re} a_{kj}(t)]v_j + \operatorname{Im} h_k(t)
\end{aligned}\right\} \tag{20}$$

for $k = 1, \ldots, n$. System (20) consists of $2n$ real equations for the $2n$ real unknown functions $u_1, \ldots, u_n, v_1, \ldots, v_n$. And the associated real initial conditions, obtained from (19), are

$$u_k(t_0) = \operatorname{Re} x_{0k}, \quad v_k(t_0) = \operatorname{Im} x_{0k} \quad \text{for} \quad k = 1, \ldots, n. \tag{21}$$

The fact that Eqs. (20) and (21) have at most one solution follows from Corollary 9-B since the coefficients in (20) are all real and continuous. This then assures that (18) and (19) have at most one complex vector-valued solution.

The required uniqueness for a complex linear nth-order scalar equation

$$x^{(n)} + a_{n-1}(t)x^{(n-1)} + \cdots + a_1(t)x' + a_0(t)x = h(t)$$

with

$$x(t_0) = b_0, \, x'(t_0) = b_1, \ldots, x^{(n-1)}(t_0) = b_{n-1}$$

can now be shown. One simply transforms this (complex) scalar equation into a system of n (complex) linear first-order equations as was done in the proof of Theorem 10-A. ∎

The complex-valued functions which will be of greatest interest to us are complex exponentials of the form $f(t) = e^{\lambda t}$ for λ complex, say $\lambda = \mu + i\omega$, where μ and ω are real. We have

$$f(t) = e^{\lambda t} = e^{\mu t}(\cos \omega t + i \sin \omega t), \tag{22}$$

so that

$$(\operatorname{Re} f)(t) = e^{\mu t} \cos \omega t \quad \text{and} \quad (\operatorname{Im} f)(t) = e^{\mu t} \sin \omega t.$$

LEMMA B

If λ is a complex constant, then the function f defined by

$$f(t) = e^{\lambda t} \quad \text{for} \quad t \in \mathbf{R}$$

is continuous and differentiable. Moreover,

$$f'(t) = \lambda e^{\lambda t}; \tag{23}$$

and, if $\lambda \neq 0$ and $a, b \in \mathbf{R}$,

$$\int_a^b e^{\lambda t} \, dt = \frac{1}{\lambda} (e^{\lambda b} - e^{\lambda a}). \tag{24}$$

The proof is left to the reader as Problem 11. For (23) one must differentiate (22). Equation (24) is then proved with the aid of (23).

Another useful result from calculus which carries over to complex-valued functions is the following.

LEMMA C

Let $f: [a, b] \to \mathbf{C}$ be continuous (or merely integrable). Then

$$\left| \int_a^b f(t) \, dt \right| \le \int_a^b |f(t)| \, dt.$$

Proof. Since f is integrable it follows that $\operatorname{Re} f$ and $\operatorname{Im} f$ are integrable, on $[a, b]$. Let

$$\int_a^b f(t) \, dt = r e^{i\theta} \quad \text{where} \quad r \ge 0.$$

Then, using Eq. (13),

$$\left| \int_a^b f(t) \, dt \right| = r = e^{-i\theta} \int_a^b f(t) \, dt = \int_a^b e^{-i\theta} f(t) \, dt.$$

But since this last integral must be real we have

$$\left| \int_a^b f(t) \, dt \right| = \operatorname{Re} \int_a^b e^{-i\theta} f(t) \, dt = \int_a^b \operatorname{Re}[e^{-i\theta} f(t)] \, dt \le \int_a^b |e^{-i\theta} f(t)| \, dt.$$

Then, since $|e^{-i\theta}| = (\cos^2 \theta + \sin^2 \theta)^{1/2} = 1$,

$$\left| \int_a^b f(t) \, dt \right| \le \int_a^b |f(t)| \, dt. \quad \blacksquare$$

PROBLEMS

1. Reduce to the form $a + ib$ where a and b are real:
   (a) $(4 - i) - (2 + 3i)(-1 + 5i)$.
   (b) $\dfrac{2 - 7i}{3 + 2i}$.
   (c) $(1 - i)^2 + i^3 - 5i^{17} + 2i^{130}$.

2. Find $\bar{z}$ if $z = (2 + i)^3$.

3. Verify the identities in (1).

4. Verify the properties (2). The proof of the triangle inequality $|z_1 + z_2| \le |z_1| + |z_2|$ begins as follows:

$$|z_1 + z_2|^2 = (z_1 + z_2)(\bar{z}_1 + \bar{z}_2) = |z_1|^2 + z_1\bar{z}_2 + \bar{z}_1 z_2 + |z_2|^2$$

$$= |z_1|^2 + 2 \operatorname{Re}(z_1 \bar{z}_2) + |z_2|^2 \le ?$$

5. Find all roots of the equation $z^4 - 3z^3 + 2z^2 + 2z - 4 = 0$ given that $1 + i$ is a root.

6. Verify the identity (7).

7. Verify properties (9).

8. Find the cube roots of -1.

9. Verify Eqs. (11), (12), and (13).

10. Prove that (14) implies (15).

11. Prove Lemma B.

12. (Important for future reference.) Derive the identities

$$\cos \theta = \frac{e^{i\theta} + e^{-i\theta}}{2} \quad \text{and} \quad \sin \theta = \frac{e^{i\theta} - e^{-i\theta}}{2i}$$

for θ in $\mathbf{R}$.

14. DIFFERENTIAL OPERATORS AND SECOND-ORDER EQUATIONS

This chapter is primarily devoted to the nth-order linear equation with constant coefficients

$$a_n x^{(n)} + a_{n-1} x^{(n-1)} + \cdots + a_1 x' + a_0 x = h(t), \tag{1}$$

where $a_n \neq 0$. We shall often seek a solution of (1) which also satisfies the initial conditions

$$x(t_0) = b_0, \quad x'(t_0) = b_1, \quad \ldots, \quad x^{(n-1)}(t_0) = b_{n-1}. \tag{2}$$

In the special case when $h(t) \equiv 0$ Eq. (1) is said to be a (linear) *homogeneous* equation. (Note that this usage of the word " homogeneous " is different from that in Section 6.) In this case we will sometimes call the unknown function y and write

$$a_n y^{(n)} + a_{n-1} y^{(n-1)} + \cdots + a_1 y' + a_0 y = 0. \tag{3}$$

A useful tool for the systematic discovery of solutions of (1) or (3) is the set of " polynomials " in the differential operator $D = \dfrac{d}{dt}$. Given an ordinary polynomial

$$p(\lambda) = a_n \lambda^n + a_{n-1} \lambda^{n-1} + \cdots + a_1 \lambda + a_0,$$

we define the symbolic *polynomial operator* in D,

$$p(D) = a_n D^n + a_{n-1} D^{n-1} + \cdots + a_1 D + a_0 I,$$

by the requirement that

$$p(D)z = a_n D^n z + a_{n-1} D^{n-1} z + \cdots + a_1 Dz + a_0 z \tag{4}$$

for every function $z = z(t)$ having n derivatives. (The symbol " I " used above stands for the *identity operator* defined by $Iz \equiv z$.) It will be simpler, and quite adequate for our purposes, if we assume henceforth that the functions z used in the definition of $p(D)$ have derivatives of all orders.

Note that with the above notation we can express Eqs. (1) and (3) respectively as

$$p(D)x = h(t) \quad \text{and} \quad p(D)y = 0.$$

It is important to observe that the definition of $p(D)$ makes sense if the coefficients, $a_n, \ldots, a_1$, and a_0, and the functions z are assumed to be complex-valued. Henceforth, let us assume them to be complex-valued unless stated otherwise. The reason for this will soon become clear.

A basic property of $p(D)$ is its *linearity*. That is if z_1 and z_2 are functions (possibly complex-valued) and c_1 and c_2 are constants (possibly complex), then

$$p(D)(c_1 z_1 + c_2 z_2) = c_1 p(D)z_1 + c_2 p(D)z_2 . \tag{5}$$

This follows easily from (4) and the linearity of each D^j,

$$D^j(c_1 z_1 + c_2 z_2) = c_1 D^j z_1 + c_2 D^j z_2 .$$

(Problem 1.) Now assume we have another polynomial, $q(\lambda) = b_m \lambda^m + b_{m-1}\lambda^{m-1} + \cdots + b_1\lambda + b_0$, with complex coefficients, so that the effect of

$$q(D) = b_m D^m + b_{m-1} D^{m-1} + \cdots + h_1 D + b_0 I$$

is also well defined on every (sufficiently differentiable) complex-valued z. We shall then define *addition* and *multiplication* of the operators $p(D)$ and $q(D)$ by the equations

$$[p(D) + q(D)]z = p(D)z + q(D)z$$

and $\tag{6}$

$$[p(D)q(D)]z = p(D)[q(D)z]$$

for *every* infinitely differentiable function z.

The principal usefulness of the polynomial operator notation lies in the following results. Let $r(\lambda)$ and $s(\lambda)$ be the ordinary polynomials defined by

$$r(\lambda) \equiv p(\lambda) + q(\lambda) \quad \text{and} \quad s(\lambda) \equiv p(\lambda)q(\lambda).$$

Then it can be shown that

$$\left.\begin{aligned} p(D) + q(D) &= r(D) \\[2ex] p(D)q(D) &= s(D). \end{aligned}\right\} \tag{7}$$

and

(Problem 2.) These equations should not be considered obvious. What they assert is that, for every infinitely differentiable function z,

$$\left.\begin{aligned} p(D)z + q(D)z &= r(D)z \\[2ex] p(D)[q(D)z] &= s(D)z. \end{aligned}\right\} \tag{7'}$$

and, more subtly,

From Eqs. (7) it follows that addition and multiplication of polynomial operators in D satisfy the usual associative, commutative, and distributive laws. This implies that *a polynomial in D can be factored just like an ordinary polynomial.* In particular, let

$$p(D) = a_n D^n + a_{n-1} D^{n-1} + \cdots + a_1 D + a_0 I$$

and let $\lambda_1, \lambda_2, \ldots, \lambda_n$ be the n roots (not necessarily distinct) of $p(\lambda) = 0$. Then, as we know,

$$p(\lambda) = a_n(\lambda - \lambda_1)(\lambda - \lambda_2) \cdots (\lambda - \lambda_n).$$

Hence,

$$p(D) = a_n(D - \lambda_1 I)(D - \lambda_2 I) \cdots (D - \lambda_n I). \tag{8}$$

Here is the reason for emphasizing that the various results about polynomial operators are valid in the case of complex coefficients: Even though we will usually start with an Equation (1) or (3) having real coefficients, it is quite likely that some roots of $p(\lambda) = 0$ will be complex, and hence some factors of $p(D)$ will have complex coefficients.

Example 1. Let us return once more to the second-order linear differential equation

$$mx'' + bx' + kx = 0 \tag{9}$$

with real constant coefficients and with $m \neq 0$. The initial conditions, when needed, will be

$$x(t_0) = x_0 \quad \text{and} \quad x'(t_0) = v_0. \tag{10}$$

Equation (9) is equivalent to

$$p(D)x = (mD^2 + bD + kI)x = 0.$$

If λ_1 and λ_2 are the roots of $m\lambda^2 + b\lambda + k = 0$, then, by (8), Eq. (9) is equivalent to

$$m(D - \lambda_1 I)(D - \lambda_2 I)x = 0.$$

This holds regardless of whether λ_1 and λ_2 are real or complex. Introducing $x_1 \equiv (D - \lambda_2 I)x$, we have $(D - \lambda_1 I)x_1 = 0$, or

$$x_1' - \lambda_1 x_1 = 0,$$

which implies $x_1(t) = c_1 e^{\lambda_1 t}$, where c_1 is a constant. Note that this too is true for real *or* complex λ_1 by virtue of the results of Section 13. To find x, it remains to solve $(D - \lambda_2 I)x = x_1$, that is

$$x' - \lambda_2 x = c_1 e^{\lambda_1 t}. \tag{11}$$

To this point, we have not said whether λ_1 and λ_2 are distinct or identical numbers. But now, in solving Eq. (11), it becomes necessary to distinguish between those two cases.

If $\lambda_2 \neq \lambda_1$ (that is, if $b^2 \neq 4mk$), use of the integrating factor $e^{-\lambda_2 t}$ leads to the solution of (11),

$$x(t) = \frac{c_1}{\lambda_1 - \lambda_2} e^{\lambda_1 t} + c_2 e^{\lambda_2 t}.$$

(Problem 3(a).) By renaming the constant c_1, we can rewrite this in the form

$$x(t) = c_1 e^{\lambda_1 t} + c_2 e^{\lambda_2 t}. \tag{12}$$

If λ_1 and λ_2 are real, this is exactly what we found in Example 10-1, and in that case we saw that (12) is the general solution of Eq. (9). If, on the other hand, λ_1 and λ_2 are complex numbers, then, with the aid of Lemma 13-B, the reader should verify by direct substitution that x as defined by (12) does indeed satisfy Eq. (9). We will show soon that it is the general solution.

In case of a double root, $\lambda_2 = \lambda_1 = \lambda$ (that is, if $b^2 = 4mk$), the same procedure gives, as the solution of (11),

$$x(t) = c_1 t e^{\lambda t} + c_2 e^{\lambda t}. \tag{13}$$

(Problem 3(b).) The reader is asked to prove that the constants c_1 and c_2 in (13) will always be (uniquely) determined by the initial conditions (10) for any t_0. (Problem 4.)

Let us now assume that m and k are positive numbers and that $b \geq 0$, this being the natural case in physical applications. Then the behavior of solutions can be classified in terms of the magnitude of the "damping coefficient" b.

(a) If $b^2 < 4mk$, the system is said to be *underdamped*. We then have $\lambda_1 = \mu + i\omega$ and $\lambda_2 = \mu - i\omega$, a pair of complex conjugate numbers, where

$$\mu = \frac{-b}{2m} < 0 \quad \text{and} \quad \omega = \frac{(4mk - b^2)^{1/2}}{2m} > 0. \tag{14}$$

In this case let us rewrite Eq. (12) as

$$x(t) = c_1 e^{\mu t}(\cos \omega t + i \sin \omega t) + c_2 e^{\mu t}(\cos \omega t - i \sin \omega t),$$

where the constants c_1 and c_2 may be real or complex. If we introduce new constants

$$C_1 \equiv c_1 + c_2 \quad \text{and} \quad C_2 \equiv ic_1 - ic_2,$$

the solution takes the form

$$x(t) = C_1 e^{\mu t} \cos \omega t + C_2 e^{\mu t} \sin \omega t. \tag{15}$$

Again it appears that C_1 and C_2 may be real or complex numbers.

Now let us consider Eq. (9) in conjunction with the initial conditions (10) where x_0 and v_0 are real. Is it possible to choose C_1 and C_2 so that $x(t)$ as defined by (15) satisfies (10)?

Substitution of (15) into Eqs. (10) leads to two conditions on the constants C_1 and C_2:

$$
\left.
\begin{aligned}
C_1 e^{\mu t_0} \cos \omega t_0 + C_2\, e^{\mu t_0} \sin \omega t_0 &= x_0, \\
C_1(\mu e^{\mu t_0} \cos \omega t_0 - \omega e^{\mu t_0} \sin \omega t_0) & \\
+ C_2(\mu e^{\mu t_0} \sin \omega t_0 + \omega e^{\mu t_0} \cos \omega t_0) &= v_0.
\end{aligned}
\right\} \tag{16}
$$

To determine whether these equations are solvable for C_1 and C_2, we calculate the determinant of the coefficients

$$
\begin{vmatrix}
e^{\mu t_0} \cos \omega t_0 & e^{\mu t_0} \sin \omega t_0 \\
(\mu e^{\mu t_0} \cos \omega t_0 - \omega e^{\mu t_0} \sin \omega t_0) & (\mu e^{\mu t_0} \sin \omega t_0 + \omega e^{\mu t_0} \cos \omega t_0)
\end{vmatrix}
= \omega e^{2\mu t_0}.
$$

Since $\omega e^{2\mu t_0} \neq 0$, regardless of the value of t_0, it follows that Eqs. (16) can *always* be solved for C_1 and C_2. Putting the resulting values of C_1 and C_2 into (15) we obtain *a* function, x, on **R** which satisfies (9) and (10). And Corollary 10-B assures us that this is *the* unique solution of (9) and (10).

Note that C_1 and C_2 obtained from Eqs. (16) will be *real* numbers because x_0, v_0, μ, ω, and t_0 are all real. Hence the unique solution (15) of Eqs. (9) and (10) will be a real-valued function, as of course it must be if it is to represent the position of an object bouncing on the end of a spring.

Equation (15) represents damped oscillatory motion as illustrated in Figure 1. The solution is contained in an "envelope" between $Ae^{\mu t}$ and $-Ae^{\mu t}$ for $A = (C_1^2 + C_2^2)^{1/2}$; and the solution has infinitely many zeros spaced π/ω units apart. [Problem 5(a).]

(b) If $b^2 > 4mk$, Eq. (9) is said to be *overdamped*, and the solution is given by (12) where λ_1 and λ_2 are distinct negative numbers. The shape of the solution curve now depends on the initial conditions (10) as illustrated in Figure 2. It can be shown that, unless it is identically zero, the solution can become zero for at most one value of $t > t_0$. [Problem 5(b).]

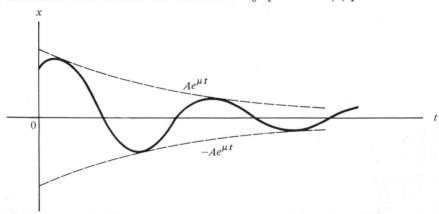

FIGURE 1. **Underdamped motion.**

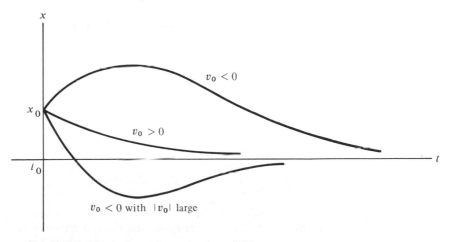

FIGURE 2. **Overdamped motion.**

(c) If $b^2 = 4mk$, Eq. (9) is said to be *critically damped*, and the solution is given by (13) with $\lambda = -b/2m < 0$. In this case the possible types of solution resemble those in Figure 2 for the overdamped case. [Problem 5(c).]

Note that in the case of overdamping, the solution (12) contains one exponential, say $e^{\lambda_1 t}$, which decays more slowly than in either the underdamped or critically damped cases, that is, $\lambda_1 > -b/2m$. Thus it is possible that the solution in this case may "die out" more slowly than in the case of a smaller damping coefficient, such as in critical damping.

In an automobile suspension system, for example, the shock absorbers provide damping. It is desirable that the system have critical damping or slight overdamping so that the vehicle does not keep on bouncing after hitting a bump. In fact, if after hitting an isolated bump or rut your car bounces through the "equilibrium level" more than once, then your shock absorbers are bad.

PROBLEMS

1. Prove that $p(D)$ as defined by Eq. (4) is linear, that is, satisfies Eq. (5).
2. Prove Eqs. (7), that is, (7'),

   (a) for the special case $p(D) = a_1 D + a_0 I$ and $q(D) = b_1 D + b_0 I$,
   *(b) in general.

3. Carry out the solution of Eq. (11) and verify your answers by direct substitution into (9) in case

   (a) $\lambda_2 \neq \lambda_1$ (real or complex), and
   (b) $\lambda_2 = \lambda_1 = \lambda$ (real).

4. Prove that the constants c_1 and c_2 in (13) are always (uniquely) determined if initial conditions (10) are given.

5. Assume $m > 0$, $b \geq 0$, $k > 0$ in Eq. (9) and x_0 and v_0 real in Eq. (10).

   (a) For the underdamped case, $b^2 < 4mk$, show that (15) can be rewritten as

   $$x(t) = Ae^{\mu t} \sin(\omega t + \phi),$$

   where $A = (C_1^2 + C_2^2)^{1/2}$, $\sin \phi = C_1/A$, and $\cos \phi = C_2/A$. Use this to verify the assertions about Figure 1.

   (b) For the overdamped case, $b^2 > 4mk$, verify that solutions can take the three shapes indicated in Figure 2. Specifically, show that if $x(t_0) = x_0$ and $x'(t_0) = v_0$, then

   $$x(t) = \frac{\lambda_2 x_0 - v_0}{\lambda_2 - \lambda_1} e^{\lambda_1(t - t_0)} + \frac{v_0 - \lambda_1 x_0}{\lambda_2 - \lambda_1} e^{\lambda_2(t - t_0)}.$$

   From this prove that (unless x is identically zero) $x(t)$ has at most one zero for $t > t_0$; and in order that $x(t)$ have a zero for some $t > t_0$ one must have v_0 and x_0 of opposite sign with $v_0/x_0 < \lambda_2$, (i.e., $|v_0|$ sufficiently large), where $\lambda_2 < \lambda_1 < 0$.

   (c) For the critically damped case, $b^2 = 4mk$, $\lambda_1 = \lambda_2 = \lambda = -b/2m < 0$. Show that the solution is now

   $$x(t) = (v_0 - \lambda x_0)(t - t_0)e^{\lambda(t - t_0)} + x_0 e^{\lambda(t - t_0)}.$$

   Again show that (if x is not identically zero) $x(t)$ can be zero at most once for $t > t_0$ and this occurs if $v_0/x_0 < \lambda$. Why can you be sure that $x(t) \to 0$ as $t \to \infty$?

6. Find the general solution of the nonhomogeneous equation $x'' - 4x' + 4x = e^{3t}$ by imitating the calculations in Example 1.

*7. The method of solution illustrated in Example 1 can sometimes be applied to equations with variable coefficients.

   (a) Show that the equation

   $$t^2 x'' + (t^3 + t)x' + (t^2 - 1)x = 2t^3$$

   is equivalent to

   $$(D + tI)(D + t^{-1}I)x = 2t \quad \text{as long as} \quad t \neq 0,$$

   where the product of operators is defined as in Eq. (6).

   (b) Solve the equation in part (a).

   (c) Show that $(D + tI)(D + t^{-1}I) \neq (D + t^{-1}I)(D + tI)$, so that multiplication of operators with variable coefficients is *not commutative*, in general.

*8. Find the general solution of $x'' + (1 - t^2)x = 0$. [*Hint:* First verify that the operator factors into $(D - tI)(D + tI)$.]

*9. If y is any solution of $y'' + a_1(t)y' + a_0(t)y = 0$ with $y(t) \neq 0$ for $t \in (\alpha, \beta)$, show that the operator $D^2 + a_1(t)D + a_0(t)I$ can be factored as follows:

   $$D^2 + a_1(t)D + a_0(t)I = [D + a_1(t)I + b(t)I][D - b(t)I]$$

   where $b(t) = y'(t)/y(t)$.

*10. Apply Problem 9 to find another factorization (besides the one given) of the operator occurring in Problem 7(a):

   $$D^2 + (t + t^{-1})D + (1 - t^{-2})I.$$

15. CONSTANT COEFFICIENTS (THE HOMOGENEOUS CASE)

Any polynomial in D with constant coefficients can be written as a product of first-order factors. Thus, in principle, we could use the method of Section 14 to solve any linear equation with constant coefficients,

$$a_n x^{(n)} + a_{n-1} x^{(n-1)} + \cdots + a_1 x' + a_0 x = h(t), \tag{1}$$

where $a_n \neq 0$. This would involve n integrations, each of which would contribute an arbitrary constant of integration. Thus one could hope to also satisfy n initial conditions

$$x(t_0) = b_0, \quad x'(t_0) = b_1, \quad \ldots, \quad x^{(n-1)}(t_0) = b_{n-1}. \tag{2}$$

But the actual computations using this method would become unwieldy unless the order of the equation were small and the right-hand side, $h(t)$, were quite simple. We shall develop easier ways of solving such an equation.

In this section we concentrate on the homogeneous equation associated with (1),

$$a_n y^{(n)} + a_{n-1} y^{(n-1)} + \cdots + a_1 y' + a_0 y = 0. \tag{3}$$

We shall begin, as in Example 10-1, by seeking exponential solutions of the form $e^{\lambda t}$. Substituting $y(t) = e^{\lambda t}$ into Eq. (3) one finds

$$(a_n \lambda^n + a_{n-1} \lambda^{n-1} + \cdots + a_1 \lambda + a_0)e^{\lambda t} = 0.$$

Since $e^{\lambda t}$ is never zero, we must require

$$a_n \lambda^n + a_{n-1} \lambda^{n-1} + \cdots + a_1 \lambda + a_0 = 0. \tag{4}$$

This nth-degree polynomial equation is called the *characteristic equation* for (3) or (1). For each root, $\lambda_1, \lambda_2, \ldots,$ or λ_n of (4),

$$y_j(t) = e^{\lambda_j t} \quad (j = 1, \ldots, \quad \text{or} \quad n)$$

defines a solution of (3). And it follows from the linearity of $p(D)$ that an arbitrary linear combination of these,

$$y(t) = c_1 e^{\lambda_1 t} + c_2 e^{\lambda_2 t} + \cdots + c_n e^{\lambda_n t}, \tag{5}$$

also defines a solution. We thus obtain a solution with n arbitrary constants, and we can *hope* that it will always be possible to choose $c_1, \ldots, c_n$ to satisfy any given initial conditions of the form

$$y(t_0) = b_0, \quad y'(t_0) = b_1, \quad \ldots, \quad y^{(n-1)}(t_0) = b_{n-1}. \tag{6}$$

But how shall we proceed if Eq. (4) has multiple roots, so that there are only, say, $q < n$ distinct roots, $\lambda_1, \lambda_2, \ldots, \lambda_q$? In this case Eq. (5) would effectively involve no more than q arbitrary constants.

Such a situation was encountered in Example 14-1 when $b^2 = 4mk$. Then the characteristic equation $m\lambda^2 + b\lambda + k = 0$ had only one root, $\lambda = -b/2m$.

This was, of course, a double root (i.e., a root of multiplicity 2). In that case we found that the differential equation, which was equivalent to

$$(D - \lambda I)^2 y = 0,$$

had the solution

$$y(t) = c_1 t e^{\lambda t} + c_2 e^{\lambda t},$$

for all values of the constants c_1 and c_2. That is, in addition to $e^{\lambda t}$, another solution was defined by $t e^{\lambda t}$.

We shall find that this is a prototype for the general case. Lemmas A and B will establish that if Eq. (4) has a root λ of multiplicity m, then $e^{\lambda t}$, $t e^{\lambda t}$, $t^2 e^{\lambda t}$, ..., and $t^{m-1} e^{\lambda t}$ are all solutions of (3).

LEMMA A

Let $\lambda \in \mathbf{C}$, and let m be a positive integer. Then $(D - \lambda I)^m (t^k e^{\lambda t}) = 0$ for $k = 0, 1, \ldots, m - 1$.

Proof. Whether λ is real or complex,

$$(D - \lambda I)(t^k e^{\lambda t}) = k t^{k-1} e^{\lambda t} + \lambda t^k e^{\lambda t} - \lambda t^k e^{\lambda t} = k t^{k-1} e^{\lambda t}.$$

In case $k = 0$ the assertion of the lemma follows at once.

If $k \geq 1$, it follows by induction on m that

$$(D - \lambda I)^m (t^k e^{\lambda t}) = k(k - 1) \cdots (k - m + 1) t^{k-m} e^{\lambda t} \quad \text{for} \quad m = 1, \ldots, k.$$

In particular we have

$$(D - \lambda I)^k (t^k e^{\lambda t}) = k! e^{\lambda t}.$$

So

$$(D - \lambda I)^m (t^k e^{\lambda t}) = 0 \quad \text{for each} \quad m > k. \quad \blacksquare$$

LEMMA B

Let λ_j with multiplicity m_j for $j = 1, \ldots, q$ be the roots of (4). (Thus $m_1 + m_2 + \cdots + m_q = n$.) Then the function defined by

$$y(t) = P_1(t) e^{\lambda_1 t} + P_2(t) e^{\lambda_2 t} + \cdots + P_q(t) e^{\lambda_q t}, \tag{7}$$

where P_j is an arbitrary polynomial of degree $m_j - 1$ or lower, is a solution of (3) on $\mathbf{R}$.

Proof. Since we can express the "characteristic polynomial," $p(\lambda) = a_n \lambda^n + a_{n-1} \lambda^{n-1} + \cdots + a_1 \lambda + a_0$, in factored form as

$$p(\lambda) = a_n (\lambda - \lambda_1)^{m_1} (\lambda - \lambda_2)^{m_2} \cdots (\lambda - \lambda_q)^{m_q},$$

it follows that $p(D)$ can be factored similarly. So the differential equation (3) can be rewritten as

$$a_n (D - \lambda_1 I)^{m_1} (D - \lambda_2 I)^{m_2} \cdots (D - \lambda_q I)^{m_q} y = 0. \tag{3'}$$

Moreover, the factors in this product of operators can be written in any order. Thus it follows from Lemma A that each function $t^k e^{\lambda_j t}$, for $k = 0, 1, \ldots,$ $m_j - 1$ and $j = 1, \ldots, q$, defines a solution of Eq. (3). But y as defined in (7) is just a linear combination of these functions, and hence is also a solution of (3) thanks to the linearity of $p(D)$. ∎

Example 2. Let us apply Lemma B to find solutions of

$$y^{(4)} - 6y^{(3)} + 12y'' - 8y' = 0.$$

The characteristic equation,

$$\lambda^4 - 6\lambda^3 + 12\lambda^2 - 8\lambda = 0,$$

has solutions $\lambda = 0, 2, 2, 2$. Thus solutions of the homogeneous equation are defined by

$$y(t) = c_1 + c_2 e^{2t} + c_3 te^{2t} + c_4 t^2 e^{2t},$$

for arbitrary constants $c_1, c_2, c_3,$ and c_4.

Remark. Usually, the coefficients in Eq. (3) will be real numbers. Then the coefficients in the characteristic equation (4) will be real, and it follows that any complex roots must occur in conjugate pairs with the same multiplicities. Thus, if $\lambda_j = \mu + i\omega$ is a complex root with multiplicity $m_j = m$, for some $j = 1, \ldots,$ or q, then for *some other* index k we must have $\lambda_k = \bar{\lambda}_j = \mu - i\omega$ and $m_k = m$. In this case it will be convenient to replace the corresponding terms in (7),

$$P_j(t)e^{\lambda_j t} + P_k(t)e^{\lambda_k t},$$

where P_j and P_k are polynomials of degree $m - 1$ or lower, with

$$Q(t)e^{\mu t} \cos \omega t + R(t)e^{\mu t} \sin \omega t, \tag{8}$$

where Q and R are polynomials of degree $m - 1$ or lower. The reader should verify (Problem 1) that this is always possible and that

$$Q(t) = P_j(t) + P_k(t) \quad \text{and} \quad R(t) = iP_j(t) - iP_k(t). \tag{9}$$

The solution of Eq. (3) defined by (7) involves $m_1 + m_2 + \cdots + m_q = n$ arbitrary constants. And we can expect, from the uniqueness theorems and from the examples considered thus far, that the "general solution" of Eq. (3) will be a function containing n arbitrary constants. But one must be careful not to jump to conclusions too quickly from this.

The fact that a solution of (3) contains n arbitrary constants does *not* prove that it is the general solution.

Example 3. It is easily verified that the equation

$$y''' + 4y' = 0$$

is satisfied by

$$y(t) = c_1 + c_2 \cos 2t + c_3 \sin^2 t,$$

where c_1, c_2, and c_3 are arbitrary constants.

But is this the general solution? In other words, can every possible solution be obtained from this expression by appropriate choices of c_1, c_2, and c_3?

The answer in this case is "no" because we do not have three *independent* constants at our disposal. Since $\sin^2 t = \frac{1}{2} - \frac{1}{2} \cos 2t$, our solution can be rewritten as

$$y(t) = (c_1 + \tfrac{1}{2}c_3) + (c_2 - \tfrac{1}{2}c_3)\cos 2t.$$

Thus there are really only two arbitrary constants, $(c_1 + \tfrac{1}{2}c_3)$ and $(c_2 - \tfrac{1}{2}c_3)$. To be specific, a solution of $y''' + 4y' = 0$ which is *not* covered by the above is $y(t) = \sin 2t$.

The next theorem asserts that the difficulty encountered in Example 3 will not occur if the solution of Eq. (3) with n constants is obtained via Lemma B (modified according to (8) if Eq. (4) has complex roots).

THEOREM C

Let a_n, ..., a_0 be real constants with $a_n \neq 0$ and let the roots of the characteristic equation (4) be the complex conjugate pairs

$$\mu_j \pm i\omega_j \quad \text{with multiplicity} \quad m_j \quad \text{for} \quad j = 1, \ldots, r,$$

and the real roots

$$\lambda_j \quad \text{with multiplicity} \quad m_j \quad \text{for} \quad j = 2r + 1, \ldots, q.$$

(Thus $\sum_{j=1}^{r} 2m_j + \sum_{j=2r+1}^{q} m_j = n$.) Then the general solution of Eq. (1) is given, for all t, by

$$y(t) = \sum_{j=1}^{r} [Q_j(t) \cos \omega_j t + R_j(t) \sin \omega_j t]e^{\mu_j t} + \sum_{j=2r+1}^{q} P_j(t)e^{\lambda_j t}, \qquad (10)$$

where each P_j, Q_j, and R_j is an arbitrary polynomial of degree $\leq m_j - 1$.

The proof of this theorem will be given in Section 16.

Example 4. Find the general solution of the 11'th order equation

$$(D + 3I)(D - 2I)^4(D^2 + 4D + 5I)(D^2 + 4I)^2 y = 0.$$

The roots of the characteristic equation are

$$\lambda = -3, \text{ with multiplicity } m = 1,$$
$$\lambda = 2, \text{ with multiplicity } m = 4,$$
$$\lambda = -2 \pm i, \text{ each with multiplicity } m = 1,$$
$$\lambda = \pm 2i, \text{ each with multiplicity } m = 2.$$

Thus the general solution (with 11 arbitrary constants) is

$$y(t) = c_1 e^{-3t} + (c_2 + c_3 t + c_4 t^2 + c_5 t)^3 e^{2t} + c_6 e^{-2t} \cos t + c_7 e^{-2t} \sin t$$
$$+ (c_8 + c_9 t)\cos 2t + (c_{10} + c_{11} t)\sin 2t.$$

PROBLEMS

1. If λ_k and λ_j are complex conjugate roots of Eq. (4) each having multiplicity m, verify that the corresponding terms in (7) can be rewritten as in (8).
2. Find the general solution in the form (10) for each of the following equations.

   (a) $y'' - 9y = 0.$
   (b) $y'' + 9y = 0.$
   (c) $y'' - 6y' + 9y = 0.$
   (d) $y'' + 4y' + 9y = 0.$
   (e) $y''' + 3y'' + 3y' + y = 0.$
   (f) $y^{(4)} - 2y''' + 2y'' - 2y' + y = 0.$
   (g) $y^{(4)} + 2y'' + y = 0.$

3. Solve the following equations of Problem 2 with the indicated initial conditions:

   (a) with $y(0) - 0,\ y'(0) = 6.$
   (b) with $y(\pi/2) = 2,\ y'(\pi/2) = -1.$
   (c) with $y(1) = -1,\ y'(1) = 2.$

16. SUPERPOSITION AND LINEAR INDEPENDENCE

The previous section developed a procedure for finding a solution involving n arbitrary constants, $c_1, \ldots, c_n$, of a linear homogeneous nth-order differential equation with constant coefficients. The present section will provide a proof that this procedure gives the general solution.

The concepts, methods, and theorems in this section actually apply to a more general linear differential equation with variable coefficients,

$$x^{(n)} + a_{n-1}(t)x^{(n-1)} + \cdots + a_1(t)x' + a_0(t)x = h(t). \tag{1}$$

The assumption made here that the coefficient of $x^{(n)}$ is "one" is merely a matter of convenience. In case the leading term is $a_n(t)x^{(n)}$ where $a_n(t) \neq 0$ one could divide the equation through by $a_n(t)$ to put it into the form of Eq. (1).

Our ultimate objective is to find a (unique) solution of Eq. (1) which satisfies given initial conditions

$$x(t_0) = b_0, \quad x'(t_0) = b_1, \quad \ldots, \quad x^{(n-1)}(t_0) = b_{n-1}, \tag{2}$$

where $t_0, b_0, \ldots,$ and b_{n-1} are given.

We shall soon find that the analysis of Eq. (1) is related to the study of the associated homogeneous equation

$$y^{(n)} + a_{n-1}(t)y^{(n-1)} + \cdots + a_1(t)y' + a_0(t)y = 0. \tag{3}$$

The basic feature of the linear homogeneous equation (3) is that any linear combination of solutions is again a solution. Furthermore, any solution of (3) can be added to any solution of (1) to produce another solution of (1). These properties, and more, are covered by the following "superposition" theorem. It is really nothing more than an assertion of the linearity of a "polynomial operator" in D with variable coefficients. [Compare Eq. 14-(5) for the constant coefficient case.]

THEOREM A (SUPERPOSITION)

Let $h(t) = c_1 h_1(t) + \cdots + c_k h_k(t)$ where $c_1, \ldots, c_k$ are constants, and for $j = 1, \ldots, k$ let x_j be a (particular) solution of

$$x^{(n)} + a_{n-1}(t)x^{(n-1)} + \cdots + a_0(t)x = h_j(t)$$

on an interval $J = (\alpha, \beta)$. Then

$$x \equiv c_1 x_1 + \cdots + c_k x_k$$

is a solution of (1) on J. In case some h_j's are identically zero the corresponding x_j's can be any solutions of Eq. (3).

Proof. Substituting x into the left-hand side of (1), we find

$$x^{(n)} + \cdots + a_0(t)x = \sum_{j=1}^{k} c_j[x_j^{(n)} + \cdots + a_0(t)x_j]$$

$$= \sum_{j=1}^{k} c_j h_j(t) = h(t). \quad \blacksquare$$

We *expect* the general solution of Eq. (1) or (3) to involve n arbitrary constants. But, as Example 15-3 showed, we *cannot* conclude that just because a solution contains n arbitrary constants it must be the general solution.

To make matters precise we introduce the following concepts from linear algebra.

DEFINITIONS

A set of functions, $v_1, \ldots, v_k$ (possibly complex valued) on an interval J is said to be *linearly dependent* if there exist constants $c_1, \ldots, c_k$, not all zero, such that

$$c_1 v_1(t) + \cdots + c_k v_k(t) = 0 \quad \text{for all} \quad t \quad \text{in} \quad J.$$

Otherwise the functions are said to be *linearly independent*. In other words, $v_1, \ldots, v_k$ is a linearly independent set if $c_1 v_1 + \cdots + c_k v_k$ is the zero function only when $c_1 = c_2 = \cdots = c_k = 0$.

Example 1. The three functions $v_1(t) = 1$, $v_2(t) = \cos 2t$, and $v_3(t) = \sin^2 t$ (encountered in Example 15-3) are linearly dependent on any interval J since

$$1v_1(t) + (-1)v_2(t) + (-2)v_3(t) = 0 \quad \text{for all} \quad t \quad \text{in} \quad \mathbf{R}.$$

Example 2. Define v_1 and v_2 on $(-1, 1)$ by

$$v_1(t) = \begin{cases} t^3 & \text{for} \quad -1 < t \le 0 \\ 0 & \text{for} \quad 0 < t < 1 \end{cases} \quad \text{and} \quad v_2(t) = \begin{cases} 0 & \text{for} \quad -1 < t \le 0 \\ t^3 & \text{for} \quad 0 < t < 1. \end{cases}$$

Then v_1 and v_2 are linearly independent. For if $c_1 v_1(t) + c_2 v_2(t) = 0$ for $-1 < t < 1$ we have only to set $t = -\frac{1}{2}$ to conclude that we must have $c_1 = 0$ and set $t = \frac{1}{2}$ to see that $c_2 = 0$ also.

A particular determinant function which is closely related to the concept of linear independence is given a special name.

DEFINITION

If $v_1, \ldots, v_n$ are (real- or complex-valued) functions on an interval J and each has $n - 1$ derivatives, their *Wronskian* is the function defined on J by

$$W(v_1, \ldots, v_n) = \begin{vmatrix} v_1 & \cdots & v_n \\ v_1' & \cdots & v_n' \\ \vdots & & \vdots \\ v_1^{(n-1)} & \cdots & v_n^{(n-1)} \end{vmatrix}.$$

The value of this function at t will be denoted by $W(v_1, \ldots, v_n)(t)$ or sometimes simply by $W(t)$.

We have already encountered the Wronskian in special cases in Example 10-1 and 14-1. The following theorems extend the method used in those examples.

THEOREM B

(i) Let $y_1, \ldots, y_n$ be functions on $J = (\alpha, \beta)$ each having $n - 1$ derivatives. If $W(t_1) = W(y_1, \ldots, y_n)(t_1) \ne 0$ for some t_1 in J, then $y_1, \ldots, y_n$ are linearly independent functions. (The converse is false.)

(ii) More specifically, let $y_1, \ldots, y_n$ be solutions of Eq. (3) on J, where $a_{n-1}, \ldots, a_1$, and a_0 are continuous functions. Then $y_1, \ldots, y_n$ are linearly independent if and only if $W(t) \ne 0$ for every t in J.

Proof. (i) Let $W(t_1) \ne 0$ for some t_1 in J, and assume that for some choice of constants $c_1, \ldots, c_n$ one has

$$c_1 y_1(t) + \cdots + c_n y_n(t) = 0 \quad \text{for all} \quad t \quad \text{in} \quad J.$$

Then it follows also that

$$c_1 y_1'(t) + \cdots + c_n y_n'(t) = 0,$$
$$\vdots$$
$$c_1 y_1^{(n-1)}(t) + \cdots + c_n y_n^{(n-1)}(t) = 0.$$

Replacing t by t_1, the above becomes a system of n linear algebraic equations for the numbers $c_1, \ldots, c_n$; and the determinant of the coefficients is $W(t_1) \neq 0$. Thus $c_1 = c_2 = \cdots = c_n = 0$ is the only solution.

To see that the converse of (i) is false, it suffices to consider $n = 2$ with $y_1 = v_1$ and $y_2 = v_2$, the functions of Example 2. Then it is easy to verify that $W(t) = W(v_1, v_2)(t) = 0$ for $-1 < t < 1$. However, as shown in Example 2, v_1 and v_2 are linearly independent.

(ii) Now consider the special case when $y_1, \ldots, y_n$ are solutions of some nth-order linear homogeneous differential equation (3) with continuous coefficients. The "if" part of assertion (ii) is covered by (i).

To prove the "only if" assertion, let us assume that $W(t_1) = 0$ for *some* t_1 in J. Then a theorem from algebra asserts that the system of equations for $c_1, \ldots, c_n$,

$$c_1 y_1(t_1) + \cdots + c_n y_n(t_1) = 0,$$

$$c_1 y_1'(t_1) + \cdots + c_n y_n'(t_1) = 0,$$

$$\vdots$$

$$c_1 y_1^{(n-1)}(t_1) + \cdots + c_n y_n^{(n-1)}(t_1) = 0,$$

has a nontrivial solution, that is, a solution for $c_1, \ldots, c_n$ not all zero. Using these values of $c_1, \ldots, c_n$, consider the solution of Eq. (3) defined by

$$y \equiv c_1 y_1 + \cdots + c_n y_n \quad \text{on} \quad J.$$

Since $y(t_1) = 0$, $y'(t_1) = 0$, $\ldots$, and $y^{(n-1)}(t_1) = 0$, the uniqueness theorem asserts that $y(t) \equiv 0$. In other words, the functions $y_1, \ldots, y_n$ are *linearly dependent*. Thus $y_1, \ldots, y_n$ can be linearly independent only if their Wronskian is nonzero for all t in J. ∎

COROLLARY C

Let $y_1, \ldots, y_n$ be solutions of Eq. (3) on an interval J where the coefficient functions $a_{n-1}, \ldots, a_1$, and a_0 are continuous. Then $W(t) = W(y_1, \ldots, y_n)(t)$ is either zero for all t in J or different from zero for all t in J.

Proof. If $W(t) \neq 0$ for some t in J, then $y_1, \ldots, y_n$ are linearly independent. But this implies that $W(t) \neq 0$ for all t in J. ∎

We are now ready to describe the form of the general solution of Eq. (1).

THEOREM D

If Eq. (3) with continuous coefficients has n linearly independent solutions $y_1, \ldots, y_n$ on an interval J and if $\tilde{x}$ is a (particular) solution of Eq. (1) on J, then every solution of Eq. (1) on J is the sum of $\tilde{x}$ and some linear combination of $y_1, \ldots, y_n$. That is, the general solution of Eq. (1) on J is

$$x = \tilde{x} + c_1 y_1 + \cdots + c_n y_n, \tag{4}$$

where $c_1, \ldots, c_n$ are unspecified complex constants. Moreover, the constants $c_1, \ldots, c_n$ are determined (uniquely) if initial conditions (2) are specified using any point $(t_0, b_0, \ldots, b_{n-1}) \in J \times \mathbf{R}^n$.

Remark. Nothing we have said thus far proves that n linearly independent solutions of Eq. (3) *exist*. This will eventually be proved under the same assumptions (continuous coefficients) in Section 28. For the present we will be able to apply Theorem D only to those special cases in which we can exhibit the desired set of n linearly independent solutions. The most important such special case will be the case of constant coefficients.

Proof of Theorem D. Theorem A asserts that (4) is a solution of (1) for every choice of the constants $c_1, \ldots, c_n$.

It remains to show that if the initial values (2) are specified for any t_0 in J, then a set of constants $c_1, \ldots, c_n$ can be determined (uniquely) so that (4) represents *a* solution, and hence, by uniqueness, *the* solution. We require

$$\left.\begin{aligned}
c_1 y_1(t_0) + \cdots + c_n y_n(t_0) &= b_0 - \tilde{x}(t_0) \\
c_1 y_1'(t_0) + \cdots + c_n y_n'(t_0) &= b_1 - \tilde{x}'(t_0), \\
&\vdots \\
c_1 y_1^{(n-1)}(t_0) + \cdots + c_n y_n^{(n-1)}(t_0) &= b_{n-1} - \tilde{x}^{(n-1)}(t_0).
\end{aligned}\right\} \quad (5)$$

This is a system of n linear algebraic equations in n unknowns, $c_1, \ldots, c_n$. Since the determinant $W(y_1, \ldots, y_n)(t_0) \neq 0$ by Theorem B (ii), it follows that one can solve for $c_1, \ldots, c_n$ (uniquely). And the resulting function (4) satisfies Eqs. (1) and (2).

Finally, to show that (4) gives the general solution of Eq. (1), let $\hat{x}$ be *any* solution on J. Choose some t_0 in J and reconsider Eqs. (5) with $b_0, b_1, \ldots,$ and b_{n-1} replaced by the numbers $\hat{x}(t_0), \hat{x}'(t_0), \ldots,$ and $\hat{x}^{(n-1)}(t_0)$, respectively. Solve for $c_1, \ldots, c_n$ and put these numbers into Eq. (4). Then the resulting solution x must satisfy the same initial conditions (2) as does $\hat{x}$. Hence, by uniqueness, x must be identical with $\hat{x}$. ∎

COROLLARY E

Equation (3) with continuous coefficients cannot have more than n linearly independent solutions on an interval J.

Proof. Suppose (for contradiction) that $y_1, \ldots, y_{n+1}$ are linearly independent solutions of (3) on J. Then clearly, $y_1, \ldots, y_n$ are linearly independent. Applying Theorem D, we conclude that

$$y_{n+1} = c_1 y_1 + \cdots + c_n y_n$$

for some appropriate choice of constants $c_1, \ldots, c_n$. This contradicts the linear independence of $y_1, \ldots, y_{n+1}$. Why? ∎

Our main application of Theorem D will be to the nth-order linear equation with constant coefficients. In the homogeneous case this can be written as

$$y^{(n)} + a_{n-1}y^{(n-1)} + \cdots + a_1 y' + a_0 y = 0, \tag{3'}$$

where a_0, a_1, ..., and a_{n-1} are constants.

Section 15 gave a procedure for finding n solutions of Eq. (3') on $J = \mathbf{R}$, say y_1, ..., y_n, in terms of the roots of the characteristic equation

$$\lambda^n + a_{n-1}\lambda^{n-1} + \cdots + a_1\lambda + a_0 = 0. \tag{6}$$

If we can show that the n solutions of (3') are linearly independent, then according to Theorem D the general solution of (3') can be expressed in the form

$$y = c_1 y_1 + \cdots + c_n y_n$$

Let us begin by considering a special case.

Example 3. Let $n = 3$ and assume that Eq. (6) has three *distinct* solutions λ_1, λ_2, and λ_3. Then

$$y_1(t) = e^{\lambda_1 t}, \quad y_2(t) = e^{\lambda_2 t}, \quad \text{and} \quad y_3(t) = e^{\lambda_3 t}$$

are solutions of Eq. (3') and we shall prove that they are linearly independent. One way to proceed would be to calculate the Wronskian of these three functions. (Problem 11.) An alternate simpler proof, avoiding the need to evaluate a determinant, is as follows. Suppose that on some interval,

$$c_1 e^{\lambda_1 t} + c_2 e^{\lambda_2 t} + c_3 e^{\lambda_3 t} \equiv 0.$$

(To establish the desired linear independence we must show that this implies $c_1 = c_2 = c_3 = 0$.) Dividing by $e^{\lambda_1 t}$, we have

$$c_1 + c_2 e^{(\lambda_2 - \lambda_1)t} + c_3 e^{(\lambda_3 - \lambda_1)t} = 0.$$

Differentiation now gives

$$c_2(\lambda_2 - \lambda_1)e^{(\lambda_2 - \lambda_1)t} + c_3(\lambda_3 - \lambda_1)e^{(\lambda_3 - \lambda_1)t} = 0.$$

Now divide by $e^{(\lambda_2 - \lambda_1)t}$ to get

$$c_2(\lambda_2 - \lambda_1) + c_3(\lambda_3 - \lambda_1)e^{(\lambda_3 - \lambda_2)t} = 0$$

or, differentiating,

$$c_3(\lambda_3 - \lambda_1)(\lambda_3 - \lambda_2)e^{(\lambda_3 - \lambda_2)t} = 0,$$

which can hold only if $c_3 = 0$. Similarly, one finds that $c_2 = 0$ and $c_1 = 0$.

We shall now adapt the method of Example 3 to prove the following general result.

LEMMA F

Let the roots of the characteristic equation (6) be λ_j with multiplicity m_j for $j = 1, \ldots, q$. (Thus $m_1 + \cdots + m_q = n$.) Then Eq. (3′) has n linearly independent solutions of the form

$$y_{jk}(t) = t^k e^{\lambda_j t} \quad \text{for} \quad k = 0, 1, \ldots, m_j - 1 \quad \text{and} \quad j = 1, \ldots, q.$$

Proof. We already know from Lemma 15-B that each of the functions y_{jk} is a solution of (3′). It remains to prove their linear independence.

For simplicity we shall treat the case $q = 3$ so that the *distinct* roots of Eq. (6) are λ_1, λ_2, and λ_3. The reader will see how to generalize the proof to arbitrary q.

Suppose (for contradiction) that the n functions y_{jk} are linearly dependent. Then

$$P_1(t)e^{\lambda_1 t} + P_2(t)e^{\lambda_2 t} + P_3(t)e^{\lambda_3 t} = 0, \tag{7}$$

where P_1, P_2, and P_3 are some polynomials, not all zero, with the degree of each P_j no greater than $m_j - 1$. Without loss of generality we can assume $P_3(t) \not\equiv 0$, for otherwise we merely relabel the λ_j's. If P_3 has degree l, then

$$P_3(t) = ct^l + \text{lower-order terms},$$

with $c \neq 0$. Multiply Eq. (7) by $e^{-\lambda_1 t}$ to obtain

$$P_1(t) + P_2(t)e^{(\lambda_2 - \lambda_1)t} + P_3(t)e^{(\lambda_3 - \lambda_1)t} = 0.$$

Then, since P_1 is a polynomial of degree at most $m_1 - 1$, differentiating this identity m_1 times annihilates P_1 and leaves a new identity

$$Q_2(t)e^{(\lambda_2 - \lambda_1)t} + Q_3(t)e^{(\lambda_3 - \lambda_1)t} = 0. \tag{8}$$

Moreover, each Q_j is a polynomial of the same degree as P_j (at most $m_j - 1$), and in particular

$$Q_3(t) = (\lambda_3 - \lambda_1)^{m_1} ct^l + \text{lower-order terms}.$$

Now multiply Eq. (8) by $e^{(\lambda_1 - \lambda_2)t}$ and differentiate the resulting identity m_2 times to annihilate $Q_2(t)$. This leaves

$$R_3(t)e^{(\lambda_3 - \lambda_2)t} = 0, \tag{9}$$

where R_3 is the polynomial

$$R_3(t) = (\lambda_3 - \lambda_1)^{m_1}(\lambda_3 - \lambda_2)^{m_2} ct^l + \text{lower-order terms}.$$

Since the exponential in Eq. (9) is not zero, we must have $R_3(t) \equiv 0$ and in particular $c = 0$—a contradiction. ∎

Generally speaking, in practical problems the constant coefficients $a_{n-1}, \ldots, a_0$ in Eq. (3′) will be real. However, as we know, it is entirely possible to have some or all λ_j complex, with the complex roots always occurring in conjugate pairs. In this case the following form of Lemma F is more useful.

THEOREM G

Let $a_{n-1}, \ldots, a_0$ be real constants and let the roots of the characteristic equation (6) be the complex conjugate pairs

$$\mu_j \pm i\omega_j \quad \text{with multiplicity} \quad m_j \quad \text{for} \quad j = 1, \ldots, r,$$

and the real roots

$$\lambda_j \quad \text{with multiplicity} \quad m_j \quad \text{for} \quad j = 2r + 1, \ldots, q.$$

(Thus $\sum_{j=1}^{r} 2m_j + \sum_{j=2r+1}^{q} m_j = n$.) Then Eq. (3') has n linearly independent solutions of the form

$$t^k e^{\mu_j t} \cos \omega_j t, \quad t^k e^{\mu_j t} \sin \omega_j t \quad \text{for} \quad k = 0, 1, \ldots, m_j - 1, \quad j = 1, \ldots, r$$

and

$$t^k e^{\lambda_j t} \quad \text{for} \quad k = 0, 1, \ldots, m_j - 1, \quad j = 2r + 1, \ldots, q.$$

This together with Theorem D proves Theorem 15-C.

*Proof. Let $\mu \pm i\omega$ be a pair of complex conjugate solutions of (4) with multiplicity m. Then we can write (according to Problem 13-12)

$$e^{\mu t} \cos \omega t = \frac{e^{(\mu + i\omega)t} + e^{(\mu - i\omega)t}}{2}$$

and

$$e^{\mu t} \sin \omega t = \frac{e^{(\mu + i\omega)t} - e^{(\mu - i\omega)t}}{2i}.$$

Thus it follows that, for each $k = 0, 1, \ldots, m - 1$,

$$t^k e^{\mu t} \cos \omega t \quad \text{and} \quad t^k e^{\mu t} \sin \omega t$$

define solutions of Eq. (3').

There remains the question of linear independence to complete the proof. Suppose that some linear combination of

$$t^k e^{\mu_j t} \cos \omega_j t, \quad t^k e^{\mu_j t} \sin \omega_j t$$

(for $k = 0, \ldots, m_j - 1; j = 1, \ldots, r$) and

$$t^k e^{\lambda_j t}$$

(for $k = 0, \ldots, m_j - 1; j = 2r + 1, \ldots, q$) vanishes identically. Say

$$\sum_{j=1}^{r} [Q_j(t)e^{\mu_j t} \cos \omega_j t + R_j(t)e^{\mu_j t} \sin \omega_j t] + \sum_{j=2r+1}^{q} P_j(t)e^{\lambda_j t} = 0, \quad (10)$$

where $P_j(t)$, $Q_j(t)$, and $R_j(t)$ are polynomials of degree $m_j - 1$ or lower. Equation (10) can be rewritten as

$$\sum_{j=1}^{r} \left[\frac{Q_j(t)}{2} + \frac{R_j(t)}{2i} \right] e^{(\mu_j + i\omega_j)t} + \sum_{j=1}^{r} \left[\frac{Q_j(t)}{2} - \frac{R_j(t)}{2i} \right] e^{(\mu_j - i\omega_j)t}$$

$$+ \sum_{j=2r+1}^{q} P_j(t) e^{\mu_j t} = 0,$$

or simply

$$\sum_{j=1}^{q} P_j(t) e^{\lambda_j t} = 0,$$

with natural definitions of λ_j and $P_j(t)$ for $j = 1, \ldots, 2r$. Now, by Lemma F, each $P_j(t)$ must be identically zero. In particular each

$$\frac{Q_j(t)}{2} + \frac{R_j(t)}{2i} \equiv 0 \quad \text{and} \quad \frac{Q_j(t)}{2} - \frac{R_j(t)}{2i} \equiv 0,$$

which imply

$$Q_j(t) = 0 \quad \text{and} \quad R_j(t) = 0.$$

So all coefficients in (10) must be zero, proving the required linear independence. ∎

PROBLEMS

1. Determine whether the sets of functions defined as follows are linearly dependent or linearly independent.

    (a) $v_1(t) = \sin t$, $v_2(t) = \cos t$ for $0 < t < \pi$.
    (b) $v_1(t) = \sin t$, $v_2(t) = \cos t$, $v_3(t) = 0$ for $0 < t < \pi$.
    (c) $v_1(t) = t^2$, $v_2(t) = 1 + t^2$ for $-1 < t < 1$.
    (d) $v_1(t) = t$, $v_2(t) = e^t$, $v_3(t) = \sin 2t$ for $0 \le t \le 1$.
    (e) $v_1(t) = e^{-2t}$, $v_2(t) = e^{2t}$, $v_3(t) = \cosh 2t$ for $1 \le t \le 2$.

2. If the functions v_1 and v_2 defined in Example 2 were both solutions of some second-order linear homogeneous equation (3) with continuous coefficients, then Theorem B would be contradicted. Show directly that *neither* v_1 nor v_2 can be a solution of such an equation.

3. The linear homogeneous equation $y'' + 2y' + y = 0$ has (by Theorem G) linearly independent solutions $y_1(t) = e^{-t}$ and $y_2(t) = te^{-t}$.

    (a) Prove that another pair of linearly independent solutions of this equation is defined by $y_1(t) = (1 - t)e^{-t}$ and $y_2(t) = (\pi + 17t)e^{-t}$. Thus the latter pair is equally acceptable for use in Theorem D.
    (b) Find still another pair of linearly independent solutions of the equation.

4. Use the results of Problem 3 to find the general solution of $x'' + 2x' + x = t$. [*Hint:* A particular solution can be found in the form $\tilde{x}(t) = A + Bt$ by an appropriate choice of A and B.]

5. Suppose that $e^{(\mu+i\omega)t}$ and $e^{(\mu-i\omega)t}$ (where μ and ω are real constants with $\omega \neq 0$) both define solutions of some linear homogeneous ordinary differential equation. (We do not specify the order of the equation nor whether its coefficients are constants.) Prove that $y_1(t) = e^{\mu t} \cos \omega t$ and $y_2(t) = e^{\mu t} \sin \omega t$ are necessarily linearly independent solutions of that same equation.

6. Suppose that $e^{(\mu+i\omega)t}$ (where μ and ω are real constants and $\omega \neq 0$) defines a solution of some linear homogeneous ordinary differential equation with *real-valued* coefficients. Then prove that y_1 and y_2, as defined in Problem 5, are linearly independent solutions of that same equation.

7. Find the general solution, valid for $t > 0$, of

$$t^2 y'' + b_1 t y' + b_0 y = 0,$$

where b_1 and b_0 are real constants such that $(b_1 - 1)^2 > 4b_0$. This is an example of "Euler's equation." [*Hint:* Try to find a solution of the form $y(t) = t^\lambda$ for constant λ.]

8. Define $y_1(t) = t^2$ and $y_2(t) = t + t^4$ for $-1 < t < 1$. Prove that y_1 and y_2 are linearly independent and yet $W(y_1, y_2)(0) = 0$. Verify that y_1 and y_2 both satisfy the linear homogeneous equation.

$$y''' - 8t^2 y'' + 40ty' - 64y = 0 \quad \text{for} \quad -1 < t < 1.$$

Why does this not contradict Theorem B (and Corollary C)?

9. Prove that there cannot exist any third-order linear homogeneous equation of the form (3) with continuous coefficients, $a_2(t)$, $a_1(t)$, $a_0(t)$, which is satisfied by $y(t) = t^3$ on $(-1, 1)$.

10. Consider Eq. (3) with $n = 2$. Show that if y_1 and y_2 are any two solutions on $J = (\alpha, \beta)$, then

$$\frac{d}{dt} W(y_1, y_2)(t) = -a_1(t) W(y_1, y_2)(t),$$

so that

$$W(y_1, y_2)(t) = W(y_1, y_2)(t_0) e^{-\int_{t_0}^t a_1(s)ds} \quad \text{for all} \quad t, t_0 \quad \text{in} \quad J.$$

This result (known as Abel's formula) provides another proof of Corollary C for $n = 2$. How? (An analogous result holds for arbitrary n, and it can be obtained using a little knowledge of the theory of determinants.)

11. If λ_1, λ_2, and λ_3 are distinct (real or complex) numbers, prove that $e^{\lambda_1 t}$, $e^{\lambda_2 t}$, and $e^{\lambda_3 t}$ are linearly independent on any interval by computing their Wronskian. How does the work involved compare to that in Example 3?

17. CONSTANT COEFFICIENTS
(GENERAL SOLUTION FOR SPECIAL h)

We now consider the linear equation with constant coefficients

$$x^{(n)} + a_{n-1}x^{(n-1)} + \cdots + a_1 x' + a_0 x = h(t), \tag{1}$$

where the function h has a certain special form. We may ask for the general solution of Eq. (1) or we may seek that particular solution which also satisfies given initial conditions

$$x(t_0) = b_0, \qquad x'(t_0) = b_1, \quad \ldots, \quad x^{(n-1)}(t_0) = b_{n-1}. \qquad (2)$$

The method described in Section 15 (and justified by the Theorems of Section 16) provides the general solution of the associated homogeneous equation

$$y^{(n)} + a_{n-1} y^{(n-1)} + \cdots + a_1 y' + a_0 y = 0. \qquad (3)$$

Having found the general solution of (3), we can then invoke Theorem 16-D to find the general solution of Eq. (1) *if* we can find *some* particular solution $\tilde{x}$ of Eq. (1). A fairly general method for finding an $\tilde{x}$ will be presented in Section 19. For the present we restrict our attention to the case when h belongs to an important special class—namely, to the case when h is itself a solution of some homogeneous linear differential equation with constant coefficients.

The work of Section 15 shows that if $h(t)$ is a linear combination of terms

$$t^k e^{\lambda t}, \quad t^k e^{\mu t} \cos \omega t, \quad \text{and} \quad t^k e^{\mu t} \sin \omega t \qquad (4)$$

(for arbitrary real constants λ, μ, and ω, and integers $k \geq 0$), then h itself will satisfy some homogeneous linear equation with constant coefficients. This equation can be conveniently written $q(D)h = 0$, where $q(D)$ is a polynomial with constant coefficients in the operator $D = d/dt$. Then we say that the operator $q(D)$ *annihilates* h.

Example 1. Find the simplest linear homogeneous differential equation with (real) constant coefficients which has as *one of its solutions* the function

$$h(t) = 2e^t + \sin 2t - 3t \sin 2t.$$

Note that the operator $(D - I)$ annihilates the term $2e^t$, that is,

$$(D - I)2e^t = 0 \quad \text{for all} \quad t.$$

To annihilate the function $\sin 2t = (e^{i2t} - e^{-i2t})/2i$ we can use the operator $(D^2 + 4I) = (D + 2iI)(D - 2iI)$, while both the function $\sin 2t$ *and* $-3t \sin 2t$ are annihilated by $(D^2 + 4I)^2$. So, using the properties of polynomial operators in D with constant coefficients, we see that h satisfies

$$(D - I)(D^2 + 4I)^2 h = 0.$$

In this example h is annihilated by $q(D) = (D - I)(D^2 + 4I)^2$.

After these preliminaries we are ready to illustrate the method for solving Eq. (1) when h has the desired special form.

Example 2. Find the general solution of

$$x'' + x' - 2x = 3e^{-2t} + te^{-2t} + 5 \cos 3t. \qquad (5)$$

As usual, we begin by considering the characteristic equation $\lambda^2 + \lambda - 2 = 0$. This has roots $\lambda_1 = 1$ and $\lambda_2 = -2$ each of multiplicity one. Thus the general solution of Eq. (5) is given by

$$x(t) = \tilde{x}(t) + c_1 e^t + c_2 e^{-2t},$$

where $\tilde{x}$ is "any particular solution."

But now observe that the terms on the right side of Eq. (5),

$$h(t) = 3e^{-2t} + te^{-2t} + 5 \cos 3t,$$

are all terms of the type listed in (4). Thus, there must be some linear *homogeneous* differential equation with constant coefficients for which h itself is a *solution*. Indeed, we find that

$$(D + 2I)^2(D + 3iI)(D - 3iI)h = 0.$$

Thus the operator $q(D) = (D + 2I)^2(D^2 + 9I)$ annihilates h.

If we now rewrite Eq. (5) as

$$(D - I)(D + 2I)x = h(t),$$

then "multiplication" from the left by $q(D)$ gives

$$(D + 2I)^2(D^2 + 9I)(D - I)(D + 2I)x = 0.$$

In other words, the general solution of the nonhomogeneous second-order equation (5) must also be *a* solution of the new homogeneous sixth-order equation

$$(D - I)(D + 2I)^3(D^2 + 9I)x = 0.$$

Thus

$$x(t) = c_1 e^t + c_2 e^{-2t} + c_3 te^{-2t} + c_4 t^2 e^{-2t} + c_5 \cos 3t + c_6 \sin 3t$$

for some choice of the constants $c_1, c_2, c_3, c_4, c_5,$ and c_6.

But we know already that, regardless of the values of c_1 and c_2, the first two terms will always satisfy the homogeneous equation $y'' + y' - 2y = 0$. So, to find a particular solution of (5), we should seek constants $A, B, C,$ and D such that

$$\tilde{x}(t) = Ate^{-2t} + Bt^2 e^{-2t} + C \cos 3t + D \sin 3t$$

satisfies Eq. (5). For this we compute

$$\tilde{x}'(t) = A(e^{-2t} - 2te^{-2t}) + B(2te^{-2t} - 2t^2 e^{-2t}) - 3C \sin 3t + 3D \cos 3t,$$

and

$$\tilde{x}''(t) = A(-4e^{-2t} + 4te^{-2t}) + B(2e^{-2t} - 8te^{-2t} + 4t^2 e^{-2t})$$
$$- 9C \cos 3t - 9D \sin 3t.$$

Substitution into (5) thus leads to the condition

$$3e^{-2t} + te^{-2t} + 5 \cos 3t = -3Ae^{-2t} + 2Be^{-2t} - 6Bte^{-2t}$$
$$+ (-2C + 3D - 9C)\cos 3t + (-2D - 3C - 9D)\sin 3t,$$

or

$$-3A + 2B = 3, \quad -6B = 1, \quad -11C + 3D = 5, \quad -3C - 11D = 0.$$

Solving these simultaneous algebraic equations, we find

$$A = -\frac{10}{9}, \quad B = -\frac{1}{6}, \quad C = -\frac{11}{26}, \quad D = \frac{3}{26}.$$

Finally then, the general solution of Eq. (5) is

$$x(t) = -\frac{10}{9} te^{-2t} - \frac{1}{6} t^2 e^{-2t} - \frac{11}{26} \cos 3t + \frac{3}{26} \sin 3t + c_1 e^t + c_2 e^{-2t}.$$

The above method for finding $\tilde{x}$ is sometimes called the method of "undetermined coefficients." The coefficients which were, at first, undetermined were A, B, C, and D. We now give a precise description of the procedure as a theorem.

THEOREM A

Rewrite Eq. (1) in the form $p(D)x = h(t)$. Assume there exists another polynomial operator, $q(D)$, with constant coefficients such that $q(D)h = 0$. [This will be the case if and only if h is equivalent to a sum of terms of the types in (4).] Then the general solution of the new homogeneous equation

$$q(D)p(D)x = 0$$

will include terms which represent the general solution of Eq. (3) plus terms which, with appropriate coefficients, give a particular solution of Eq. (1).

The proof would look very similar to Example 2.

Example 3. Find the general solution of

$$x'' + 4x = \sin 2t. \tag{6}$$

The characteristic equation, $\lambda^2 + 4 = 0$, has roots $\lambda = \pm 2i$. Thus, the general solution of the homogeneous equation is

$$y(t) = c_1 \cos 2t + c_2 \sin 2t.$$

But since $(D^2 + 4I)$ annihilates $h(t) = \sin 2t$, every solution of Eq. (6), $(D^2 + 4I)x = h(t)$, must also be a solution of the new homogeneous equation

$$(D^2 + 4I)^2 x = 0.$$

Thus,

$$x(t) = c_1 \cos 2t + c_2 \sin 2t + At \cos 2t + Bt \sin 2t,$$

for appropriate choices of the undetermined coefficients, A and B. Hence, a particular solution of the nonhomogeneous equation can be found in the form

$$\tilde{x}(t) = At \cos 2t + Bt \sin 2t.$$

Substitution leads us to set $A = \frac{1}{4}$, $B = 0$. Thus, the general solution is

$$x(t) = -\tfrac{1}{4}t \cos 2t + c_1 \cos 2t + c_2 \sin 2t. \tag{7}$$

This example illustrates the phenomenon of "resonance." The homogeneous equation $y'' + 4y = 0$ has solutions $\cos 2t$ and $\sin 2t$ which oscillate at the "frequency" $1/\pi$ cycles per unit time, say $1/\pi$ cycles per second. (Note that $\cos \omega t$ and $\sin \omega t$ describe periodic functions with period $2\pi/\omega$, and hence with frequency $\omega/2\pi$ cycles per unit of time.)

The function $h(t) = \sin 2t$ on the right-hand side of the equation can be thought of as a "forcing" or "driving" function. Example 3 illustrates the fact that when the driving function is periodic with the same frequency as the "natural frequency" of the equation, then we get a solution which oscillates with ever growing and unbounded amplitude.

In more refined models of physical systems the oscillations do not grow without bound, but are limited by "friction" or "resistance." See Section 18.

PROBLEMS

1. Find the general solutions for each of the following:

   (a) $x''' + 6x'' + 12x' + 8x = 0$.
   (b) $x'' + x = \cos t$.
   (c) $x' + x = \sin t$ (Problem 2-13, but now much easier).
   (d) $x'' - x' - 2x = 1 + \sin 2t + e^t$.
   (e) $x''' - 4x'' + 4x' = e^t$.
   (f) $x''' - 4x'' + 4x' = e^{2t}$.
   (g) $x''' + x' - 10x = t^2$.
   (h) $x''' + x'' - 10x' = t^2$.
   (i) $x'' - x' - 2x = te^{-t} + 2e^{-t} - e^{2t}$.
   (j) $x'' + 4x' + 5x = e^{-2t} \cos t$. Describe the behavior of the (general) solution as $t \to \infty$.
   (k) $x'' + 3x' + 2x = \sin t \cos t$.
   (l) $x'' + 4x = \sin^2 t$.

2. Find the (unique) solutions of the first three equations of Problem 1 which satisfy the following initial conditions:

   (a) with $x(0) = 2$, $x'(0) = 0$, $x''(0) = -10$.
   (b) with $x(\pi/2) = 0$, $x'(\pi/2) = 1$.
   (c) with $x(\pi) = 2$.

3. Find the simplest form of *a* particular solution which you know will work for each of the following. Do not evaluate the coefficients.

(a) $x'' + x = t \sin t$.
(b) $x''' + 9x'' + 27x' + 27x = t + t^2 e^{-3t}$.

18. RESONANCE

If a second-order linear differential equation represents some real physical system, one does not expect to encounter unbounded oscillations as in Example 17-3. In a physical system "energy" is lost through friction (or damping), often represented by an x' term on the left-hand side of the differential equation.

Nevertheless a phenomenon called resonance still occurs. We begin by indicating two physical systems which are modeled by a linear second-order nonhomogeneous differential equation.

Forced Oscillations of a Mass and Spring

Let a mass m hang from a spring as in Example 10-1 except that now the spring is suspended from a support which moves vertically. At time t let the support be $f(t)$ units below some stationary reference point as in Figure 1.

Let $y(t)$ be the position of the mass below the reference point at time t, let l be the natural unstretched length of the spring, let k be the spring constant, let g be the acceleration of gravity, and let b be the coefficient of air resistance.

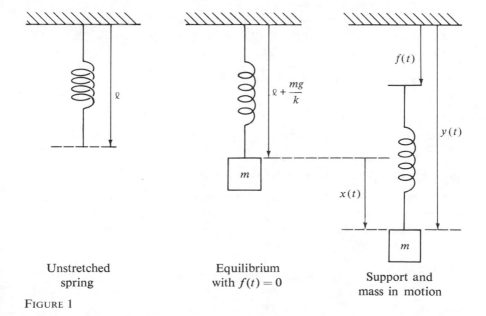

Unstretched
spring

Equilibrium
with $f(t) = 0$

Support and
mass in motion

FIGURE 1

Then

$$my''(t) = \text{total downward force on the mass}$$

$$= mg - k[y(t) - f(t) - l] - by'(t).$$

The equation of motion will be simplified if we change our point of reference. The spring with mass m attached will hang at rest from a stationary support when the downward force of gravity mg is exactly balanced by the upward force of the spring, that is, when the spring is stretched an amount mg/k. Let us use this stretched length $l + mg/k$ to locate a new reference point—the equilibrium position when $f(t) \equiv 0$. Thus we introduce

$$x = y - \left(l + \frac{mg}{k}\right)$$

as in Figure 1. Then the equation of motion becomes

$$mx'' = mg - k\left[x + l + \frac{mg}{k} - f(t) - l\right] - bx'$$

or

$$mx'' + bx' + kx = kf(t). \tag{1}$$

Example 1. A 2-kg mass when hung from a certain spring stretches that spring 4 cm (or 0.04 m) to an equilibrium position. Air resistance to the vertical motion of the system is $56v$ kg where v is the velocity in m/sec. If the upper end of the spring is 0.2 cos 7t m below a fixed reference point at time t sec, find the equation of motion.

We must first determine the spring constant k. Since $g = 9.8$ m/sec^2, gravity exerts a force of $2(9.8) = 19.6$ newtons on the mass of 2 kg. Thus $0.04k = 19.6$, or $k = 490$ kg/m, and Eq. (1) becomes

$$2x'' + 56x' + 490x = 490(0.2 \cos 7t)$$

or

$$x'' + 28x' + 245x = 49 \cos 7t,$$

where $x(t)$ is the distance below the equilibrium point as in Figure 1.

Forced Oscillations in an *RLC* Circuit

Let a potential source $V(t)$ (volts) be connected to a series electrical circuit containing resistance R (ohms), capacitance C (farads), and inductance L (henries), each of which is a positive constant (Figure 2). Then the flow of current $x(t)$ (amps) is governed by the equation

$$Lx'(t) + Rx(t) + \frac{1}{C}\left[q_0 + \int_{t_0}^{t} x(s)\, ds\right] = V(t),$$

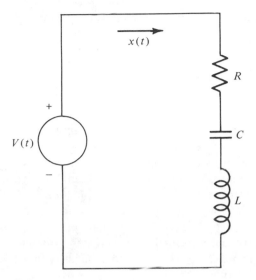

FIGURE 2

where q_0 is a constant representing the charge on C at the instant t_0. Time is measured in seconds. Differentiating this equation we obtain the second-order differential equation

$$Lx'' + Rx' + \frac{1}{C}x = V'(t) = h(t). \tag{2}$$

Since this equation is equivalent to Eq. (1) we can study both the mechanical and the electrical system simultaneously. We shall concentrate on Eq. (2).

Let us take $V(t) = a \sin \omega_1 t$, where a and ω_1 are real constants with $\omega_1 > 0$ and let us assume $0 < R^2 < 4L/C$. We shall proceed to find the general solution of the resulting Eq. (2):

$$Lx'' + Rx' + \frac{1}{C}x = a\omega_1 \cos \omega_1 t. \tag{3}$$

Setting

$$\mu = -\frac{R}{2L} \quad \text{and} \quad \omega = \left(\frac{1}{LC} - \frac{R^2}{4L^2}\right)^{1/2}$$

we find (Problem 1) the general solution of the homogeneous equation associated with (3) to be

$$y(t) = c_1 e^{\mu t} \cos \omega t + c_2 e^{\mu t} \sin \omega t. \tag{4}$$

This represents damped (or decaying) "free oscillations" with the natural frequency $\omega/2\pi$. See Figure 14-1. Now seek a particular solution of the nonhomogeneous equation (3) in the form

$$\tilde{x}(t) = A \cos \omega_1 t + B \sin \omega_1 t.$$

Evaluating A and B by substitution into Eq. (3) we find (Problem 2) the general solution $x = \tilde{x} + y$ to be

$$x(t) = \frac{a}{(\omega_1 L - 1/\omega_1 C)^2 + R^2}\left[-\left(\omega_1 L - \frac{1}{\omega_1 C}\right)\cos \omega_1 t + R \sin \omega_1 t\right]$$

$$+ c_1 e^{\mu t}\cos \omega t + c_2 e^{\mu t}\sin \omega t. \quad (5)$$

The constants c_1 and c_2 would be determined as usual if appropriate initial conditions were specified. Note, however, that *regardless of the initial conditions* the contribution from the terms involving c_1 and c_2 dies out as t increases since $\mu < 0$. These are therefore referred to as "transient" terms. [When, as in this case, every solution of the homogeneous equation, $Ly'' + Ry' + (1/C)y = 0$, tends to zero as $t \to \infty$ one says that the trivial solution ($y \equiv 0$) of the homogeneous equation is "asymptotically stable."]

The periodic terms which persist undiminished as t increases represent the "steady-state" current $x_{ss}(t)$. [In Eq. (5) x_{ss} happens to coincide with $\tilde{x}$.] To examine x_{ss} it is convenient to introduce

$$\phi = \text{Arctan}\left(\frac{(\omega_1 L - 1/\omega_1 C)}{R}\right). \quad (6)$$

We can then write the steady-state current (Problem 3) as

$$x_{ss}(t) = \frac{a}{[(\omega_1 L - 1/\omega_1 C)^2 + R^2]^{1/2}}\sin(\omega_1 t - \phi). \quad (7)$$

Note that the steady-state current oscillates with the same frequency as the driving voltage, but with a "phase shift" ϕ. The amplitude of the oscillations of x_{ss},

$$\frac{a}{[(\omega_1 L - 1/\omega_1 C)^2 + R^2]^{1/2}},$$

depends strongly on the driving frequency $\omega_1/2\pi$.

This leads to a more realistic discussion of the phenomenon of resonance than that given in Example 17-3. For given a, R, L, and C, the steady-state current amplitude will be maximized when $\omega_1 = 1/\sqrt{LC}$. Why? We note that the driving frequency which gives the maximum amplitude for x_{ss} is somewhat greater than the natural frequency of the system since

$$\omega_1 = \frac{1}{\sqrt{LC}} > \frac{1}{\sqrt{LC}}\left(1 - \frac{R^2 C}{4L}\right)^{1/2} = \omega.$$

However, if $R^2 C/L \ll 1$, then these frequencies will be approximately the same.

Perhaps the most common everyday use of resonance occurs in the tuning

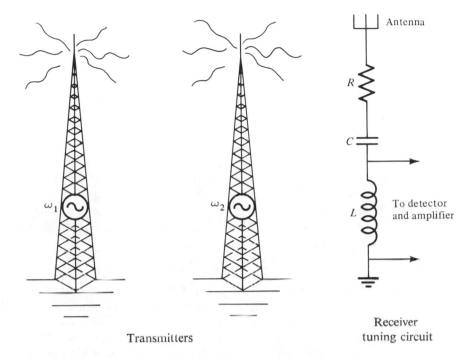

Transmitters   Receiver
                                tuning circuit

FIGURE 3

of a radio or TV set. In fact the circuit of Figure 2 can be regarded as a simplified model of a radio tuning circuit which we redraw in Figure 3.

The antenna of a radio receiver is influenced by the signals from 10,000 commercial AM, FM, and TV broadcasting stations around the world plus the signals from millions of amateur, CB, police, navigational and micro-wave transmitters, and microwave ovens (not to mention the disorganized electromagnetic emissions from electric motors, automobile ignition systems, and natural atmospheric phenomena). At any particular antenna, most of these signals are very weak. But nevertheless, with appropriate atmospheric conditions, a typical medium-priced radio would be capable of converting into sound the signals from several hundred commercial stations.

How then is it possible for the listener to select that single station he wants to hear, and reject (or virtually reject) all other signals?

The commercial transmitters in a given locality all transmit with different basic frequencies. In Figure 3 the multitude of transmitters is represented by two, with frequencies of $\omega_1/2\pi$ and $\omega_2/2\pi$.

The RLC circuit of the receiver can then be considered to be as in Figure 2 with

$$V(t) = a_1 \sin \omega_1 t + a_2 \sin \omega_2 t. \tag{8}$$

Putting $V'(t)$ from this on the right-hand side of Eq. (2) and solving, we find the steady-state current

$$x_{ss}(t) = \frac{a_1 \sin(\omega_1 t - \phi_1)}{[(\omega_1 L - 1/\omega_1 C)^2 + R^2]^{1/2}} + \frac{a_2 \sin(\omega_2 t - \phi_2)}{[(\omega_2 L - 1/\omega_2 C)^2 + R^2]^{1/2}}. \quad (9)$$

Now the tuning knob of the radio is connected to the capacitor C, and tuning consists of changing the value of C. In order to detect the signal of transmitter 1 and essentially exclude that of 2 we adjust the magnitude of C ("tune the radio") until $1/\sqrt{LC} = \omega_1$.

Example 2. In an AM radio we might have $L = 250 \times 10^{-6}$ henries and $R = 2$ ohms. Suppose that the signals from two transmitters at frequencies of 636 kilohertz (kHz) (kilocycles per second) and 796 kHz are reaching the antenna of such a radio with equal strength. This means that in Eqs. (8) and (9) we have $a_1 = a_2 = a$, and

$$\omega_1 = 636,000 \times 2\pi = 4 \times 10^6, \qquad \omega_2 = 796,000 \times 2\pi = 5 \times 10^6.$$

If we wanted to "receive" the broadcast at 636 kHz, we would "tune the radio" by adjusting the value of C so that $\omega_1 L - 1/\omega_1 C = 0$—in this case to $C = 250 \times 10^{-12}$ farads—in order to maximize the desired component of x_{ss} in (9). Then we would find that x_{ss} has two terms—the one from the 636-kHz transmitter having amplitude $a/R = a/2$ and that from the 796-kHz transmitter having an amplitude of only

$$\frac{a}{[(\omega_2 L - 1/\omega_2 C)^2 + R^2]^{1/2}} = \frac{a}{450}.$$

Thus we would "receive" the desired broadcast and "reject" the other.

Resonance can be interpreted similarly in the mechanical system modeled by Eq. (1). If we move the support up and down with sinusoidal motion at a frequency of $\omega_1/2\pi$ cycles per second we will get the strongest steady-state oscillations (of the mass m) in case $\omega_1 = \sqrt{k/m}$ (approximately) if $b^2/m \ll k$. See Problem 7.

A familiar day-to-day application of mechanical resonance occurs when trying to extricate a car from mud, sand, or snow. The car has a natural frequency of oscillation in the ruts in which it is stuck. Thus you should try to "rock" the car back and forth at approximately that same frequency (determined experimentally) in order to maximize your oscillations and increase your chances of getting out of the rut.

Resonance is a liability to the designer of a bridge or an airplane. He must try to avoid building into his structure any natural frequencies which are likely to be "driven" by traffic crossing the bridge, or by the engines of the airplane. If serious resonance occurs, the resulting oscillations can be disastrous for the structure.

Another example of resonance is observed in the "sympathetic vibrations" of a piano string when a sound of an appropriate frequency is produced nearby.

PROBLEMS

1. Verify that (4) gives the general solution of $Ly'' + Ry' + \frac{1}{C}y = 0$.

2. Verify that Eq. (5) gives the general solution of Eq. (3).
3. Verify that $x_{ss}(t)$ as expressed in Eq. (7) is equivalent to the first two terms on the right-hand side of Eq. (5).
4. Verify that the voltage $V(t)$ in (8) leads to the steady-state current $x_{ss}(t)$ in (9).
5. With respect to Example 2:

   (a) Verify the appropriateness of the choice of C to best receive the 636-kHz signal, and verify the amplitudes of the two terms in x_{ss} in this case. Using the same value of C, find the value of ω corresponding to free oscillations in the circuit as in Eq. (4).
   (b) Find the value of C for best reception of the 796-kHz signal. Using this value of C, find the amplitudes of the two terms in x_{ss}, and find the value of ω for free oscillations in the circuit.

6. Find the general solution of $x'' + 28x' + 245x = 49 \cos 7t$, the equation of Example 1.
7. If in Eq. (1) we have $f(t) = a \cos \omega_1 t$, find the amplitude of the steady-state oscillations. What value of ω_1 maximizes this amplitude?
*8. Why is the maximum amplitude in Problem 7 *not* obtained when $\omega_1 = \sqrt{k/m}$ corresponding to the value of $1/\sqrt{LC}$ in the electrical case?

19. VARIATION OF PARAMETERS

Once again consider the linear equation with variable coefficients

$$x^{(n)} + a_{n-1}(t)x^{(n-1)} + \cdots + a_1(t)x' + a_0(t)x = h(t), \tag{1}$$

perhaps with initial conditions

$$x(t_0) = b_0, \quad x'(t_0) = b_1, \quad \ldots, \quad x^{(n-1)}(t_0) = b_{n-1}, \tag{2}$$

and the associated homogeneous equation

$$y^{(n)} + a_{n-1}(t)y^{(n-1)} + \cdots + a_1(t)y' + a_0(t)y = 0. \tag{3}$$

The functions $a_{n-1}, \ldots, a_1, a_0$, and h are assumed to be continuous on an interval $J = (\alpha, \beta)$.

In Section 17 we studied Eq. (1) in the special case when the coefficients $a_{n-1}, \ldots, a_0$ were constants and the function h was itself a solution of some linear homogeneous differential equation with constant coefficients. For that important special case, Section 17 provided a systematic method for the complete solution of Eq. (1).

Unfortunately, no such simple general method is available for solving linear equations with variable coefficients (when $n \geq 2$). The present section introduces a technique for solving Eq. (1), or at least reducing it to a simpler problem, *provided* one knows some solution(s) of the homogeneous Eq. (3) to begin with. This method is central in the study of differential equations, and it will be encountered again later in the study of systems of equations. Here we do not require that h have any special properties beyond continuity.

The simplest illustration of the method uses a familiar equation from Section 2.

Example 1. Consider the nonhomogeneous linear first-order equation

$$x' + a(t)x = h(t), \tag{4}$$

where a and h are continuous functions on J, together with the initial condition

$$x(t_0) = x_0, \tag{5}$$

where $t_0 \in J$. Forgetting the integrating factor method of Section 2, we shall present an alternative method of solution which rests upon our first finding the general solution of the associated homogeneous equation

$$y' + a(t)y = 0. \tag{6}$$

Our uniqueness theorems show that if y is a solution of (6), then either $y(t) = 0$ for all t in J or else $y(t) \neq 0$ for all t in J. Why? If $y(t) \neq 0$ for all t, we can solve Eq. (6) by the method of separation of variables. This gives $\ln |y(t)| + \int_{t_0}^{t} a(s)\, ds = c_1$, or

$$y(t) = ce^{-\int_{t_0}^{t} a(s)\, ds} \equiv cy_1(t).$$

Considering c as an arbitrary constant, we have the general solution of Eq. (6).

Now the idea in the method called "variation of parameters" (or "variation of constants") is this. Let us seek a solution of the nonhomogeneous Eq. (4) in the form

$$x(t) = y_1(t)p(t), \tag{7}$$

where the arbitrary constant (or parameter) c in the solution of Eq. (6) has been replaced by an unknown function p—a "varying parameter." This expression for x must be substituted into Eq. (4) in order to discover the condition(s) on p. The calculations for this substitution are conveniently displayed as follows:

$$
\begin{array}{c|l}
a(t) & x(t) \;= y_1(t)p(t) \\
1 & x'(t) = y_1'(t)p(t) + y_1(t)p'(t) \\
\hline
0 & \quad + y_1(t)p'(t) = h(t). \tag{8}
\end{array}
$$

The quantities $a(t)$ and 1 at the far left are the appropriate multipliers for x and x' respectively before the columns are totaled to complete the substitution into Eq. (4), yielding (8). Note that with the columns aligned as above, the first column to the right of the equal sign corresponds to substitution of $y_1(t)p$ into the left-hand side of Eq. (4) or (6) *with p regarded as a constant.* Since y_1 satisfies Eq. (6), *this column must total zero.*

From (8) we find $p'(t) = h(t)/y_1(t)$. Hence $p(t) = c + \int_{t_0}^t [h(s)/y_1(s)]\, ds$, or

$$x(t) = p(t)y_1(t) = cy_1(t) + \int_{t_0}^t h(s) \left[\frac{y_1(t)}{y_1(s)}\right] ds.$$

Since $y_1(t_0) = 1$, $x(t_0) = c$. So $c = x_0$ and

$$x(t) = x_0\, e^{-\int_{t_0}^t a(s)ds} \int_{t_0}^t h(s)e^{-\int_s^t a(u)du}\, ds \tag{9}$$

—precisely the solution obtained in Section 2 (Problem 2-2).

Remarks. In the calculation leading to Eq. (8), the decision to write $x(t) = y_1(t)p(t)$ instead of the more explicit form $x(t) = e^{\int_{t_0}^t a(s)ds} p(t)$ was an arbitrary choice of convenience. If this is not clear, rewrite the calculation using the more explicit form and its derivative to see that the outcome is precisely the same.

The next example shows how the same trick used in Example 1 may also work for a second-order linear equation.

Example 2. Find the general solution of the " Euler equation "

$$t^2 x'' + 3tx' + x = 2t. \tag{10}$$

Observe that this equation can be put into the form of Eq. (1) on the interval $(0, \infty)$ or on the interval $(-\infty, 0)$, but *not* on any interval which contains the point 0. Thus we restrict ourselves to either $(0, \infty)$ or $(-\infty, 0)$.

First consider the associated homogeneous equation

$$t^2 y'' + 3ty' + y = 0, \tag{11}$$

and, as in Problem 16-7, seek a solution of the form $y(t) = t^\lambda$ for $t > 0$, where λ is a constant. [For $t < 0$ one could use $y(t) = (-t)^\lambda$.] Substitution into Eq. (10) gives $\lambda(\lambda - 1)t^\lambda + 3\lambda t^\lambda + t^\lambda = 0$, and, since $t \neq 0$, this implies

$$\lambda(\lambda - 1) + 3\lambda + 1 = 0,$$

or $\lambda^2 + 2\lambda + 1 = 0$. This quadratic equation has the double root $\lambda_1 = \lambda_2 = -1$. Thus $y_1(t) = t^{-1}$ is a solution of the homogeneous Eq. (11).

Now let us try to find a solution of the nonhomogeneous Eq. (10) in the form $x = y_1 p$. The calculations involved in substituting $x = y_1 p$ into (10) are displayed below in a format analogous to that used in Example 1. This time it is more economical to use the specific form, t^{-1}, of our solution of the homogeneous equation rather than $y_1(t)$.

1	$x = t^{-1}p$
$3t$	$x' = -t^{-2}p + t^{-1}p'$
t^2	$x'' = 2t^{-3}p - 2t^{-2}p' + t^{-1}p''$

$$0 + \quad p' + \quad tp'' = 2t. \tag{12}$$

The result is a first-order linear equation for the unknown $q = p'$, namely,

$$q' + t^{-1}q = 2.$$

This is solved (with the aid of an integrating factor) to give $tq = t^2 + c$, or

$$p'(t) = q(t) = t + \frac{c}{t}.$$

Hence $p(t) = t^2/2 + c \ln|t| + c_1$ and

$$x(t) = t^{-1}p(t) = \frac{t}{2} + c\frac{1}{t}\ln|t| + c_1\frac{1}{t}.$$

For any constants, c and c_1, this x is a solution of the original differential equation (10) on $(0, \infty)$. It happens also to be a solution on $(-\infty, 0)$.

To prove that it is the general solution, we would have to show that

$$y_1(t) = \frac{1}{t} \quad \text{and} \quad y_2(t) = \frac{1}{t}\ln|t|$$

represent two linearly independent solutions of the homogeneous Eq. (11). This is left to the reader as Problem 1.

In problems at the end of this section the reader will encounter further examples of Euler's equation

$$t^n x^{(n)} + a_{n-1}t^{n-1}x^{(n-1)} + \cdots + a_1 tx' + a_0 x = h(t), \tag{13}$$

where $a_{n-1}, \ldots, a_1, a_0$ are constants.

In its most general form, the method of "variation of parameters" aims to study a nonhomogeneous equation (1) of order n with the aid of k linearly independent solutions of the corresponding homogeneous equation (3) where $1 \le k \le n$. The most important case is that in which $k = n$.

When $k < n$, as in Example 2, the method has the effect of producing a differential equation of lower order than the given equation. So in this case, one sometimes refers to the method as "reduction of order."

The Method

If $y_1, \ldots, y_k$ are linearly independent solutions of Eq. (3) with $1 \le k \le n$, we shall seek a solution of Eq. (1) in the form

$$x = y_1 p_1 + \cdots + y_k p_k, \tag{14}$$

where $p_1, \ldots, p_k$ are functions yet to be defined. [We know that if $p_1, \ldots, p_k$ were all constants, then (14) would represent just another solution of Eq. (3) and, of course, it would not even be linearly independent of $y_1, \ldots, y_k$.]

Now, as will be seen later, the single requirement that x satisfy Eq. (1) will *not* uniquely determine $p_1, \ldots, p_k$ in (14). In other words, if there is one choice of functions $p_1, \ldots, p_k$ which provide a solution of Eq. (1), then there are other choices which would also work. This (unproved) assertion implies a freedom to impose certain other conditions on $p_1, \ldots, p_k$. The further conditions will be chosen in a way that simplifies the calculations.

The substitution of (14) into Eq. (1) requires a computation of successive derivatives of x as follows:

$$x' = y_1'p_1 + \cdots + y_k'p_k + y_1p_1' + \cdots + y_kp_k',$$

$$x'' = y_1''p_1 + \cdots + y_k''p_k + 2y_1'p_1' + \cdots + 2y_k'p_k' + y_1p_1'' + \cdots + y_kp_k'',$$

and so on. Clearly these get ever more cumbersome. But notice how much simpler this would become if one imposed the following $k - 1$ conditions on the first derivatives, $p_1', \ldots, p_k'$, of the k unknown functions:

$$\left. \begin{array}{r} y_1p_1' + \cdots + y_kp_k' = 0, \\ y_1'p_1' + \cdots + y_k'p_k' = 0, \\ \vdots \\ y_1^{(k-2)}p_1' + \cdots + y_k^{(k-2)}p_k' = 0. \end{array} \right\} \tag{15}$$

These conditions and the corresponding calculation of the derivatives of (14) for substitution into Eq. (1) are conveniently remembered and recorded in the following format:

$$\begin{array}{c|l} a_0(t) & x = y_1p_1 + \cdots + y_kp_k, \\ a_1(t) & x' = y_1'p_1 + \cdots + y_k'p_k + \underline{y_1p_1' + \cdots + y_kp_k' = 0,} \\ a_2(t) & x'' = y_1''p_1 + \cdots + y_k''p_k + \underline{y_1'p_1' + \cdots + y_k'p_k' = 0,} \end{array}$$

and so on.

Note that if $k = 1$, as it was in Examples 1 and 2, then (15) is vacuous, that is, no extra conditions are imposed. If $k = 2$, we impose only the first condition in (15), ..., and if $k = n$, we impose $n - 1$ conditions.

It is *not at all obvious* that one can find functions $p_1, \ldots, p_k$ which satisfy the $k - 1$ conditions (15) and such that (14) satisfies Eq. (1). Further calculations in specific examples will determine whether this is possible.

Example 3 (with $k = n = 2$). Find the general solution of

$$tx'' - (t + 2)x' + 2x = t^3 \quad \text{for} \quad t > 0, \tag{16}$$

using the fact (which is easily verified) that $y_1(t) = e^t$ and $y_2(t) = t^2 + 2t + 2$ are two linearly independent solutions of the associated homogeneous equation. We seek a solution in the form $x = y_1 p_1 + y_2 p_2$. And, since $k = 2$, we impose the single condition (15) $y_1 p_1' + y_2 p_2' = 0$. The calculations are written as follows:

$$\begin{array}{c|l}
2 & x = y_1 p_1 + y_2 p_2 \\
-t - 2 & x' = y_1' p_1 + y_2' p_2 + \boxed{y_1 p_1' + y_2 p_2' = 0} \\
t & x'' = y_1'' p_1 + y_2'' p_2 + y_1' p_1' + y_2' p_2'
\end{array}$$

$$0 + \quad 0 \quad + t y_1' p_1' + t y_2' p_2' = t^3,$$

where the last line is the result of substitution into Eq. (16). So there are two *algebraic* equations for p_1' and p_2':

$$y_1 p_1' + y_2 p_2' = 0,$$

$$y_1' p_1' + y_2' p_2' = t^2.$$

From these,

$$p_1'(t) = \frac{\begin{vmatrix} 0 & y_2(t) \\ t^2 & y_2'(t) \end{vmatrix}}{W(y_1, y_2)(t)} = (t^2 + 2t + 2)e^{-t},$$

$$p_2'(t) = \frac{\begin{vmatrix} y_1(t) & 0 \\ y_1'(t) & t^2 \end{vmatrix}}{W(y_1, y_2)(t)} = -1.$$

[Is it anything more than a fortunate coincidence that $W(y_1, y_2)(t) \neq 0$?] Integration gives

$$p_1(t) = -t^2 e^{-t} - 4t e^{-t} - 6e^{-t} + c_1 \quad \text{and} \quad p_2(t) = -t + c_2,$$

so that

$$x(t) = -t^3 - 3t^2 - 6t - 6 + c_1 e^t + c_2(t^2 + 2t + 2),$$

or, more compactly,

$$x(t) = -t^3 + c_1 e^t + c_3(t^2 + 2t + 2),$$

where $c_3 = c_2 - 3$.

The general theorem for which Example 3 is a prototype is as follows.

THEOREM A (VARIATION OF PARAMETERS)

Let $y_1, \ldots, y_n$ be linearly independent solutions of Eq. (3) on J. Then the general solution of Eq. (1) is given by

$$x(t) = \sum_{j=1}^{n} y_j(t) \int^t \frac{(-1)^{n+j} W_j(s) h(s)}{W(y_1, \ldots, y_n)(s)} \, ds, \tag{17}$$

where W_j is the Wronskian of $y_1, \ldots, y_{j-1}, y_{j+1}, \ldots, y_n$ [see Eq. (20) below], and each $\int^t$ indicates an indefinite integral with arbitrary constant of integration.

Proof. We seek a solution of Eq. (1) in the form

$$x = y_1 p_1 + \cdots + y_n p_n \tag{18}$$

with conditions (15) imposed for $k = n$. These $n - 1$ conditions plus the nth condition imposed by Eq. (1) itself form a system of n linear algebraic equations for the unknowns $p_1', \ldots, p_n'$:

$$
\begin{aligned}
y_1 p_1' + \cdots + y_n p_n' &= 0, \\
y_1' p_1' + \cdots + y_n' p_n' &= 0, \\
&\vdots \\
y_1^{(n-2)} p_1' + \cdots + y_n^{(n-2)} p_n' &= 0, \\
y_1^{(n-1)} p_1' + \cdots + y_n^{(n-1)} p_n' &= h.
\end{aligned}
\tag{19}
$$

Verify this. Moreoever, by Theorem 16-B(ii), the determinant of the coefficients is $W(y_1, \ldots, y_n)(t) \neq 0$ for all t in J. So Eqs. (19) have a unique solution for $p_1'(t), \ldots, p_n'(t)$ for each t in J. To express this, let us denote by $W_j(t)$ (for each $j = 1, \ldots, n$) the Wronskian of $y_1, \ldots, y_{j-1}, y_{j+1}, \ldots, y_n$, that is, the $(n-1) \times (n-1)$ determinant

$$
W_j(t) \equiv
\begin{vmatrix}
y_1 & \cdots & y_{j-1} & y_{j+1} & \cdots & y_n \\
y_1' & \cdots & y_{j-1}' & y_{j+1}' & \cdots & y_n' \\
\vdots & & \vdots & \vdots & & \vdots \\
y_1^{(n-2)} & \cdots & y_{j-1}^{(n-2)} & y_{j+1}^{(n-2)} & \cdots & y_n^{(n-2)}
\end{vmatrix}.
\tag{20}
$$

Then, by Cramer's rule, the unique solution of system (19) is

$$p_j'(t) = \frac{(-1)^{n+j} W_j(t) h(t)}{W(y_1, \ldots, y_n)(t)}, \qquad j = 1, \ldots, n.$$

Integrating and substituting into Eq. (18), we obtain (17). And we note that if each $\int^t$ has an additive arbitrary constant of integration then, by Theorem 16-D, (17) is the general solution of Eq. (1). ∎

Observe that Theorem A gives a very complete result in the case of constant coefficients. For if the coefficients $a_{n-1}, \ldots, a_1, a_0$ in Eq. (1) are constants, then we know from Theorem 16-G that there *do* indeed exist n linearly independent solutions of Eq. (3). Thus, since h is continuous, Eq. (1) does have a general solution, which can be made to match any initial conditions of the form (2).

In solving the problems below, you are urged to use the *method* of variation of parameters rather than attempting to use Eq. (17) as a formula for the solution.

PROBLEMS

1. Complete Example 2 by verifying that $y_1(t) = t^{-1}$ and $y_2(t) = t^{-1} \ln|t|$ are linearly independent solutions of Eq. (11).

2. If λ_1, λ_2, and λ_3 are three distinct real numbers, prove that the functions defined by $y_1(t) = t^{\lambda_1}$, $y_2(t) = t^{\lambda_2}$, and $y_3(t) = t^{\lambda_3}$ are linearly independent on $(0, \infty)$

   (a) by computing the Wronskian, and
   (b) by adapting the method used in Example 16-3.

3. Find general solutions of each of the following equations and state where they are valid. The first four are examples of Euler's equation.

   (a) $4t^2 x'' + 4tx' - x = 0$.
   (b) $4t^2 x'' + 8tx' + x = 0$.
   (c) $t^3 x''' - t^2 x'' + tx' = 0$.
   (d) $t^3 x''' + tx' - x = 0$.
   (e) $tx'' - (t+2)x' + 2x = t^3$ for $t > 0$. This is the equation of Example 3. But now solve it *using only* the observation that $y(t) = e^t$ is a solution of the associated homogeneous equation.
   (f) $tx'' - x' + t^3 x = 0$. [*Hint:* $x(t) = \sin(t^2/2)$ is one solution.]

4. Another approach to the solution of Euler's equations begins with the introduction of new variables, $s = \ln|t|$ and $w(s) = x(e^s) = x(t)$ in case $t > 0$ or $w(s) = x(-e^s) = x(t)$ in case $t < 0$. Apply this trick to the equations of Problems 3(a), 3(b), 3(c), and 3(d).

5. The method indicated in Problem 4 is especially appropriate in case our first method for Euler's equation leads to complex values of λ. Find the general solution of

   (a) $t^2 x'' + 5tx' + 5x = 0$.
   (b) $t^3 x''' + 3t^2 x'' + 7tx' - 20x = 0$.

6. Use the *method* of variation of parameters to find general solutions of the following equations, and wherever possible compare your answer with that obtained using the method of undetermined coefficients (Section 17).

   (a) $x'' + 3x' + 2x = 1 + 2t + e^{-t}$.
   (b) $x' + x = t$.
   (c) $x'' + x = \sin t$.
   (d) $x'' + x = e^t$.
   (e) $x'' + x = \tan t$.
   (f) $x'' + x = \sec t$.
   (g) $x'' + x = \sec^2 t$.
   (h) $x''' + 3x'' + 2x' = e^{-t}$.
   (i) $t^2 x'' - 2tx' + 2x = t^3 e^t$.
   (j) $t^2 x'' + 3tx' + x = t^{-1}$.

*7. Find the general solution of

$$t^5 y''' + t^2 y'' - 2ty' + 2y = 0$$

after observing that $y_1(t) = t$ and $y_2(t) = t^2$ give two particular solutions.

8. Find, by the *method* of variation of parameters, the general solution of each of the following equations when h is assumed to be a given continuous function. Check your answers by substitution.

(a) $x'' + 4x' - 5x = h(t)$.
(b) $x''' + 2x'' + 4x' + 8x = h(t)$.
(c) $x'' + x = h(t)$.
(d) $x'' + 4x' + 5x = h(t)$.

9. Let $x''(t) = h(t)$ on J where h is a continuous function. Use the method of variation of parameters to derive the result

$$x(t) = x(t_0) + x'(t_0)(t - t_0) + \int_{t_0}^{t} (t - s)h(s) \, ds,$$

where t_0 is an arbitrary point in J. Verify the formula by differentiation, also checking the values of $x(t_0)$ and $x'(t_0)$. Compare this result with Taylor's theorem (Theorem A2-B) for the case $m = 2$. Can you prove that the above formula is equivalent to the one given by Eq. A2-(1) with $m = 2$?

*10. The more ambitious reader may want to repeat Problem 9 for $x^{(n)}(t) = h(t)$, obtaining

$$x(t) = \sum_{k=0}^{n-1} \frac{x^{(k)}(t_0)}{k!} (t - t_0)^k + \frac{1}{(n-1)!} \int_{t_0}^{t} (t - s)^{n-1} h(s) \, ds.$$

For an alternative easier problem, take $n = 3$.

11. Find the general solution of $mx'' + bx' + kx = 0$ when $b^2 = 4mk$ by first finding an exponential solution $y_1(t) = e^{\lambda t}$, and then applying the method of this section. (Of course, the answer is already known from Example 14-1.)

12. It is sometimes useful when studying an equation of the form

$$x'' + a_1(t)x' + a_0(t)x = h(t) \tag{21}$$

to introduce a new unknown function z defined by

$$z(t) = x(t)e^{\frac{1}{2}\int_{t_0}^{t} a_1(s)ds}, \tag{22}$$

using some convenient t_0. Then

$$x(t) = z(t)e^{-\frac{1}{2}\int_{t_0}^{t} a_1(s)ds}.$$

(a) Assuming that a_1 is differentiable, substitute this into Eq. (21) to find the new differential equation satisfied by z,

$$z'' + [a_0(t) - \tfrac{1}{4}a_1^2(t) - \tfrac{1}{2}a_1'(t)]z = h(t)e^{\frac{1}{2}\int_{t_0}^{t} a_1(s)ds}. \tag{23}$$

Notice that, in contrast to Eq. (21), Eq. (23) contains no first-derivative term.

(b) Apply this change of variables to the "Bessel's equation of order $\tfrac{1}{2}$,"

$$t^2 y'' + t y' + (t^2 - \tfrac{1}{4})y = 0 \quad \text{for} \quad t > 0. \tag{24}$$

You should then be able to find the general solution of Eq. (24),

$$y(t) = c_1 t^{-1/2} \cos t + c_2 t^{-1/2} \sin t.$$

*20. PLANETS AND SATELLITES AND CERTAIN OTHER SPACE TRAVELERS

We conclude this chapter by considering an old but important application—the equations for the motion of a planet about the sun. The model to be used was previously introduced in Example 10-3. It consists of a pair of decidedly nonlinear second-order equations, representing Newton's law of gravitation. Certain clever changes of variable, which date back to Newton, will enable us to reduce the problem to a linear second-order equation.

The same equations which we use to describe the motion of the earth under the sole influence of the sun are also used to describe the motion of the earth's natural satellite, the moon, or an artificial satellite. They also describe the motion of any "free-falling" object, such as a rocket after burn-out, which is influenced solely by the gravitational attraction of the earth. The various assumptions involved in this model (such as spherical symmetry of the central attracting mass, no atmospheric resistance, and no influence from other planets) were discussed following Example 10-3.

Without further discussion then, we consider the pair of second-order equations

$$x_1'' = -\frac{kM}{(x_1^2 + x_2^2)^{3/2}} x_1 \quad \text{and} \quad x_2'' = -\frac{kM}{(x_1^2 + x_2^2)^{3/2}} x_2, \tag{1}$$

where $k > 0$ is a universal constant of gravitation and M is the mass of the central attracting body. Thus if we are considering the motion of the satellite earth under the influence of the sun, then M is the mass of the sun (Figure 1). If we are considering the motion of the moon, or a man-made earth satellite, or a burned-out rocket under the influence of the earth, then M is the mass of the earth. It is assumed in Eqs. (1) that $x_1^2(t) + x_2^2(t)$ does not become zero.

We now introduce polar coordinates $r(t) > 0$ and $\theta(t)$ by setting

$$r(t) = [x_1^2(t) + x_2^2(t)]^{1/2} \tag{2}$$

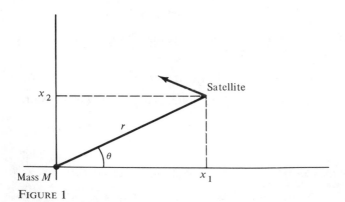

Mass M

FIGURE 1

and then choosing $\theta(t)$ such that

$$x_1(t) = r(t) \cos \theta(t) \quad \text{and} \quad x_2(t) = r(t) \sin \theta(t). \tag{3}$$

Since x_1 and x_2, in order to be solutions of Eq. (1), must be continuous, it follows that r is continuous, and hence θ must also be continuous if θ is chosen in a natural way (Problem 1). Since x_1 and x_2 must be twice differentiable, it now follows from Eqs. (2) and (3) that both r and θ must also be twice differentiable.

Differentiation of Eqs. (3) gives

$$x_1' = r' \cos \theta - r\theta' \sin \theta \quad \text{and} \quad x_2' = r' \sin \theta + r\theta' \cos \theta.$$

Hence,

$$(x_1')^2 + (x_2')^2 = (r')^2 + r^2(\theta')^2 \tag{4}$$

and

$$x_1 x_2' - x_2 x_1' = r^2 \theta'. \tag{5}$$

Now differentiate $r^2 = x_1^2 + x_2^2$ to get

$$rr' = x_1 x_1' + x_2 x_2',$$

and differentiate again to find

$$rr'' + (r')^2 = x_1 x_1'' + (x_1')^2 + x_2 x_2'' + (x_2')^2.$$

Substituting (4) into this we find the "radial component of the acceleration,"

$$a_r \equiv r'' - r(\theta')^2 = \frac{x_1 x_1'' + x_2 x_2''}{r}. \tag{6}$$

We next differentiate Eq. (5) to get

$$\frac{d}{dt}(r^2 \theta') = r^2 \theta'' + 2rr'\theta' = x_1 x_2'' - x_2 x_1'',$$

and hence find the "θ component of the acceleration,"

$$a_\theta \equiv r\theta'' + 2r'\theta' = \frac{x_1 x_2'' - x_2 x_1''}{r}. \tag{7}$$

It is now a simple matter to substitute Eqs. (1) into the right-hand sides of Eqs. (6) and (7) to obtain the equations of motion in polar form:

$$r'' - r(\theta')^2 = -\frac{kM}{r^2} \tag{8}$$

and

$$r\theta'' + 2r'\theta' = 0. \tag{9}$$

These equations will be considered together with initial conditions of the form

$$r(t_0) = r_0, \quad r'(t_0) = v_{0r}, \quad \theta(t_0) = \theta_0, \quad \theta'(t_0) = \frac{v_{0\theta}}{r_0}. \tag{10}$$

Before solving Eqs. (8) and (9) let us note some elementary solutions which are often discussed during a first introduction to planetary motion. If we assume $r'' \equiv 0$, $r' \equiv 0$, and $\theta'' \equiv 0$, then we will have a circular orbit with constant angular velocity θ' provided

$$r \cdot (\theta')^2 = \frac{kM}{r^2}.$$

This assumes that the satellite travels in a circle at just the right constant radius and just the right constant angular velocity so that the "centrifugal force" balances the "gravitational force." Now, if we wanted to launch a satellite into such a circular orbit we would have to be sure that when the final rocket burned out, say at $t = t_0$, we would have an *exactly*-appropriate combination of the initial values (10). Specifically, we would need $v_{0r} = 0$ and $r_0 v_{0\theta}^2 = kM$. Such precision is, of course, impossible to achieve.

Does this mean, then, that our satellite will either spiral back down to earth or else travel out into space forever? Is a periodic orbit impossible to achieve because a circular orbit is impossible to achieve?

The solution of Eqs. (8) and (9) will show that, if we assume M to be a point mass, then in fact it is "easy" to achieve a repeating orbit. That is, almost all initial conditions lead to closed orbital trajectories unless the initial velocity is very large. The orbits, as we shall discover, are ellipses—not circles.

To begin solving Eqs. (8) and (9), let us recall how (7) was obtained. The left-hand side of (7) is $(r^2\theta')'/r$. Thus, we see that Eq. (9) can be rewritten as

$$\frac{1}{r}\frac{d}{dt}(r^2\theta') = 0,$$

from which

$$r^2(t)\theta'(t) = r_0 v_{0\theta} = h, \tag{11}$$

a constant, for all t. Equation (11) says that the line segment from the satellite to the central mass sweeps out equal areas in equal times—*Kepler's second law of planetary motion*. (Problem 2.) Note that this assertion follows solely from the fact that $a_\theta = 0$ and does not depend on the expression for a_r. Equation (11) can also be interpreted as saying that the "angular momentum" of the satellite about the origin is constant.

In case $h = 0$ we must have $\theta'(t) \equiv 0$, and then Eq. (8) reduces to that of Example 7-2.

We shall now treat the much more interesting case, $h \neq 0$. Without loss of generality we assume $h > 0$. Then $\theta'(t) > 0$ for all t, and hence the inverse of the function θ, call it η, must exist. This enables us to treat θ as the independent variable instead of t. Thus we can write $t = \eta(\theta)$. Now introduce a new unknown function, u, by defining

$$u(\theta) = \frac{1}{r(\eta(\theta))}.$$

Then, with $t = \eta(\theta)$,

$$r(t) = \frac{1}{u(\theta)},$$

Differentiating, with the aid of the chain rule, we find

$$r'(t) = -\frac{1}{u^2(\theta)} u'(\theta)\theta'(t) = -hu'(\theta),$$

where we have used the fact from Eq. (11) that $\theta'(t) = hu^2(\theta)$. Furthermore,

$$r''(t) = -hu''(\theta)\theta'(t) = -h^2u^2(\theta)u''(\theta).$$

Substituting the last expression and $\theta'(t) = hu^2(\theta)$ into Eq. (8), and regarding θ as the independent variable, we get

$$-h^2u^2(\theta)u''(\theta) - h^2u^3(\theta) = -kMu^2(\theta),$$

or

$$u''(\theta) + u(\theta) = \frac{kM}{h^2}. \tag{12}$$

This is a simple *linear* equation with constant coefficients. We see at once that the general solution of Eq. (12) is

$$u(\theta) = \frac{kM}{h^2} + c_1 \cos \theta + c_2 \sin \theta \tag{13}$$

For convenience we introduce a constant

$$e = \frac{h^2}{kM} (c_1^2 + c_2^2)^{1/2} \tag{14}$$

and a constant ϕ such that

$$c_1 = \frac{kM}{h^2} e \cos \phi \quad \text{and} \quad c_2 = \frac{kM}{h^2} e \sin \phi.$$

Then (13) can be rewritten as

$$u(\theta) = \frac{kM}{h^2} [1 + e \cos(\theta - \phi)]. \tag{15}$$

Now take reciprocals of both sides to find

$$r = \frac{h^2/kM}{1 + e\cos(\theta - \phi)}. \tag{16}$$

Equation (16) shows that the orbit is a conic section with eccentricity e. If $e < 1$ an ellipse with one focus at the origin results, representing the orbit of a planet or satellite—*Kepler's first law of planetary motion*. If $e = 1$ we get a parabola and if $e > 1$ we get one branch of a hyperbola, representing, for example, the path of a nonreturning comet moving under the sole influence of the sun.

We asserted earlier that it is "easy" to put an object into a repeating orbit. What we meant, more precisely, is that it is easy to choose initial conditions for Eqs. (1) such that the trajectory will be an ellipse. Let us now examine the role played by the initial conditions (10) in determining the value of e. The reader should verify (Problem 3) that these initial conditions translate into

$$u(\theta_0) = \frac{1}{r_0} \quad \text{and} \quad u'(\theta_0) = -\frac{v_{0r}}{r_0 v_{0\theta}} \tag{17}$$

and hence, using Eq. (15) and its derivative,

$$e^2 = \left(\frac{r_0 v_{0\theta}^2}{kM} - 1\right)^2 + \left(\frac{r_0 v_{0r} v_{0\theta}}{kM}\right)^2. \tag{18}$$

From this one finds that $e \geq 1$ if and only if for the initial speed we have

$$(v_{0r}^2 + v_{0\theta}^2)^{1/2} \geq \sqrt{\frac{2kM}{r_0}}. \tag{19}$$

Thus if the initial speed is less than $(2kM/r_0)^{1/2}$ and if $v_{0\theta} \neq 0$, an elliptical orbit will result. But if the initial speed is $\geq (2kM/r_0)^{1/2}$, the "satellite" will escape and hence not be a satellite.

Is something wrong? If it is so easy to achieve an elliptical orbit, why do we use such powerful and expensive rockets in order to launch a satellite? (Problem 5.)

For the case $h = 0$ considered in Example 7-2 it was not convenient to solve for r as a function of t since nonelementary integrals would be involved. In the present case, with $h > 0$, we could now combine Eqs. (8) and (16) to get a second-order differential equation for θ as a function of t. However, this equation would be nonelementary.

Nevertheless, using only algebraic manipulations, we can at least deduce an expression for the period T of an elliptical orbit.

Let $A(t)$ be the area swept out during time t by the line segment from the origin to the satellite. Then, by Eq. (11),

$$A'(t) = \tfrac{1}{2}r^2(t)\theta'(t) = \tfrac{1}{2}h. \tag{20}$$

Assume $e < 1$, so that the orbit is elliptical, and let a and b be the lengths of the semimajor and semiminor axes, respectively. Then the area of the ellipse is πab. Thus, if T is the period of the orbit, Eq. (20) gives

$$\tfrac{1}{2}hT = \pi ab.$$

But from Eq. (16) we can compute

$$2a = r_{max} + r_{min}$$

$$= \frac{h^2}{kM}\left(\frac{1}{1-e} + \frac{1}{1+e}\right) = \frac{h^2}{kM}\frac{2}{1-e^2}.$$

Thus, since $b^2 = a^2(1 - e^2)$, we find

$$T^2 = \left(\frac{2\pi ab}{h}\right)^2 = 4\pi^2 \frac{a}{h^2} a^3(1 - e^2) = \frac{4\pi^2}{kM}a^3. \tag{21}$$

If M is the mass of the sun, we conclude from Eq. (21) that the squares of the periods of the various planets are proportional to the cubes of the lengths of their orbits' major axes—*Kepler's third law of planetary motion.*

Historically the study of planetary motion was quite the opposite of that presented here. Johannes Kepler (1571–1630), over a period of 20 years, painstakingly analyzed the observational data accumulated by his former teacher Tycho Brahe and other astronomers of earlier centuries. On the basis of this observational data Kepler finally put forth his three laws of planetary motion. From these three laws, Isaac Newton (1642–1727) *deduced* the equations of motion listed here as Eqs. (1).

PROBLEMS

1. What do we mean by saying, after Eqs. (3), that θ should be chosen in a " natural way"?
2. Verify the assertion made following Eq. (11) regarding the sweeping out of equal areas in equal times.
3. Verify Eqs. (17) and (18) and condition (19).
4. Compare condition (19) with the escape condition found in Example 7-2 for the case $h = 0$.
5. What is the principal defect in our model which causes us to erroneously conclude that it is easy to put a satellite into an elliptical orbit? How would you correct this defect?
6. Use the mean distance from the earth to the sun (150×10^6 km) as an approximation to the semimajor axis of the earth's orbit, and use the mean distance from earth to the moon (384×10^3 km) as an approximation to the semimajor axis of the moon's orbit about the earth. Then compute, using Eq. (21), an approximate value for the ratio M_s/M_e, the mass of the sun to the mass of the earth. How would you compute the relative mass of Jupiter?
7. Our model for planetary motion assumes that the sun is not affected by the presence of the much smaller planet Earth. Remove this assumption by following the procedure suggested in Problem 10-10. What effect does this have on the conclusions drawn from Eqs. (16), (19), and (21)?

Chapter Four

LINEAR ORDINARY DIFFERENTIAL SYSTEMS

The study of a system of linear differential equations (or a linear differential system) of the form

$$x_i' = \sum_{j=1}^{n} a_{ij}(t)x_j + h_i(t) \qquad (i = 1, \ldots, n)$$

is generally simplified by the use of matrix notation and matrix theory.

Sections 21 and 22 introduce matrix notation and basic properties. If this material is familiar, by all means pass over it.

The rest of the chapter applies these ideas to differential systems. Much of this is analogous to what was done for nth-order scalar equations in the previous chapter. The general discussion of superposition of solutions and of linear independence is similar to the previous case. Then methods are presented for finding exact and complete solutions of homogeneous systems with constant coefficients. The discussion of oscillations and damping given here is strongly oriented toward applications, as was the earlier discussion of "resonance." And the final section of this chapter presents a method of "variation of parameters" which, as in the scalar case, yields the solution of a nonhomogeneous system in terms of the solutions of the associated homogeneous system.

21. VECTOR AND MATRIX PRELIMINARIES

Recall the definitions and notation introduced in Section 9 for column vectors of n components. In general, the components of such a column n-vector will be considered to be real *or* complex numbers.

The components of an n-vector ξ are denoted by subscripts as $\xi_1, \xi_2, \ldots, \xi_n$, so that

$$\xi = \text{col}\,(\xi_1, \xi_2, \ldots, \xi_n).$$

If several n-vectors are under consideration simultaneously it will be convenient to distinguish them by the use of subscripts in parentheses, say $\xi_{(1)}, \ldots, \xi_{(k)}$. With this notation, the ith component of vector $\xi_{(j)}$ is $\xi_{(j)i}$.

The following concepts should now be added to those already given in Section 9.

DEFINITIONS

Vectors $\xi_{(1)}, \ldots, \xi_{(k)}$ are said to be *linearly dependent* if there exist scalars (that is, real or complex numbers) $c_1, \ldots, c_k$, not all zero, such that

$$c_1 \xi_{(1)} + c_2 \xi_{(2)} + \cdots + c_k \xi_{(k)} = \mathbf{0},$$

the zero vector. Otherwise the set of vectors $\{\xi_{(1)}, \ldots, \xi_{(k)}\}$ is said to be *linearly independent*.

Note the similarity of the above definitions to the ones given in Section 16 for linear dependence and linear independence of a set of functions. The reader who has studied linear algebra should recognize these definitions as special cases of the definitions of linear dependence and linear independence in an arbitrary "linear vector space."

Example 1. Taking $n = 3$ and $k = 3$, determine whether the vectors

$$\xi_{(1)} = \begin{pmatrix} 1 \\ 2 \\ 3 \end{pmatrix}, \quad \xi_{(2)} = \begin{pmatrix} 0 \\ 1 \\ 1 \end{pmatrix}, \quad \text{and} \quad \xi_{(3)} = \begin{pmatrix} -1 \\ 0 \\ -1 \end{pmatrix}$$

are linearly dependent or linearly independent. We must consider the equation

$$c_1 \xi_{(1)} + c_2 \xi_{(2)} + c_3 \xi_{(3)} = \mathbf{0}.$$

This is equivalent to the three equations

$$c_1 \qquad\quad - c_3 = 0,$$
$$2c_1 + c_2 \qquad = 0,$$
$$3c_1 + c_2 - c_3 = 0,$$

for the unknowns c_1, c_2, c_3. If the determinant of the coefficients of this system were nonzero, it would follow that c_1, c_2, and c_3 would all be zero. However, in this case the determinant *is zero*,

$$\begin{vmatrix} 1 & 0 & -1 \\ 2 & 1 & 0 \\ 3 & 1 & -1 \end{vmatrix} = 0,$$

so we cannot conclude that c_1, c_2, and c_3 are all zero. In fact, one easily verifies that our system will be satisfied provided $c_3 = c_1$ and $c_2 = -2c_1$. One solution, among infinitely many which are possible, is $c_1 = 1$, $c_2 = -2$, $c_3 = 1$. Thus

$$\xi_{(1)} - 2\xi_{(2)} + \xi_{(3)} = \mathbf{0},$$

which means that our three vectors are linearly dependent.

Example 2. Determine linear dependence or linear independence of the vectors

$$\xi_{(1)} = \begin{pmatrix} 1 \\ 2 \\ 3 \end{pmatrix}, \quad \xi_{(2)} = \begin{pmatrix} 0 \\ 1 \\ 1 \end{pmatrix}, \quad \text{and} \quad \xi_{(3)} = \begin{pmatrix} -1 \\ 0 \\ 1 \end{pmatrix}.$$

The equation $c_1 \xi_{(1)} + c_2 \xi_{(2)} + c_3 \xi_{(3)} = 0$ stands for

$$\begin{aligned} c_1 \qquad\quad - c_3 &= 0, \\ 2c_1 + c_2 \qquad &= 0, \\ 3c_1 + c_2 + c_3 &= 0. \end{aligned}$$

But now the determinant of the coefficients is

$$\begin{vmatrix} 1 & 0 & -1 \\ 2 & 1 & 0 \\ 3 & 1 & 1 \end{vmatrix} = 2 \neq 0.$$

Thus the only solution is the trivial solution, $c_1 = c_2 = c_3 = 0$. So the three given vectors in this example are linearly independent.

The following theorem generalizes the discussion of Examples 1 and 2.

THEOREM A

A set of n column n-vectors $\{\xi_{(1)}, \ldots, \xi_{(n)}\}$ is linearly independent if and only if the determinant

$$\begin{vmatrix} \xi_{(1)1} & \xi_{(2)1} & \cdots & \xi_{(n)1} \\ \xi_{(1)2} & \xi_{(2)2} & \cdots & \xi_{(n)2} \\ \vdots & \vdots & & \vdots \\ \xi_{(1)n} & \xi_{(2)n} & \cdots & \xi_{(n)n} \end{vmatrix} \neq 0.$$

Proof. The equation $c_1 \xi_{(1)} + \cdots + c_n \xi_{(n)} = 0$ represents the n equations

$$\begin{aligned} c_1 \xi_{(1)1} + \cdots + c_n \xi_{(n)1} &= 0, \\ &\vdots \\ c_1 \xi_{(1)n} + \cdots + c_n \xi_{(n)n} &= 0. \end{aligned} \tag{1}$$

But this, considered as a system of n equations in n unknowns, $c_1, \ldots, c_n$, has only the trivial solution $c_1 = \cdots = c_n = 0$ if and only if the determinant of the coefficients is not zero. (See Remark below.) This is the assertion of the theorem. ∎

Remark. The reader may not know the theorem of algebra, invoked above, which states that if the determinant in Theorem A *is* zero, then there *does* *exist* a nontrivial solution of the homogeneous system (1). This was previously used in the proof of Theorem 16-B. Lack of knowledge of this theorem will be of no practical concern. For, as in Example 1, whenever this situation occurs in practice, it will be possible to explicitly find such a nontrivial solution of (1).

THEOREM B

If $\xi_{(1)}, \ldots, \xi_{(k)}$ are column n-vectors with $k > n$, then the vectors must be linearly dependent.

Proof. The vector equation $c_1 \xi_{(1)} + \cdots + c_k \xi_{(k)} = 0$ stands for

$$c_1 \xi_{(1)1} + \cdots + c_k \xi_{(k)1} = 0,$$
$$\vdots$$
$$c_1 \xi_{(1)n} + \cdots + c_k \xi_{(k)n} = 0.$$

Let us consider these as n equations for the k unknown numbers $c_1, \ldots, c_k$. To this system we can adjoin $k - n$ additional trivial equations of the form

$$c_1 \cdot 0 + \cdots + c_k \cdot 0 = 0.$$

This gives a system of k equations in k unknowns *with determinant of the coefficients equal to zero.* Thus there does exist a nontrivial solution for $c_1, \ldots, c_k$. Hence, $\xi_{(1)}, \ldots, \xi_{(k)}$ are linearly dependent. ∎

A rectangular array of real or complex numbers will be called a *matrix*; for example,

$$A = \begin{pmatrix} a_{11} & a_{12} & \cdots & a_{1n} \\ a_{21} & a_{22} & \cdots & a_{2n} \\ \vdots & \vdots & & \vdots \\ a_{n1} & a_{n2} & \cdots & a_{nn} \end{pmatrix}, \tag{2}$$

where each a_{ij} $(i, j = 1, \ldots, n)$ is a number, called the ijth element of A. In general, a matrix need not have the same number of rows as columns. But ours always will. Thus, a matrix A with n rows and n columns, as illustrated above, is called an $n \times n$ (read "n-by-n") square matrix, or simply an $n \times n$ matrix.

The following definitions for matrices are completely analogous to definitions given in Section 9 for vectors.

DEFINITIONS

If

$$A = \begin{pmatrix} a_{11} & \cdots & a_{1n} \\ \vdots & & \vdots \\ a_{n1} & \cdots & a_{nn} \end{pmatrix} \quad \text{and} \quad B = \begin{pmatrix} b_{11} & \cdots & b_{1n} \\ \vdots & & \vdots \\ b_{n1} & \cdots & b_{nn} \end{pmatrix}, \tag{3}$$

then *addition* of matrices is defined by

$$A + B \equiv \begin{pmatrix} a_{11} + b_{11} & \cdots & a_{1n} + b_{1n} \\ \vdots & & \vdots \\ a_{n1} + b_{n1} & \cdots & a_{nn} + b_{nn} \end{pmatrix}, \tag{4}$$

and *multiplication of A by a scalar, c,* is defined by

$$cA \equiv \begin{pmatrix} ca_{11} & \cdots & ca_{1n} \\ \vdots & & \vdots \\ ca_{n1} & \cdots & ca_{nn} \end{pmatrix}, \tag{5}$$

that is, $A + B$ and cA are new matrices whose ijth elements are $a_{ij} + b_{ij}$ and ca_{ij}, respectively. If c is a scalar, we shall sometimes write Ac instead of cA. Let us agree that these are synonymous.

Subtraction of matrices is defined by

$$A - B \equiv A + (-1)B.$$

We say $A = B$ if and only if

$$a_{ij} = b_{ij} \quad \text{for all} \quad i, j = 1, \ldots, n.$$

Example 3. If

$$A = \begin{pmatrix} 1 & 2 \\ -2 & 3 \end{pmatrix} \quad \text{and} \quad B = \begin{pmatrix} 0 & -5 \\ 1 & 2 \end{pmatrix}, \quad \text{then} \quad 2A - 3B = \begin{pmatrix} 2 & 19 \\ -7 & 0 \end{pmatrix}.$$

As far as the above definitions are concerned the matrices might as well be considered as vectors with n^2 components each. We could say that the n^2 elements just happen to be written in an $n \times n$ square array rather than in a column of length n^2. Thus, the following properties of addition and multiplication by scalars follow at once from the corresponding properties in Section 9.

THEOREM C

If A, B, and C are $n \times n$ matrices and c_1 and c_2 are scalars, then

(i) $A + B = B + A$,
(ii) $(A + B) + C = A + (B + C)$,
(iii) there exists a unique $n \times n$ matrix, 0, called the zero matrix, such that $A + 0 = A$ for every $n \times n$ matrix A,
(iv) for each $n \times n$ matrix A there exists a unique $n \times n$ matrix, $-A$, such that $A + (-A) = 0$,
(v) $c_1(c_2 A) = (c_1 c_2)A$,
(vi) $(c_1 + c_2)A = c_1 A + c_2 A$,
(vii) $c_1(A + B) = c_1 A + c_1 B$,
(viii) $1A = A$.

In conditions (iii) and (iv) the required matrices 0 and $-A$ are, respectively

$$\begin{pmatrix} 0 & \cdots & 0 \\ \vdots & & \vdots \\ 0 & \cdots & 0 \end{pmatrix} \quad \text{and} \quad \begin{pmatrix} -a_{11} & \cdots & -a_{1n} \\ \vdots & & \vdots \\ -a_{n1} & \cdots & -a_{nn} \end{pmatrix},$$

where A is as in (3). Notice that the symbol 0 will sometimes stand for the number zero and sometimes for the zero matrix. The context should always make clear which it is.

In order to gain anything from the use of matrix notation we must also introduce the following definitions for multiplication.

DEFINITION

If A is the $n \times n$ matrix (2) and if ξ is a column n-vector with components $\xi_1, \ldots, \xi_n$, we shall define the *product* $A\xi$ by

$$A\xi = \begin{pmatrix} a_{11} & \cdots & a_{1n} \\ a_{21} & \cdots & a_{2n} \\ \vdots & & \vdots \\ a_{n1} & \cdots & a_{nn} \end{pmatrix} \begin{pmatrix} \xi_1 \\ \xi_2 \\ \vdots \\ \xi_n \end{pmatrix} = \begin{pmatrix} a_{11}\xi_1 + \cdots + a_{1n}\xi_n \\ a_{21}\xi_1 + \cdots + a_{2n}\xi_n \\ \vdots \\ a_{n1}\xi_1 + \cdots + a_{nn}\xi_n \end{pmatrix}. \tag{6}$$

In other words, $A\xi$ is a new n-vector whose ith component is

$$a_{i1}\xi_1 + \cdots + a_{in}\xi_n = \sum_{j=1}^{n} a_{ij}\xi_j.$$

Example 4.

$$\begin{pmatrix} 1 & 2 & 0 \\ -2 & 3 & 2 \\ 0 & 1 & \sqrt{2} \end{pmatrix} \begin{pmatrix} 2 \\ 3 \\ -1 \end{pmatrix} = \begin{pmatrix} 8 \\ 3 \\ 3 - \sqrt{2} \end{pmatrix}.$$

As one application of this new notation we remark that a system of n linear algebraic equations in n unknowns $\xi_1, \ldots, \xi_n$,

$$a_{11}\xi_1 + \cdots + a_{1n}\xi_n = b_1,$$
$$a_{21}\xi_1 + \cdots + a_{2n}\xi_n = b_2,$$
$$\vdots$$
$$a_{n1}\xi_1 + \cdots + a_{nn}\xi_n = b_n,$$

can now be written very compactly as

$$A\xi = \mathbf{b},$$

where A and ξ are as in Eq. (6) and $\mathbf{b} = \text{col}(b_1, \ldots, b_n)$.

If ξ and η are two n-vectors, A is an $n \times n$ matrix, and c_1 and c_2 are scalars, then it is easy to verify that

$$A(c_1\xi + c_2\eta) = c_1 A\xi + c_2 A\eta \tag{7}$$

(Problem 2). Thus A is a "linear operator."

DEFINITION

If A and B are the $n \times n$ matrices in (3), we define the *product* AB by

$$AB = \begin{pmatrix} \sum\limits_{k=1}^{n} a_{1k}b_{k1} & \cdots & \sum\limits_{k=1}^{n} a_{1k}b_{kn} \\ \vdots & & \vdots \\ \sum\limits_{k=1}^{n} a_{nk}b_{k1} & \cdots & \sum\limits_{k=1}^{n} a_{nk}b_{kn} \end{pmatrix}, \tag{8}$$

that is, AB is the $n \times n$ matrix whose ijth element is

$$\sum_{k=1}^{n} a_{ik}b_{kj} = a_{i1}b_{1j} + a_{i2}b_{2j} + \cdots + a_{in}b_{nj}.$$

Another way of phrasing this definition is as follows: Let $\mathbf{b}_{(j)}$ stand for the jth column of B considered as a column n-vector. Then AB is the new matrix whose jth column is the vector $A\mathbf{b}_{(j)}$ defined as in Eq. (6).

Example 5. Let A and B be the matrices of Example 3. Then

$$AB = \begin{pmatrix} 2 & -1 \\ 3 & 16 \end{pmatrix} \quad \text{and} \quad BA = \begin{pmatrix} 10 & -15 \\ -3 & 8 \end{pmatrix}.$$

This shows that matrix multiplication is not, in general, commutative, that is, $AB \neq BA$ in general. Nevertheless, matrix multiplication does have many properties which are familiar from ordinary arithmetic.

THEOREM D

Let A, B, and C be any $n \times n$ matrices. Then

  (i) $(A + B)C = AC + BC$,
 (ii) $C(A + B) = CA + CB$,
(iii) for any scalar, c, $A(cB) = (cA)B = c(AB)$,
(iv) $(AB)C = A(BC)$,
 (v) there exists a unique $n \times n$ matrix, I, called the identity matrix, such that $IA = A = AI$ for all A.

Proof. Let A and B be as in (3) and let

$$C = \begin{pmatrix} c_{11} & \cdots & c_{1n} \\ \vdots & & \vdots \\ c_{n1} & \cdots & c_{nn} \end{pmatrix}.$$

Then to prove (i) we note that the ijth element of $(A + B)C$ is

$$\sum_{k=1}^{n} (a_{ik} + b_{ik})c_{kj} = \sum_{k=1}^{n} a_{ik}c_{kj} + \sum_{k=1}^{n} b_{ik}c_{kj},$$

which is the ijth element of $AC + BC$. The proof of (ii) is similar.
The proof of (iii) is straightforward and is left as Problem 5.
To prove (iv) we note that the ijth element of $(AB)C$ is

$$\sum_{k=1}^{n} \left(\sum_{l=1}^{n} a_{il}b_{lk} \right) c_{kj} = \sum_{k=1}^{n} \left(\sum_{l=1}^{n} a_{il}b_{lk}c_{kj} \right)$$

$$= \sum_{l=1}^{n} \left(\sum_{k=1}^{n} a_{il}b_{lk}c_{kj} \right) = \sum_{l=1}^{n} a_{il} \left(\sum_{k=1}^{n} b_{lk}c_{kj} \right).$$

But the last form is exactly the ijth element of $A(BC)$.
A matrix which has the property attributed to I in (v) is the matrix

$$I \equiv \begin{pmatrix} 1 & 0 & \cdots & 0 \\ 0 & 1 & \cdots & 0 \\ \vdots & \vdots & & \vdots \\ 0 & 0 & \cdots & 1 \end{pmatrix},$$

an $n \times n$ matrix with ones on the "main diagonal" and zeros elsewhere.
More precisely, I is the $n \times n$ matrix whose ijth element is 1 when $i = j$, and
0 when $i \neq j$. The reader should convince himself that $IA = A = AI$ for every
$n \times n$ matrix A. It remains to show that no other $n \times n$ matrix has this
property.
Let I' be any other $n \times n$ matrix with the property that $I'A = A = AI'$ for
every $n \times n$ matrix A. Then it follows that

$$I' = I'I = I. \quad \blacksquare$$

DEFINITION

If A and B are $n \times n$ matrices with the property that

$$AB = I = BA,$$

we shall say that B is the *inverse* of A and write

$$A^{-1} = B.$$

We shall also say A is *invertible*.

To justify the use of the expression "the" inverse of A we must prove the
following result.

LEMMA E

If $AB = I = BA$ and $AC = I = CA$, then $B = C$. In fact, if only $AC = I$
$= BA$, then $B = C$.

Proof. $B = BI = B(AC) = (BA)C = IC = C.$ ∎

It also follows at once from the definition that if B is the inverse of A, then A is also the inverse of B. This can be stated as follows: If A^{-1} exists, then so does $(A^{-1})^{-1}$, and

$$(A^{-1})^{-1} = A.$$

Example 6. If $A = \begin{pmatrix} 1 & 2 \\ -2 & 3 \end{pmatrix}$, then $A^{-1} = \begin{pmatrix} \frac{3}{7} & -\frac{2}{7} \\ \frac{2}{7} & \frac{1}{7} \end{pmatrix}$.

For

$$\begin{pmatrix} 1 & 2 \\ -2 & 3 \end{pmatrix}\begin{pmatrix} \frac{3}{7} & -\frac{2}{7} \\ \frac{2}{7} & \frac{1}{7} \end{pmatrix} = \begin{pmatrix} 1 & 0 \\ 0 & 1 \end{pmatrix} = \begin{pmatrix} \frac{3}{7} & -\frac{2}{7} \\ \frac{2}{7} & \frac{1}{7} \end{pmatrix}\begin{pmatrix} 1 & 2 \\ -2 & 3 \end{pmatrix}.$$

DEFINITION

If A is any given matrix, the new matrix $A^T = B$ obtained by setting $b_{ij} = a_{ji}$ for all i, j is called the *transpose* of A. Thus the jth column of A^T is simply the jth row of A.

Example 7. If $A = \begin{pmatrix} 1 & 2 & 3 \\ 4 & 5 & 6 \\ 7 & 8 & 9 \end{pmatrix}$, then $A^T = \begin{pmatrix} 1 & 4 & 7 \\ 2 & 5 & 8 \\ 3 & 6 & 9 \end{pmatrix}$.

THEOREM F

An $n \times n$ matrix A, as in (1) has an inverse if and only if the determinant of A (denoted by $\det A$) is not zero, and then

$$A^{-1} = \frac{1}{\det A}\begin{pmatrix} \det A_{11} & -\det A_{21} & \cdots & (-1)^{n+1}\det A_{n1} \\ -\det A_{12} & \det A_{22} & \cdots & (-1)^{n+2}\det A_{n2} \\ \vdots & \vdots & & \vdots \\ (-1)^{1+n}\det A_{1n} & (-1)^{2+n}\det A_{2n} & \cdots & \det A_{nn} \end{pmatrix}, \quad (9)$$

where A_{ij} is the $(n-1) \times (n-1)$ matrix obtained from A by deletion of the ith row and the jth column. (The matrix on the right-hand side of Eq. (9) is the transpose of the matrix of the "cofactors" of the elements of A.)

*Proof. (Assuming some knowledge of determinants.) Let $\det A \neq 0$ and let B be the right-hand side of Eq. (9). Then

$$AB = \frac{1}{\det A}\begin{pmatrix} \det A & 0 & \cdots & 0 \\ 0 & \det A & \cdots & 0 \\ \vdots & \vdots & & \vdots \\ 0 & 0 & \cdots & \det A \end{pmatrix} = I.$$

Similarly, one also finds $BA = I$. Thus $B = A^{-1}$.

Now assume A^{-1} exists and consider the system of n equations represented by

$$A\xi = 0, \tag{10}$$

where $\xi_1, \ldots, \xi_n$, the components of ξ, are the unknowns. Multiply Eq. (10) from the left by the matrix A^{-1} to find

$$A^{-1}(A\xi) = 0,$$

or

$$\xi = I\xi = (A^{-1}A)\xi = A^{-1}(A\xi) = 0.$$

In other words, the only solution of Eq. (10) is $\xi_1 = \xi_2 = \cdots = \xi_n = 0$. This implies that det $A \neq 0$. ∎

Example 8. If

$$A = \begin{pmatrix} 1 & 2 & 0 \\ 0 & -1 & 4 \\ 3 & 1 & -3 \end{pmatrix},$$

then

$$\det A - 1 \det \begin{pmatrix} -1 & 4 \\ 1 & -3 \end{pmatrix} - 2 \det \begin{pmatrix} 0 & 4 \\ 3 & -3 \end{pmatrix} = 23.$$

Thus, by Theorem F, A^{-1} exists and

$$A^{-1} = \frac{1}{23} \begin{pmatrix} -1 & 6 & 8 \\ 12 & -3 & -4 \\ 3 & 5 & -1 \end{pmatrix}.$$

As Problem 6 the reader should verify by direct computation that, with A^{-1} defined this way, $A^{-1}A = I = AA^{-1}$.

Equation (9) produces A^{-1} quite easily if A is a 2×2 matrix. However, for larger matrices there is a much more efficient method. Equation (9) was presented here because it is relatively easy to state. If you know a better way to compute A^{-1} use it.

PROBLEMS

1. Determine whether each of the following sets of vectors is linearly dependent or linearly independent. In case of linear dependence, exhibit an appropriate linear combination $c_1\xi_{(1)} + \cdots + c_k\xi_{(k)} = 0$.

   (a) $\xi_{(1)} = \text{col } (1, 1)$, $\xi_{(2)} = \text{col } (1, -1)$.
   (b) $\xi_{(1)} = \text{col } (1, 0, 1)$, $\xi_{(2)} = \text{col } (2, 1, 0)$, $\xi_{(3)} = \text{col } (3, -1, 1)$.
   (c) $\xi_{(1)} = \text{col } (1, 0, 1)$, $\xi_{(2)} = \text{col } (0, 1, 1)$,
   $\xi_{(3)} = \text{col } (1, 1, 0)$, $\xi_{(4)} = \text{col } (1, 1, 1)$
   (d) $\xi_{(1)} = \text{col } (0, 1, 1, 2)$, $\xi_{(2)} = \text{col } (1, 2, 3, 4)$, $\xi_{(3)} = \text{col } (1, 0, 1, 0)$.

2. Verify Eq. (7).

3. If $A = \begin{pmatrix} 1 & 0 & 2 \\ 0 & 3 & -1 \\ -2 & 1 & 2 \end{pmatrix}$ and $B = \begin{pmatrix} -2 & 1 & 3 \\ 2 & 0 & 1 \\ 1 & 4 & 0 \end{pmatrix}$, compute AB, BA, and B^T.

4. If $A = \begin{pmatrix} 2 & 0 \\ 5 & -3 \end{pmatrix}$, $B = \begin{pmatrix} -1 & 3 \\ 1 & 2 \end{pmatrix}$, and $C = \begin{pmatrix} 0 & 2 \\ -4 & 1 \end{pmatrix}$, compute

   (a) $(A + B)C$ and $AC + BC$.   (b) $(AB)C$ and $A(BC)$.

5. Prove part (iii) of Theorem D.
6. Verify by direct multiplication that $A^{-1}A = I = AA^{-1}$ when A and A^{-1} are the matrices defined in Example 8.
7. Find the inverses, if they exist, of the matrices A and B of Problem 3.
8. Show that if A and B are two $n \times n$ matrices, then

$$(A^T)^T = A \quad \text{and} \quad (AB)^T = B^T A^T.$$

22. DIAGONALIZATION OF MATRICES

An extensive theory for the manipulation of matrices can be found in books on matrix analysis. This section covers just enough of that theory to yield some results which are particularly useful in the study of differential equations.

We begin our study of a constant matrix, A, by seeking special vectors ("eigenvectors") $\xi \neq 0$ such that $A\xi$ is just equal to some scalar multiple of ξ.

DEFINITION

If A is a given $n \times n$ matrix, λ is a real or complex number, and $\xi \neq 0$ is an n-vector such that

$$A\xi = \lambda\xi \quad \text{or} \quad (A - \lambda I)\xi = 0, \tag{1}$$

then we shall call λ an *eigenvalue* (or characteristic value) of A, and ξ an *eigenvector* (or characteristic vector) of A corresponding to the eigenvalue λ.

Example 1. Find the eigenvalues and eigenvectors of the matrix

$$A = \begin{pmatrix} 1 & 4 \\ 3 & 2 \end{pmatrix}.$$

In order that there exist a vector $\xi \neq 0$ such that

$$(A - \lambda I)\xi = 0$$

it is necessary (and sufficient) that

$$\det (A - \lambda I) = 0.$$

Thus we begin by solving

$$\begin{vmatrix} 1 - \lambda & 4 \\ 3 & 2 - \lambda \end{vmatrix} = \lambda^2 - 3\lambda - 10 = 0.$$

This quadratic equation has solutions $\lambda_1 = 5$ and $\lambda_2 = -2$. Let us now consider the resulting equations for ξ, $(A - 5I)\xi = 0$ and $(A + 2I)\xi = 0$.
The first of these, $(A - 5I)\xi = 0$, becomes

$$\begin{pmatrix} -4 & 4 \\ 3 & -3 \end{pmatrix} \begin{pmatrix} \xi_1 \\ \xi_2 \end{pmatrix} = 0$$

or

$$-4\xi_1 + 4\xi_2 = 0,$$
$$3\xi_1 - 3\xi_2 = 0.$$

A nonzero solution of these equations is the vector

$$\xi_{(1)} = \text{col}\,(1, 1).$$

Thus, $\xi_{(1)}$ is an eigenvector of A corresponding to the eigenvalue λ_1. Of course, any nonzero multiple of $\xi_{(1)}$ would also be an eigenvector corresponding to λ_1.
Similarly, $(A + 2I)\xi = 0$ becomes

$$3\xi_1 + 4\xi_2 = 0,$$
$$3\xi_1 + 4\xi_2 = 0,$$

so that

$$\xi_{(2)} = \text{col}\,(4, -3)$$

is an eigenvector of A corresponding to the eigenvalue λ_2.

The generalization of this example to an arbitrary $n \times n$ matrix, A, is the following.

THEOREM A
Given any $n \times n$ matrix A, the solutions of the polynomial equation

$$\det (A - \lambda I) = 0 \tag{2}$$

are the eigenvalues of A. [Equation (2) is called the *characteristic equation* for A.]

The proof, an abstraction of Example 1, is left as Problem 1.

Example 2. Consider again the matrix

$$A = \begin{pmatrix} 1 & 4 \\ 3 & 2 \end{pmatrix}.$$

Form a new matrix P whose columns are the eigenvectors of A which we found in Example 1, namely,

$$P = \begin{pmatrix} 1 & 4 \\ 1 & -3 \end{pmatrix}.$$

Applying Theorem 21-F, we find that P has an inverse, and in fact,

$$P^{-1} = -\frac{1}{7} \begin{pmatrix} -3 & -4 \\ -1 & 1 \end{pmatrix} = \frac{1}{7} \begin{pmatrix} 3 & 4 \\ 1 & -1 \end{pmatrix}.$$

Now let us compute the product

$$\Lambda = P^{-1} A P = \frac{1}{7} \begin{pmatrix} 3 & 4 \\ 1 & -1 \end{pmatrix} \begin{pmatrix} 5 & -8 \\ 5 & 6 \end{pmatrix} = \begin{pmatrix} 5 & 0 \\ 0 & -2 \end{pmatrix} = \begin{pmatrix} \lambda_1 & 0 \\ 0 & \lambda_2 \end{pmatrix}.$$

The particularly simple form of this result is no mere coincidence, as we shall soon see.

An $n \times n$ matrix Λ with components λ_{ij} is called a *diagonal matrix* if $\lambda_{ij} = 0$ whenever $i \neq j$, that is,

$$\Lambda = \begin{pmatrix} \lambda_{11} & 0 & \cdots & 0 \\ 0 & \lambda_{22} & \cdots & 0 \\ \vdots & \vdots & \ddots & \vdots \\ 0 & 0 & \cdots & \lambda_{nn} \end{pmatrix}.$$

Example 2 illustrates a rather general procedure for obtaining a diagonal matrix Λ from a given $n \times n$ matrix A. In the next theorem we show that whenever A has n *linearly independent* eigenvectors, as in Example 2, then there exists a matrix P such that $P^{-1} A P$ is diagonal. When we have been able to find a matrix P such that $P^{-1} A P$ is diagonal, it is customary to say we have *diagonalized* A.

THEOREM B

Let A be an $n \times n$ matrix which has n linearly independent eigenvectors, $\xi_{(1)}, \ldots, \xi_{(n)}$. Then the matrix

$$P \equiv \begin{pmatrix} \xi_{(1)1} & \cdots & \xi_{(n)1} \\ \vdots & & \vdots \\ \xi_{(1)n} & \cdots & \xi_{(n)n} \end{pmatrix}$$

has an inverse and

$$P^{-1}AP = \Lambda \tag{3}$$

is a diagonal matrix. The diagonal elements of Λ are $\lambda_1, \ldots, \lambda_n$, the eigenvalues to which the eigenvectors $\xi_{(1)}, \ldots, \xi_{(n)}$ correspond.

Remarks. When two matrices, A and B, are related by an equation of the form $Q^{-1}AQ = B$ [as is the case in (3)] it is said that A and B are *similar*; and the transformation which consists of multiplying A from the left by Q^{-1} and from the right by Q is called a *similarity transformation*.

In the proof of Theorem B, and on future occasions, we shall use the special *unit vectors* $\mathbf{I}_{(1)}, \ldots, \mathbf{I}_{(n)}$ defined by

$$\mathbf{I}_{(j)} = \begin{pmatrix} 0 \\ \vdots \\ 1 \\ \vdots \\ 0 \end{pmatrix} \leftarrow j\text{th row}, \tag{4}$$

a column vector with each element 0 except the jth which is 1.

Proof of Theorem B. It is given that

$$A\xi_{(j)} = \lambda_j \xi_{(j)} \quad \text{for each} \quad j = 1, \ldots, n.$$

Thus,

$$AP = \begin{pmatrix} \lambda_1 \xi_{(1)1} & \cdots & \lambda_n \xi_{(n)1} \\ \vdots & & \vdots \\ \lambda_1 \xi_{(1)n} & \cdots & \lambda_n \xi_{(n)n} \end{pmatrix}.$$

Now, by Theorems 21-A and 21-F, the linear independence of $\xi_{(1)}, \ldots, \xi_{(n)}$ guarantees the existence of P^{-1}. Moreover, for each $j = 1, \ldots, n$ we must have $\xi_{(j)} = P\mathbf{I}_{(j)}$. Why? Hence,

$$P^{-1}\xi_{(j)} = \mathbf{I}_{(j)}.$$

It follows that

$$P^{-1}AP = \begin{pmatrix} \lambda_1 & 0 & \cdots & 0 \\ 0 & \lambda_2 & \cdots & 0 \\ \vdots & & \ddots & \vdots \\ 0 & 0 & \cdots & \lambda_n \end{pmatrix} = \Lambda. \quad \blacksquare$$

The question which we have not discussed is: When does an $n \times n$ matrix A have n linearly independent eigenvectors so that we can diagonalize A? We give only a partial answer below. For the complete answer the interested reader is referred again to a book on matrix theory.

THEOREM C

If an $n \times n$ matrix A has n distinct eigenvalues, then it has n linearly independent eigenvectors.

Proof. Let $\lambda_1, \ldots, \lambda_n$ be the n distinct eigenvalues of A, and let $\xi_{(1)}, \ldots, \xi_{(n)}$, respectively, be corresponding eigenvectors. Suppose that

$$c_1 \xi_{(1)} + c_2 \xi_{(2)} + c_3 \xi_{(3)} + \cdots + c_n \xi_{(n)} = 0 \qquad (5)$$

for some scalars $c_1, c_2, \ldots, c_n$. Now multiply Eq. (5) from the left by $A - \lambda_1 I$ to obtain

$$c_1(A\xi_{(1)} - \lambda_1 \xi_{(1)}) + c_2(A\xi_{(2)} - \lambda_1 \xi_{(2)}) + c_3(A\xi_{(3)} - \lambda_1 \xi_{(3)})$$
$$+ \cdots + c_n(A\xi_{(n)} - \lambda_1 \xi_{(n)}) = 0.$$

This reduces to

$$(\lambda_2 - \lambda_1)c_2 \xi_{(2)} + (\lambda_3 - \lambda_1)c_3 \xi_{(3)} + \cdots + (\lambda_n - \lambda_1)c_n \xi_{(n)} = 0.$$

Now multiply by $A - \lambda_2 I$ to obtain

$$(\lambda_3 - \lambda_2)(\lambda_3 - \lambda_1)c_3 \xi_{(3)} + \cdots + (\lambda_n - \lambda_2)(\lambda_n - \lambda_1)c_n \xi_{(n)} = 0.$$

Proceeding similarly, we eventually reach

$$(\lambda_n - \lambda_{n-1})(\lambda_n - \lambda_{n-2}) \cdots (\lambda_n - \lambda_2)(\lambda_n - \lambda_1)c_n \xi_{(n)} = 0.$$

Since the eigenvalues are all different and since $\xi_{(n)} \neq 0$, we conclude that $c_n = 0$. Now use this information in the next previous equation to conclude that $c_{n-1} = 0$. Proceeding backwards step by step, we eventually conclude that each $c_j = 0$ ($j = 1, \ldots, n$). This means that $\xi_{(1)}, \ldots, \xi_{(n)}$ are linearly independent. ∎

Note the similarity of this proof to the proof of Lemma 16-F.

The condition that there be n distinct eigenvalues is a sufficient, but not a necessary condition, for the existence of n linearly independent eigenvectors of the $n \times n$ matrix A.

Another sufficient condition (which we shall not prove) for the existence of n linearly independent eigenvectors is that A be real and *symmetric*. This means $a_{ij} = a_{ji}$ for all $i, j = 1, \ldots, n$. (In other words, $A^T = A$.)

Actually it can be shown that a real matrix A has n linearly independent eigenvectors if it merely commutes with A^T, that is, $AA^T = A^T A$.

Example 3. Diagonalize (if possible) the matrix

$$A = \begin{pmatrix} 5 & 0 & 2 \\ 1 & 3 & 1 \\ 2 & 0 & 5 \end{pmatrix}.$$

This matrix is not symmetric and we shall find that it does not have three distinct eigenvalues. Nevertheless, it can be diagonalized.

To find the eigenvalues, we solve

$$\det (A - \lambda I) = 0,$$

that is,

$$\begin{vmatrix} 5 - \lambda & 0 & 2 \\ 1 & 3 - \lambda & 1 \\ 2 & 0 & 5 - \lambda \end{vmatrix} = (3 - \lambda)[(5 - \lambda)(5 - \lambda) - 4]$$

$$= -(\lambda - 3)(\lambda - 3)(\lambda - 7) = 0.$$

Thus, the eigenvalues are $\lambda_1 = \lambda_2 = 3$ and $\lambda_3 = 7$. Using the double root, or "double eigenvalue," $\lambda = 3$, we seek nontrivial solutions of

$$(A - 3I)\xi = \begin{pmatrix} 2 & 0 & 2 \\ 1 & 0 & 1 \\ 2 & 0 & 2 \end{pmatrix} \xi = 0.$$

This gives only the single independent equation

$$\xi_1 + \xi_3 = 0.$$

Thus, among the infinitely many eigenvectors associated with $\lambda = 3$, it is possible to choose two *linearly independent* ones: for example,

$$\xi_{(1)} = \text{col} (1, 0, -1) \quad \text{and} \quad \xi_{(2)} = \text{col} (0, 1, 0).$$

Now consider the eigenvectors corresponding to $\lambda = 7$, namely the nontrivial solutions of

$$(A - 7I)\xi = \begin{pmatrix} -2 & 0 & 2 \\ 1 & -4 & 1 \\ 2 & 0 & -2 \end{pmatrix} \xi = 0,$$

which means

$$-2\xi_1 \qquad + 2\xi_3 = 0,$$
$$\xi_1 - 4\xi_2 + \xi_3 = 0,$$
$$2\xi_1 \qquad - 2\xi_3 = 0.$$

It is not difficult to find a solution:

$$\xi_{(3)} = \text{col} (2, 1, 2).$$

But now, as the reader should verify, the three eigenvectors $\xi_{(1)}, \xi_{(2)}, \xi_{(3)}$

are linearly independent. This means that A *can* be diagonalized. The reader should also verify that if

$$P \equiv \begin{pmatrix} 1 & 0 & 2 \\ 0 & 1 & 1 \\ -1 & 0 & 2 \end{pmatrix},$$

then

$$P^{-1} = \frac{1}{4} \begin{pmatrix} 2 & 0 & -2 \\ -1 & 4 & -1 \\ 1 & 0 & 1 \end{pmatrix} \quad \text{and} \quad P^{-1}AP = \begin{pmatrix} 3 & 0 & 0 \\ 0 & 3 & 0 \\ 0 & 0 & 7 \end{pmatrix}.$$

(Problem 3.)

In a sense, it is a rare $n \times n$ matrix which cannot be diagonalized. It is *not* true, however, that *every* $n \times n$ matrix can be diagonalized. In Problems 5 and 6, the reader is asked to show that it is not possible to diagonalize the matrix

$$A = \begin{pmatrix} -1 & -1 \\ 1 & -3 \end{pmatrix},$$

Finally, on the general topic of eigenvectors, we note the following simple theorem for future reference.

THEOREM D

Let λ be an eigenvalue of A and let $\xi_{(1)}, \ldots, \xi_{(k)}$ be eigenvectors of A, all corresponding to λ. Then any nonzero linear combination of these eigen-vectors, say

$$c_1 \xi_{(1)} + \cdots + c_k \xi_{(k)},$$

is also an eigenvector corresponding to λ.

Proof. If $c_1, \ldots, c_k$ are any constant scalars, then

$$A(c_1 \xi_{(1)} + \cdots + c_k \xi_{(k)}) = c_1 A \xi_{(1)} + \cdots + c_k A \xi_{(k)}$$
$$= \lambda(c_1 \xi_{(1)} + \cdots + c_k \xi_{(k)}). \quad \blacksquare$$

Example 4. If A is the matrix in Example 3, then for any constants c_1 and c_2, not both zero, the vector

$$c_1 \xi_{(1)} + c_2 \xi_{(2)} = \text{col}(c_1, c_2, -c_1)$$

is an eigenvector of A corresponding to $\lambda = 3$.

PROBLEMS

1. Prove Theorem A.
2. (a) If A, P, and Λ are $n \times n$ matrices such that $P^{-1}AP = \Lambda$, show that $A^k = P\Lambda^k P^{-1}$ for each $k = 1, 2, \ldots$
   (b) Use this result to compute A^{117} where A is the 2×2 matrix of Example 2.
3. Verify the expressions for P^{-1} and $P^{-1}AP$ given in Example 3.
4. Diagonalize (if possible) the following matrices.

(a) $A = \begin{pmatrix} 2 & -1 \\ -2 & 3 \end{pmatrix}$,   (b) $A = \begin{pmatrix} 1 & 2 \\ -1 & -1 \end{pmatrix}$,

(c) $A = \begin{pmatrix} 1 & 0 & -3 \\ -2 & -1 & 3 \\ 2 & 0 & -4 \end{pmatrix}$,   (d) $A = \begin{pmatrix} 0 & 1 & 2 \\ 1 & 0 & 3 \\ -1 & -1 & -3 \end{pmatrix}$.

5. If $A = \begin{pmatrix} -1 & -1 \\ 1 & -3 \end{pmatrix}$, prove that the only eigenvalue is $\lambda = -2$. Then show that there cannot be any matrix P such that

$$P^{-1}AP = \begin{pmatrix} -2 & 0 \\ 0 & -2 \end{pmatrix}.$$

*6 If A is as in Problem 5, show that there cannot be any matrix P such that

$$P^{-1}AP = \begin{pmatrix} a & 0 \\ 0 & b \end{pmatrix}, \qquad \text{any diagonal matrix.}$$

[*Hint:* First show that the eigenvalues of any such matrix $P^{-1}AP$ are the same as those of A. This requires knowledge of a theorem from matrix theory which asserts that $\det(AB) = (\det A)(\det B)$ for any two $n \times n$ matrices A and B.]

7. Illustrate Theorem D for those matrices in Problem 4 which have two or more linearly independent eigenvectors corresponding to the same eigenvalue (as in Example 4).
8. If for some $n \times n$ matrix A, we have $A\xi = 0$, for every n-vector ξ prove that $A = 0$. [*Hint:* Consider, in turn, $\xi = I_{(1)}$, $\xi = I_{(2)}$, $\ldots$, and $\xi = I_{(n)}$.]

23. SUPERPOSITION AND LINEAR INDEPENDENCE

Consider the system of n equations in n unknown functions

$$x_1' = a_{11}(t)x_1 + \cdots + a_{1n}(t)x_n + h_1(t),$$
$$\vdots$$
$$x_n' = a_{n1}(t)x_1 + \cdots + a_{nn}(t)x_n + h_n(t),$$

on an interval $J = (\alpha, \beta)$. With this system we shall often associate initial conditions of the form

$$x_i(t_0) = x_{0i} \qquad (i = 1, \ldots, n),$$

where $t_0 \in J$.

The concept of an *n-vector-valued function* was introduced in Section 9. In a completely analogous manner let us now introduce an $n \times n$-*matrix-valued function*

$$A(t) = \begin{pmatrix} a_{11}(t) & \cdots & a_{1n}(t) \\ \vdots & & \vdots \\ a_{n1}(t) & \cdots & a_{nn}(t) \end{pmatrix} \quad \text{for} \quad t \quad \text{in} \quad J,$$

where each a_{ij} is a function mapping $J \to \mathbf{R}$ or $\mathbf{C}$. For brevity, we shall often use the terms "vector-valued" and "matrix-valued" instead of "n-vector-valued" and "$n \times n$-matrix-valued," respectively.

With this notation, our system becomes

$$\mathbf{x}' = A(t)\mathbf{x} + \mathbf{h}(t) \tag{1}$$

on J, and the initial conditions are written as

$$\mathbf{x}(t_0) = \mathbf{x}_0. \tag{2}$$

Here $\mathbf{x}(t) = \text{col}\,(x_1(t), \ldots, x_n(t))$, $\mathbf{h}(t) = \text{col}\,(h_1(t), \ldots, h_n(t))$, and $\mathbf{x}_0 = \text{col}\,(x_{01}, \ldots, x_{0n})$. Together with (1), we shall also consider the associated linear *homogeneous* system

$$\mathbf{y}' = A(t)\mathbf{y}, \tag{3}$$

where $\mathbf{y}$ is again a vector-valued function.

Proceeding as in Section 9 for vector-valued functions, we shall say a matrix-valued function A is *continuous* if each of its elements a_{ij} is continuous $(i, j = 1, \ldots, n)$. We shall say A is *differentiable* (or *integrable*) if each a_{ij} is differentiable (or integrable), and then the *derivative* (or *integral*) of A will be the matrix obtained by differentiating (or integrating) each element of A.

The theorems and corollaries A through to E to be given in this section for system (1) are analogs of propositions 16-A through 16-E, respectively, for the nth-order linear scalar equation.

Since x_i will now stand for the ith component of the vector-valued function $\mathbf{x}$, we shall denote several different vector-valued functions by $\mathbf{x}_{(1)}, \ldots, \mathbf{x}_{(l)}$; and, when needed, we shall denote the ith component of $\mathbf{x}_{(j)}$ by $x_{(j)i}$.

THEOREM A (SUPERPOSITION)

Let $\mathbf{h}(t) = c_1 \mathbf{h}_{(1)}(t) + \cdots + c_k \mathbf{h}_{(k)}(t)$, where $c_1, \ldots,$ and c_k are constants (possibly complex), and $\mathbf{h}_{(1)}, \ldots, \mathbf{h}_{(k)}$ are given n-vector-valued functions on J. For $j = 1, \ldots, k$, let $\mathbf{x}_{(j)}$ be a (particular) solution of $\mathbf{x}' = A(t)\mathbf{x} + \mathbf{h}_{(j)}(t)$ on J. Then

$$\mathbf{x} = c_1 \mathbf{x}_{(1)} + \cdots + c_k \mathbf{x}_{(k)}$$

is a solution of Eq. (1) on J. In case some $\mathbf{h}_{(j)}$'s are identically zero the corresponding $\mathbf{x}_{(j)}$'s can be any solutions of Eq. (3).

Proof. Using the linearity of the differentiation operator and the linearity of multiplication by the matrix $A(t)$,

$$\mathbf{x}' = \sum_{j=1}^{k} c_j \mathbf{x}'_{(j)} = \sum_{j=1}^{k} c_j [A(t)\mathbf{x}_{(j)} + \mathbf{h}_{(j)}(t)] = A(t) \sum_{j=1}^{k} c_j \mathbf{x}_{(j)} + \sum_{j=1}^{k} c_j \mathbf{h}_{(j)}(t)$$

$$= A(t)\mathbf{x} + \mathbf{h}(t). \quad \blacksquare$$

Let us now extend to vector functions the concepts of "linearly dependence" and "linear independence."

DEFINITIONS

A set of vector-valued functions, $\mathbf{v}_{(1)}, \ldots, \mathbf{v}_{(k)}$, on an interval J is said to be *linearly dependent* if there exist constants $c_1, \ldots, c_k$, not all zero, such that

$$c_1 \mathbf{v}_{(1)}(t) + \cdots + c_k \mathbf{v}_{(k)}(t) = 0 \quad \text{for all} \quad t \quad \text{in} \quad J. \tag{4}$$

Otherwise the functions are said to be *linearly independent*.

Note that linear dependence of the vector-valued functions requires that *each* of the n scalar equations represented by (4) hold for *all* t in J. Fortunately, however, in the case of functions which happen to be solutions of a linear homogeneous system (3) with a continuous coefficient matrix it suffices to examine Eq. (4) at just one value of t. This is made precise in the next theorem.

THEOREM B

(i) Let $\mathbf{y}_{(1)}, \ldots, \mathbf{y}_{(k)}$ be any n-vector-valued functions on J. Then if the constant vectors $\mathbf{y}_{(1)}(t_1), \ldots, \mathbf{y}_{(k)}(t_1)$ are linearly independent for some t_1 in J, it follows that $\mathbf{y}_{(1)}, \ldots, \mathbf{y}_{(k)}$ are linearly independent functions on J. (The converse is false.)

(ii) Let A be a continuous matrix-valued function on the interval J, and let $\mathbf{y}_{(1)}, \ldots, \mathbf{y}_{(k)}$ be solutions of the linear homogeneous equation (3) on J. Then $\mathbf{y}_{(1)}, \ldots, \mathbf{y}_{(k)}$ are linearly independent functions if and only if for each t_1 in J the constant vectors $\mathbf{y}_{(1)}(t_1), \ldots, \mathbf{y}_{(k)}(t_1)$ are linearly independent.

Proof. (i) Clearly if $\mathbf{y}_{(1)}(t_1), \ldots, \mathbf{y}_{(k)}(t_1)$ are linearly independent for some t_1, then condition (4) can only hold if $c_1 = c_2 = \cdots = c_k = 0$. Thus $\mathbf{y}_{(1)}, \ldots, \mathbf{y}_{(k)}$ are independent on J.

To see that the converse of (i) is false it suffices to consider the pair of scalar-valued functions ($n = 1$, $k = 2$) $\mathbf{y}_{(1)}(t) = t$ and $\mathbf{y}_{(2)}(t) = 1$ on $(-1, 1)$. Then $\mathbf{y}_{(1)}$ and $\mathbf{y}_{(2)}$ are linearly independent functions on $(-1, 1)$. Why? But $\mathbf{y}_{(1)}(0)$ and $\mathbf{y}_{(2)}(0)$ are linearly dependent.

(ii) Now let $\mathbf{y}_{(1)}, \ldots, \mathbf{y}_{(k)}$ be solutions of Eq. (3) on J. Then the "if" part of assertion (ii) follows from (i). To prove the "only if" part, assume that $\mathbf{y}_{(1)}(t_1), \ldots, \mathbf{y}_{(k)}(t_1)$ are linearly *dependent* for some t_1. Then there must exist constants $c_1, \ldots, c_k$, not all zero, such that $c_1 \mathbf{y}_{(1)}(t_1) + \cdots + c_k \mathbf{y}_{(k)}(t_1) = 0$.

Using these values of $c_1, \ldots, c_k$, define a new function $\mathbf{y}$ on J by

$$\mathbf{y}(t) \equiv c_1 \mathbf{y}_{(1)}(t) + \cdots + c_k \mathbf{y}_{(k)}(t).$$

Thus $\mathbf{y}$ is a solution of Eq. (3) and $\mathbf{y}(t_1) = 0$. Hence, by uniqueness, $\mathbf{y}(t) \equiv 0$. In other words, the functions $\mathbf{y}_{(1)}, \ldots, \mathbf{y}_{(k)}$ are linearly dependent on J. This means that $\mathbf{y}_{(1)}, \ldots, \mathbf{y}_{(k)}$ can be linearly independent functions on J only if $\mathbf{y}_{(1)}(t_1), \ldots, \mathbf{y}_{(k)}(t_1)$ are linearly independent vectors for each t_1 in J. ∎

Remark. Actually Theorem B is slightly more general than its counterpart, Theorem 16-B, in as much as we now allow $k \neq n$. However, Theorem 16-B could have been made entirely analogous to the above if it had been phrased in terms of the linear independence of the vectors col $(y_j, y_j', \ldots, y_j^{(n-1)})$ for $j = 1, \ldots, k$ instead of the nonvanishing of the Wronskian. The Wronskian, being a determinant, can only be defined for a square array, that is, for n columns each having n components.

COROLLARY C

Let A be a continuous matrix-valued function on J. If $\mathbf{y}_{(1)}, \ldots, \mathbf{y}_{(n)}$ are solutions of Eq. (3), then the vectors $\mathbf{y}_{(1)}(t), \ldots, \mathbf{y}_{(n)}(t)$ are either linearly dependent for each t in J or linearly independent for each t in J. Thus the determinant

$$W(t) = W(\mathbf{y}_{(1)}, \ldots, \mathbf{y}_{(n)})(t) \equiv \begin{vmatrix} y_{(1)1}(t) & \cdots & y_{(n)1}(t) \\ \vdots & & \vdots \\ y_{(1)n}(t) & \cdots & y_{(n)n}(t) \end{vmatrix} \tag{5}$$

is either zero for each t in J or different from zero for each t in J, according to whether $\mathbf{y}_{(1)}, \ldots, \mathbf{y}_{(n)}$ are linearly dependent or linearly independent, respectively.

The proof is left as Problem 2.

The scalar-valued function $W = W(\mathbf{y}_{(1)}, \ldots, \mathbf{y}_{(n)})$ defined by (5) will be called the *Wronskian* of the vector functions $\mathbf{y}_{(1)}, \ldots, \mathbf{y}_{(n)}$.

We can now describe the "general solution" of Eq. (1).

THEOREM D

Let A be a continuous matrix-valued function on J. If Eq. (3) has n linearly independent solutions, $\mathbf{y}_{(1)}, \ldots, \mathbf{y}_{(n)}$, and if $\tilde{\mathbf{x}}$ is any (particular) solution of Eq. (1) on J, then the general solution of Eq. (1) is defined by

$$\mathbf{x}(t) = \tilde{\mathbf{x}}(t) + c_1 \mathbf{y}_{(1)}(t) + \cdots + c_n \mathbf{y}_{(n)}(t) \quad \text{for } t \text{ in } J. \tag{6}$$

By this we mean that every solution of Eq. (1) can be obtained from (6) by an appropriate choice of the (constant) scalars $c_1, \ldots, c_n$.

Moreover, for each t_0 in J and each vector $\mathbf{x}_0$, there does exist a (unique) solution of Eq. (1) which satisfies the initial condition (2).

Proof. Theorem A asserts that (6) defines a solution of Eq. (1) for every choice of the constants $c_1, \ldots, c_n$.

Now let $\hat{\mathbf{x}}$ be any given solution of Eq. (1) on J. Select any t_0 in J and consider the n scalar equations represented by

$$c_1 \mathbf{y}_{(1)}(t_0) + \cdots + c_n \mathbf{y}_{(n)}(t_0) = \hat{\mathbf{x}}(t_0) - \tilde{\mathbf{x}}(t_0). \tag{7}$$

If we regard (7) as a system of n linear algebraic equations in the n unknowns, $c_1, \ldots, c_n$, then it follows from Corollary C that the determinant of the coefficients is $W(t_0) \neq 0$. Thus a solution $c_1, \ldots, c_n$ exists (and is unique). Using these values of $c_1, \ldots, c_n$, we define a function

$$\mathbf{x}(t) \equiv \tilde{\mathbf{x}}(t) + c_1 \mathbf{y}_{(1)}(t) + \cdots + c_n \mathbf{y}_{(n)}(t) \quad \text{for all } t \text{ in } J.$$

This function $\mathbf{x}$ is a solution of (1) and, moreover, $\mathbf{x}(t_0) = \hat{\mathbf{x}}(t_0)$. It follows from the uniqueness theorem that $\hat{\mathbf{x}}(t) = \mathbf{x}(t)$ for all t in J.

If t_0 in J and a vector $\mathbf{x}_0$ are given, the existence of a (unique) solution of (1) and (2) is established as above using $\mathbf{x}_0$ in place of $\hat{\mathbf{x}}(t_0)$.   ∎

Nothing we have said thus far proves that Eq. (3) with a continuous coefficient matrix A will have n linearly independent solutions. This will eventually be proved (in Section 28) so that the hypothesis in Theorem D that Eq. (3) have n linearly independent solutions will then be seen to be unnecessary. In the meantime, to give an application of Theorem D, we must use an example for which it is possible to explicitly find n linearly independent solutions of the homogeneous equation. The most natural examples are those in which A is a *constant* matrix.

Example 1. Find the general solution on $\mathbf{R}$ of the linear system with constant coefficients,

$$\begin{aligned} x_1' &= -2x_1 + 2x_2 + 5 \sin 2t, \\ x_2' &= 2x_1 - 5x_2. \end{aligned} \tag{8}$$

(This is the system of Example 9-1.)

By analogy to the method used for scalar equations, let us begin by trying to find solutions of the associated homogeneous system

$$\mathbf{y}' = A\mathbf{y}, \quad \text{where} \quad A = \begin{pmatrix} -2 & 2 \\ 2 & -5 \end{pmatrix}, \tag{9}$$

in the form $\mathbf{y}(t) = e^{\lambda t} \boldsymbol{\xi}$, that is,

$$\mathbf{y}(t) = \begin{pmatrix} \xi_1 e^{\lambda t} \\ \xi_2 e^{\lambda t} \end{pmatrix},$$

where $\lambda, \xi_1,$ and ξ_2 are constants. (We have no guarantee, at present, that this will be successful.) Substitution of $y(t) = e^{\lambda t}\xi$ into Eq. (9) leads to the vector equation

$$\lambda e^{\lambda t}\xi = Ae^{\lambda t}\xi.$$

But, since $e^{\lambda t}$ is never zero, this is equivalent to

$$(A - \lambda I)\xi = 0, \tag{10}$$

where I is the identity matrix. Now the problem of solving Eq. (10) is exactly the problem of finding the eigenvalues λ and associated eigenvectors ξ of the constant matrix A.

We ask the reader to compute (in Problem 3) the eigenvalues, $\lambda_1 = -1$ and $\lambda_2 = -6$, and respective eigenvectors col $(2, 1)$ and col $(1, -2)$. Solutions of (9) are therefore given by

$$y_{(1)}(t) = \begin{pmatrix} 2e^{-t} \\ e^{-t} \end{pmatrix} \quad \text{and} \quad y_{(2)}(t) = \begin{pmatrix} e^{-6t} \\ -2e^{-6t} \end{pmatrix}. \tag{11}$$

Once the reader verifies that these two solutions, $y_{(1)}$ and $y_{(2)}$, are linearly independent, Theorem D provides the general solution of the *homogeneous* system (9), $y = c_1 y_{(1)} + c_2 y_{(2)}$.

To now find the general solution of the original nonhomogeneous system (8), we require, in addition to $y_{(1)}$ and $y_{(2)}$, some particular solution of (8). Because of the nature of that system we might reasonably hope to find a solution of the form

$$\tilde{x}(t) = \begin{pmatrix} B_1 \cos 2t + C_1 \sin 2t \\ B_2 \cos 2t + C_2 \sin 2t \end{pmatrix},$$

where B_1, C_1, B_2, and C_2 are undetermined constants. Substituting this expression for $\tilde{x}$ into (8) we find

$$-2B_1 \sin 2t + 2C_1 \cos 2t$$

$$= -2B_1 \cos 2t - 2C_1 \sin 2t + 2B_2 \cos 2t + 2C_2 \sin 2t + 5 \sin 2t$$

and

$$-2B_2 \sin 2t + 2C_2 \cos 2t$$

$$= 2B_1 \cos 2t + 2C_1 \sin 2t - 5B_2 \cos 2t - 5C_2 \sin 2t.$$

Let us now equate the coefficients of $\sin 2t$ and $\cos 2t$ on the two sides of each of these two equations. The result is four equations in four unknowns:

$$-2B_1 + 2C_1 \qquad\quad - 2C_2 = 5,$$

$$2B_1 + 2C_1 - 2B_2 \qquad\quad = 0,$$

$$- 2C_1 - 2B_2 + 5C_2 = 0,$$

$$-2B_1 \qquad\quad + 5B_2 + 2C_2 = 0.$$

These yield

$$B_1 = -\frac{33}{20}, \quad C_1 = \frac{19}{20}, \quad B_2 = -\frac{7}{10}, \quad C_2 = \frac{1}{10}.$$

Thus, by virtue of Theorem D, the general solution of system (8) is

$$\mathbf{x}(t) = \frac{1}{20}\begin{pmatrix} -33\cos 2t + 19\sin 2t \\ -14\cos 2t + 2\sin 2t \end{pmatrix} + c_1\begin{pmatrix} 2e^{-t} \\ e^{-t} \end{pmatrix} + c_2\begin{pmatrix} e^{-6t} \\ -2e^{-6t} \end{pmatrix}.$$

The reader should realize that "real" problems seldom have answers with simple numerical coefficients. The above example was constructed so that the answer would look fairly simple.

The following is a straightforward consequence of Theorem D. Its proof is Problem 4.

COROLLARY E

If A is a continuous matrix-valued function on J, then Eq. (3) can have at most n linearly independent solutions.

PROBLEMS

1. Show that the equation in Problem 16-8 is equivalent to the system

$$y_1' = y_2,$$

$$y_2' = y_3,$$

$$y_3' = 64y_1 - 40ty_2 + 8t^2y_3.$$

Let $\mathbf{y}_{(1)}$ and $\mathbf{y}_{(2)}$ be the vector-valued solutions corresponding to the solutions given in Problem 16-8. Now test Theorem B on these solutions, taking $t_1 = 0$.

2. Prove Corollary C.

3. Complete the derivation of the solutions $\mathbf{y}_{(1)}$ and $\mathbf{y}_{(2)}$ of system (9), and verify that these vector functions $\mathbf{y}_{(1)}$ and $\mathbf{y}_{(2)}$, given in (11), are linearly independent.

4. Prove Corollary E.

5. Following the procedure used in Example 1, find the general solutions of the following systems.

(a) $\begin{aligned} y_1' &= y_1 + 4y_2, \\ y_2' &= 3y_1 + 2y_2. \end{aligned}$

(b) $\mathbf{x}' = \begin{pmatrix} 1 & 4 \\ 3 & 2 \end{pmatrix}\mathbf{x} + \begin{pmatrix} 2 \\ \cos t \end{pmatrix}.$

(c) $\mathbf{x}' = \begin{pmatrix} -1 & 0 & 1 \\ -1 & -2 & -1 \\ -2 & -2 & -3 \end{pmatrix}\mathbf{x} + \begin{pmatrix} 1+t \\ 1 \\ t \end{pmatrix}.$

6. Find the particular solutions for the first two systems of Problem 5 which satisfy the initial conditions

(a) $\mathbf{y}(0) = \mathrm{col}\,(1, -1)$, (b) $\mathbf{x}(0) = \mathrm{col}\,(1, -1)$.

*7. If $\mathbf{y}_{(1)}$ and $\mathbf{y}_{(2)}$ are solutions of Eq. (3) with $n = 2$ prove that

$$W(\mathbf{y}_{(1)}, \mathbf{y}_{(2)})(t) = W(\mathbf{y}_{(1)}, \mathbf{y}_{(2)})(t_0)\exp\left\{\int_{t_0}^t [a_{11}(s) + a_{22}(s)]ds\right\}$$

for all t, t_0 in J. Compare Problem 16-10. (This gives an alternate proof of Corollary C.) Can you generalize this to arbitrary n?

24. CONSTANT COEFFICIENTS

Just as Section 23 provided analogs to the general theorems of Section 16, the present section will provide analogs for many of the results of Sections 15, 16, and 17 for the special case of constant coefficients.

The coefficient matrix A will now be a constant and the systems considered will be

$$\mathbf{x}' = A\mathbf{x} + \mathbf{h}(t) \tag{1}$$

with initial condition

$$\mathbf{x}(t_0) = \mathbf{x}_0, \tag{2}$$

plus the associated homogeneous system

$$\mathbf{y}' = A\mathbf{y}. \tag{3}$$

The interval J can be considered to be $\mathbf{R}$.

In case the matrix A can be diagonalized, then solution of Eq. (1) is an easy matter.

Example 1. Solve the system

$$x_1' = -2x_1 + 2x_2 + 5 \sin 2t, \tag{4}$$

$$x_2' = 2x_1 - 5x_2.$$

This is system (1) with

$$A = \begin{pmatrix} -2 & 2 \\ 2 & -5 \end{pmatrix} \quad \text{and} \quad \mathbf{h}(t) = \begin{pmatrix} 5 \sin 2t \\ 0 \end{pmatrix}.$$

When solving this system in Example 23-1 we first sought solutions of the associated homogeneous equation in the form $\mathbf{y}(t) = e^{\lambda t}\xi$, where λ is a constant scalar and ξ is a constant vector. Substitution into $\mathbf{y}' = A\mathbf{y}$ led to the equation $(A - \lambda I)\xi = \mathbf{0}$, the familiar eigenvalue-eigenvector problem for the matrix A. We found two distinct eigenvalues $\lambda_1 = -1$ and $\lambda_2 = -6$ together with associated linearly independent eigenvectors $\xi_{(1)} = \text{col}(2, 1)$

and $\xi_{(2)} = \text{col}(1, -2)$. Then, by Theorem 23-D, the general solution of system (4) was seen to be

$$\mathbf{x}(t) = \tilde{\mathbf{x}}(t) + c_1 e^{-t} \begin{pmatrix} 2 \\ 1 \end{pmatrix} + c_2 e^{-6t} \begin{pmatrix} 1 \\ -2 \end{pmatrix},$$

where $\tilde{\mathbf{x}}$ was any particular solution of (4).

Another approach to system (4) involves diagonalizing the matrix A. Using the linearly independent eigenvectors $\xi_{(1)}$ and $\xi_{(2)}$ as columns, define the matrix

$$P = \begin{pmatrix} 2 & 1 \\ 1 & -2 \end{pmatrix}, \quad \text{and compute} \quad P^{-1} = \frac{1}{5} \begin{pmatrix} 2 & 1 \\ 1 & -2 \end{pmatrix}.$$

Then the matrix A is "diagonalized" by computing

$$P^{-1}AP = \begin{pmatrix} -1 & 0 \\ 0 & -6 \end{pmatrix} = \Lambda.$$

Now let us introduce a new unknown function $\mathbf{z}$ defined by

$$\mathbf{z}(t) = P^{-1}\mathbf{x}(t).$$

Then it follows that $\mathbf{x}(t) = P\mathbf{z}(t)$, and

$$\mathbf{z}' = P^{-1}\mathbf{x}' = P^{-1}A\mathbf{x} + P^{-1}\mathbf{h}(t) = P^{-1}AP\mathbf{z} + P^{-1}\mathbf{h}(t)$$

or

$$\mathbf{z}' = \Lambda\mathbf{z} + P^{-1}\mathbf{h}(t).$$

Thus, $\mathbf{z}$ satisfies a much simpler system of equations, namely,

$$z_1' = -z_1 + 2 \sin 2t, \tag{5}$$

$$z_2' = -6z_2 + \sin 2t.$$

These equations are said to be *uncoupled* since each equation involves only one component of $\mathbf{z}$, and hence can be solved quite independently of the rest of the system. We can also say we have *decoupled* the original system (4). The general solutions of the uncoupled equations (5) are now found by the methods of Chapter Three (or even of Section 2). We find

$$\mathbf{z}(t) = \begin{pmatrix} z_1(t) \\ z_2(t) \end{pmatrix} = \frac{1}{20} \begin{pmatrix} -16 \cos 2t + 8 \sin 2t \\ -\cos 2t + 3 \sin 2t \end{pmatrix} + \begin{pmatrix} c_1 e^{-t} \\ c_2 e^{-6t} \end{pmatrix}, \tag{6}$$

where c_1 and c_2 are arbitrary constants. But, since

$$\mathbf{x}(t) = P\mathbf{z}(t) = \begin{pmatrix} 2 & 1 \\ 1 & -2 \end{pmatrix} \begin{pmatrix} z_1(t) \\ z_2(t) \end{pmatrix},$$

we have

$$\mathbf{x}(t) = \frac{1}{20}\begin{pmatrix} -33\cos 2t + 19\sin 2t \\ -14\cos 2t + 2\sin 2t \end{pmatrix} + \begin{pmatrix} 2c_1 e^{-t} + c_2 e^{-6t} \\ c_1 e^{-t} - 2c_2 e^{-6t} \end{pmatrix}, \tag{7}$$

the same result as in Example 23-1.

This decoupling method is sometimes advantageous when initial conditions are also specified. For if $\mathbf{x}(t_0) = \mathbf{x}_0$, then $\mathbf{z}(t_0) = P^{-1}\mathbf{x}_0$.

For example, if together with system (4) we require

$$\mathbf{x}(0) = \mathbf{x}_0 = \mathrm{col}\,(2, -1), \tag{8}$$

then $\mathbf{z}(0) = P^{-1}\mathbf{x}_0$. In the present case, using (6),

$$\mathbf{z}(0) = \frac{1}{20}\begin{pmatrix} -16 \\ -1 \end{pmatrix} + \begin{pmatrix} c_1 \\ c_2 \end{pmatrix} = \frac{1}{5}\begin{pmatrix} 3 \\ 4 \end{pmatrix}.$$

This yields

$$c_1 = \frac{7}{5}, \qquad c_2 = \frac{17}{20}.$$

Then $\mathbf{x}(t) = P\mathbf{z}(t)$ becomes

$$\mathbf{x}(t) = \begin{pmatrix} 2 & 1 \\ 1 & -2 \end{pmatrix}\frac{1}{20}\begin{pmatrix} -16\cos 2t + 8\sin 2t + 28e^{-t} \\ -\cos 2t + 3\sin 2t + 17e^{-6t} \end{pmatrix}$$

$$= \frac{1}{20}\begin{pmatrix} -33\cos 2t + 19\sin 2t + 56e^{-t} + 17e^{-6t} \\ -14\cos 2t + 2\sin 2t + 28e^{-t} - 34e^{-6t} \end{pmatrix}. \tag{9}$$

In Problem 1 you are asked to also obtain the solution of (4) with the initial conditions (8) directly from the general solution given by Eq. (7).

The decoupling method illustrated in Example 1 is described in general as follows.

THEOREM A

Let A have n linearly independent eigenvectors so that A is diagonalizable, that is,

$$P^{-1}AP = \Lambda = \begin{pmatrix} \lambda_1 & 0 & \cdots & 0 \\ 0 & \lambda_2 & \cdots & 0 \\ \vdots & \vdots & \ddots & \vdots \\ 0 & 0 & \cdots & \lambda_n \end{pmatrix}$$

for some constant $n \times n$ matrix P. Then $\mathbf{x}$ is a solution of Eq. (1) if and only if $\mathbf{z}$, defined by

$$\mathbf{z}(t) = P^{-1}\mathbf{x}(t), \tag{10}$$

is a solution of the uncoupled system

$$\mathbf{z}' = \Lambda\mathbf{z} + P^{-1}\mathbf{h}(t). \tag{11}$$

And then $\mathbf{x}(t) = P\mathbf{z}(t)$.

Moreover, the particular solution of Eq. (1) which also satisfies the initial condition (2) is $\mathbf{x}(t) = P\mathbf{z}(t)$ where $\mathbf{z}$ is that solution of (11) which satisfies

$$\mathbf{z}(t_0) = P^{-1}\mathbf{x}_0. \tag{12}$$

The proof of this theorem involves practically nothing that was not already done in Example 1. Thus, we shall consider it clear.

In the case treated in Theorem A, the general solution of (1) can also be written in the form

$$\mathbf{x}(t) = \tilde{\mathbf{x}}(t) + c_1 e^{\lambda_1 t}\xi_{(1)} + \cdots + c_n e^{\lambda_n t}\xi_{(n)}, \tag{13}$$

where $\xi_{(1)}, \ldots, \xi_{(n)}$ are linearly independent eigenvectors of A corresponding to the eigenvalues $\lambda_1, \ldots, \lambda_n$, respectively, and $\tilde{\mathbf{x}}$ is any particular solution of (1). This follows from Theorem 23-D together with the fact that the functions defined by $\mathbf{y}_{(j)}(t) = e^{\lambda_j t}\xi_{(j)}$ $(j = 1, \ldots, n)$ are linearly independent if $\xi_{(1)}, \ldots,$ $\xi_{(n)}$ are linearly independent vectors. Verify this last assertion.

The natural question left by Example 1 and Theorem A is: How can we find the general solution of Eq. (1) or (3) if A has *fewer* than n linearly independent eigenvectors? We shall concentrate now on homogeneous systems.

Example 2. Find the general solution of

$$\mathbf{y}' = A\mathbf{y} = \begin{pmatrix} -1 & -1 \\ 1 & -3 \end{pmatrix}\mathbf{y}. \tag{14}$$

It is easily discovered that the coefficient matrix A has only one eigenvalue, $\lambda = -2$. Thus, any eigenvector ξ must satisfy

$$(A + 2I)\xi = \begin{pmatrix} 1 & -1 \\ 1 & -1 \end{pmatrix}\xi = 0.$$

This requires that $\xi = \text{col}\,(1, 1)$ or some scalar multiple of that vector. So there do not exist two linearly independent eigenvectors. (In fact, Problems 22-5 and 22-6 showed that the matrix A cannot be diagonalized.)

Of course,

$$\mathbf{y}_{(1)}(t) = e^{-2t}\xi = \begin{pmatrix} e^{-2t} \\ e^{-2t} \end{pmatrix}$$

defines a solution of system (14). But how can we find a second linearly independent solution?

Perhaps the first idea that comes to mind, by analogy with Section 15, is to try a function of the form $y(t) = te^{-2t}\eta$, where η is a constant vector. But this will not work. For substitution into Eq. (14) gives $e^{-2t}\eta - 2te^{-2t}\eta = Ate^{-2t}\eta$, or

$$(1 - 2t)\eta = At\eta; \tag{15}$$

and the only way Eq. (15) can hold for various different values of t is if $\eta = 0$.

It will be found, however, that the general solution of system (14) *can* be obtained if we seek solutions of the form

$$y(t) = e^{-2t}\eta + te^{-2t}\zeta. \tag{16}$$

Indeed, substitution of (16) into (14) gives

$$-2e^{-2t}\eta + e^{-2t}\zeta - 2te^{-2t}\zeta = e^{-2t}A\eta + te^{-2t}A\zeta,$$

or, dividing out the common factor of e^{-2t},

$$-2\eta + \zeta - 2t\zeta = A\eta + tA\zeta.$$

Now equating terms involving like powers of t we find

$$(A + 2I)\zeta = 0, \tag{17}$$

$$(A + 2I)\eta = \zeta. \tag{18}$$

The solutions ζ of Eq. (17) are multiples of the eigenvector ξ. Solutions η of (18) (if there are any), are called *generalized eigenvectors*. But, since det $(A + 2I) = 0$, it is not at all clear that Eq. (18) has any solution when $\zeta \neq 0$.

If $\zeta = \xi$, then (18) becomes

$$\begin{pmatrix} 1 & -1 \\ 1 & -1 \end{pmatrix} \eta = \begin{pmatrix} 1 \\ 1 \end{pmatrix},$$

which, we find, does have solutions

$$\eta = \text{col}\,(1, 0) + c\xi \tag{19}$$

for all values of the scalar c. Thus, for example, we can get two solutions of (14) by putting into (16)

$$\zeta = 0, \qquad \eta = \xi,$$

and then

$$\zeta = \xi, \qquad \eta = \text{col}\,(1, 0).$$

These give, respectively,

$$y_{(1)}(t) = \begin{pmatrix} e^{-2t} \\ e^{-2t} \end{pmatrix} \quad \text{and} \quad y_{(2)}(t) = \begin{pmatrix} (1 + t)e^{-2t} \\ te^{-2t} \end{pmatrix}. \tag{20}$$

The reader should verify that $y_{(1)}$ and $y_{(2)}$ are linearly independent. (Problem 2.) Thus, the general solution of (14) is $y = c_1 y_{(1)} + c_2 y_{(2)}$.

A natural extension of this last procedure will handle the case of a system of three equations having only one or two linearly independent eigenvectors. Two examples are presented to illustrate the method. It will be noted in Problem 4 that because of the special form of these systems, (21) and (26), there is actually an alternative easier way of solving them. Nevertheless, the reader is asked to follow these examples as presented to illustrate a general method of solution.

Example 3. Find three linearly independent solutions of

$$y' = \begin{pmatrix} \mu & 1 & 0 \\ 0 & \mu & 1 \\ 0 & 0 & \mu \end{pmatrix} y. \tag{21}$$

The equation $\det (A - \lambda I) = 0$ becomes $(\mu - \lambda)^3 = 0$. Thus $\lambda = \mu$ is the only eigenvalue of A; or, we might say, μ is an eigenvalue of multiplicity three. To find the eigenvectors we solve $(A - \mu I)\xi = 0$, that is,

$$\begin{pmatrix} 0 & 1 & 0 \\ 0 & 0 & 1 \\ 0 & 0 & 0 \end{pmatrix} \xi = 0.$$

A solution is

$$\xi = \mathrm{col}\,(1, 0, 0),$$

or any multiple of this. But there are no other linearly independent eigenvectors. Thus there will be a deficiency of two linearly independent solutions of system (21). All we have so far is

$$y_{(1)}(t) = e^{\mu t}\xi = \mathrm{col}\,(e^{\mu t}, 0, 0).$$

Let us then seek solutions of system (21) in the form

$$y(t) = e^{\mu t}\eta + t e^{\mu t}\zeta + t^2 e^{\mu t}\theta. \tag{22}$$

Substitution into (21) gives

$$\mu e^{\mu t}\eta + \mu t e^{\mu t}\zeta + e^{\mu t}\zeta + \mu t^2 e^{\mu t}\theta + 2t e^{\mu t}\theta = e^{\mu t} A\eta + t e^{\mu t} A\zeta + t^2 e^{\mu t} A\theta.$$

Now divide out the common factor $e^{\mu t}$ and equate coefficients of like powers of t to get

$$(A - \mu I)\theta = 0, \tag{23}$$

$$(A - \mu I)\zeta = 2\theta, \tag{24}$$

$$(A - \mu I)\eta = \zeta. \tag{25}$$

For θ we can take any multiple of ξ. However, since $\det (A - \mu I) = 0$, it is again unclear that Eq. (24) has a solution when $\theta \neq 0$. A similar doubt arises for Eq. (25). Let us try anyway. If we take $\theta = c_1 \xi$, then Eq. (24) becomes

$$\begin{pmatrix} 0 & 1 & 0 \\ 0 & 0 & 1 \\ 0 & 0 & 0 \end{pmatrix} \zeta = \begin{pmatrix} 2c_1 \\ 0 \\ 0 \end{pmatrix}$$

which does have solutions, namely, the "generalized eigenvectors"

$$\zeta = \begin{pmatrix} c_2 \\ 2c_1 \\ 0 \end{pmatrix} = \begin{pmatrix} 0 \\ 2c_1 \\ 0 \end{pmatrix} + c_2 \xi,$$

for any scalar constant, c_2. This, in turn, gives for Eq. (25),

$$\begin{pmatrix} 0 & 1 & 0 \\ 0 & 0 & 1 \\ 0 & 0 & 0 \end{pmatrix} \eta = \begin{pmatrix} c_2 \\ 2c_1 \\ 0 \end{pmatrix}.$$

Solutions are the "generalized eigenvectors"

$$\eta = \begin{pmatrix} c_3 \\ c_2 \\ 2c_1 \end{pmatrix} = \begin{pmatrix} 0 \\ c_2 \\ 2c_1 \end{pmatrix} + c_3 \xi.$$

Putting these expressions for η, ζ, and θ into (22) we get

$$y(t) = \begin{pmatrix} (c_3 + c_2 t + c_1 t^2) e^{\mu t} \\ (c_2 + 2c_1 t) e^{\mu t} \\ 2c_1 e^{\mu t} \end{pmatrix}.$$

Now using various different choices of the arbitrary constants c_1, c_2, and c_3 we obtain three solutions,

$$y_{(1)}(t) = \begin{pmatrix} e^{\mu t} \\ 0 \\ 0 \end{pmatrix}, \quad y_{(2)}(t) = \begin{pmatrix} t e^{\mu t} \\ e^{\mu t} \\ 0 \end{pmatrix}, \quad y_{(3)}(t) = \begin{pmatrix} t^2 e^{\mu t} \\ 2t e^{\mu t} \\ 2 e^{\mu t} \end{pmatrix}.$$

The reader should verify that these are linearly independent. [Problem 3(a).]

The situation can actually become more complicated when an eigenvalue of multiplicity m yields more than one but fewer than m linearly independent eigenvectors, as in the next example.

Example 4. Find the general solution of

$$\mathbf{y}' = \begin{pmatrix} \mu & 1 & 0 \\ 0 & \mu & 0 \\ 0 & 0 & \mu \end{pmatrix} \mathbf{y}. \tag{26}$$

Again one easily finds that $\lambda = \mu$ is an eigenvalue of A of multiplicity $m = 3$. The equation for the eigenvectors, $(A - \mu I)\xi = 0$, becomes

$$\begin{pmatrix} 0 & 1 & 0 \\ 0 & 0 & 0 \\ 0 & 0 & 0 \end{pmatrix} \xi = \mathbf{0}.$$

This gives two linearly independent eigenvectors, for example,

$$\xi_{(1)} = \text{col } (1, 0, 1) \quad \text{and} \quad \xi_{(2)} = \text{col } (0, 0, 1),$$

but no more than two. Thus, we immediately obtain two linearly independent solutions of system (26), defined by

$$\mathbf{y}_{(1)}(t) = e^{\mu t}\xi_{(1)} \quad \text{and} \quad \mathbf{y}_{(2)}(t) = e^{\mu t}\xi_{(2)}.$$

There is now a deficiency of only one linearly independent solution, and we might reasonably seek a third linearly independent solution in the form

$$\mathbf{y}(t) = e^{\mu t}\boldsymbol{\eta} + t e^{\mu t}\boldsymbol{\zeta}. \tag{27}$$

Substitution of (27) into (26) leads to

$$(A - \mu I)\boldsymbol{\zeta} = \mathbf{0}, \tag{28}$$

$$(A - \mu I)\boldsymbol{\eta} = \boldsymbol{\zeta}. \tag{29}$$

(Putting $\boldsymbol{\zeta} = \mathbf{0}$ gives again the two solutions, $\mathbf{y}_{(1)}$ and $\mathbf{y}_{(2)}$, already obtained.) To find a third linearly independent solution, let us put $\boldsymbol{\zeta} = c\xi_{(1)}$ and try to solve Eq. (29) for $\boldsymbol{\eta}$:

$$(A - \mu I)\boldsymbol{\eta} = \begin{pmatrix} 0 & 1 & 0 \\ 0 & 0 & 0 \\ 0 & 0 & 0 \end{pmatrix} \boldsymbol{\eta} = \begin{pmatrix} c \\ 0 \\ c \end{pmatrix}.$$

But this system has no solution when $c \neq 0$!
We try again, taking $\boldsymbol{\zeta} = c\xi_{(2)}$. Then (29) becomes

$$\begin{pmatrix} 0 & 1 & 0 \\ 0 & 0 & 0 \\ 0 & 0 & 0 \end{pmatrix} \boldsymbol{\eta} = \begin{pmatrix} 0 \\ 0 \\ c \end{pmatrix},$$

and again, there is no solution if $c \neq 0$.

What next?

We have not exhausted all possibilities, because we can use for ζ *any* eigenvector of A. In other words, we can try to satisfy Eq. (29) by using for ζ any linear combination of $\xi_{(1)}$ and $\xi_{(2)}$, say

$$\zeta = c_1 \xi_{(1)} + c_2 \xi_{(2)} = \text{col}\,(c_1, 0, c_1 + c_2).$$

(Recall Theorem 22-D.) Then (29) becomes

$$\begin{pmatrix} 0 & 1 & 0 \\ 0 & 0 & 0 \\ 0 & 0 & 0 \end{pmatrix} \eta = \begin{pmatrix} c_1 \\ 0 \\ c_1 + c_2 \end{pmatrix}.$$

A little study of this equation shows that a solution exists if and only if $c_1 + c_2 = 0$. Thus, we can take $c_1 = 1$ and $c_2 = -1$ to get

$$\zeta = \text{col}\,(1, 0, 0) \quad \text{and} \quad \eta = \text{col}\,(0, 1, 0).$$

This gives, as a third solution of system (26),

$$\mathbf{y}_{(3)}(t) = e^{\mu t}\eta + te^{\mu t}\zeta = \begin{pmatrix} te^{\mu t} \\ e^{\mu t} \\ 0 \end{pmatrix}.$$

The reader should verify the linear independence of $\mathbf{y}_{(1)}$, $\mathbf{y}_{(2)}$, and $\mathbf{y}_{(3)}$. [Problem 3(b).]

The following theorem is presented to complete a prescription for solving the homogeneous system (3) with constant coefficients. We give the proof only for the case $n = 2$.

THEOREM B

Let the characteristic equation $\det(A - \lambda I) = 0$ have a root λ of multiplicity m. Let k be the number of linearly independent eigenvectors associated with the eigenvalue λ. Then one can find m linearly independent solutions of (3), of the form

$$\mathbf{y}(t) = e^{\lambda t}\eta + te^{\lambda t}\zeta + \cdots + t^{m-k}e^{\lambda t}\theta,$$

for some appropriate vectors η, ζ, ..., θ. Moreover, if these solutions are found for each distinct eigenvalue, the result will be a total of n linearly independent solutions of (3).

*Proof for $n = 2$. If the 2×2 matrix A has two linearly independent eigenvectors, $\xi_{(1)}$ and $\xi_{(2)}$, we know that the general solution of (3) has the form

$$\mathbf{y}(t) = c_1 e^{\lambda_1 t}\xi_{(1)} + c_2 e^{\lambda_2 t}\xi_{(2)}.$$

In case A has only one linearly independent eigenvector ξ, and hence a double eigenvalue λ, proceed as follows. (This will explain why the method worked in Example 2.) One solution is given by $e^{\lambda t}\xi$. We seek a second linearly independent solution in the form

$$e^{\lambda t}\eta + te^{\lambda t}\zeta, \tag{30}$$

as in Eq. (16). Of course, for linear independence, we must have $\zeta \neq 0$. As in Example 2, the expression (30) satisfies Eq. (3) if and only if

$$(A - \lambda I)\zeta = 0 \tag{31}$$

and

$$(A - \lambda I)\eta = \zeta. \tag{32}$$

To construct a nontrivial solution of (31) and (32), let η be *any* nonzero vector which is *not an eigenvector*. Then *define* ζ by Eq. (32). It follows that $\zeta \neq 0$. Why? We will find that ζ is an eigenvector and hence satisfies Eq. (31).

We first show that ζ and η are linearly independent. Suppose they were not. Then we would have $\zeta = c\eta$ for some $c \neq 0$. Why? Hence, from (32),

$$(A - \lambda I - cI)\eta = 0.$$

But this would say that η was an eigenvector and $\lambda + c$ an eigenvalue for A, which is a contradiction.

The fact that ζ and η are linearly independent means that

$$\begin{vmatrix} \zeta_1 & \eta_1 \\ \zeta_2 & \eta_2 \end{vmatrix} \neq 0.$$

Hence we can solve, for c_1 and c_2, the equations

$$c_1\zeta_1 + c_2\eta_1 = \xi_1,$$

$$c_1\zeta_2 + c_2\eta_2 = \xi_2,$$

where ξ is our original given eigenvector. Then

$$\xi = c_1\zeta + c_2\eta. \tag{33}$$

Since η is not an eigenvector, it follows that $c_1 \neq 0$ in (33). Multiplying Eq. (33) by $A - \lambda I$ one now finds

$$0 = c_1(A - \lambda I)\zeta + c_2\zeta$$

or

$$\left(A - \lambda I + \frac{c_2}{c_1}I\right)\zeta = 0.$$

This says that $\lambda - c_2/c_1$ is an eigenvalue of A. Hence, $c_2 = 0$ and, from (33), $\xi = c_1\zeta$ so that ζ is an eigenvector. Thus (31) and (32) are both satisfied. ∎

Deeper results of matrix analysis, which are beyond the scope of this text, would be of great value if we wanted to study systems of high order. But most systems encountered in applications are of second or third order (or occasionally fourth). And then the methods indicated in the examples of this section should be adequate for the homogeneous system with constant coefficients.

For a nonhomogeneous system (1), in case A is diagonalizable, Theorem A shows how to decouple the system into n independent linear equations of first order. Thus, in that case the problem is completely solved, in principle.

When A is not diagonalizable we have concentrated on the homogeneous system (3). Having solved the homogeneous system we can then solve certain nonhomogeneous systems by the "method of undetermined coefficients." This procedure for finding a particular solution was illustrated in Example 23-1 and Problem 23-5 and will be needed again in Problem 5(c) below. It depends on successfully guessing an appropriate form for a particular solution of the nonhomogeneous system. This will usually be possible if the vector function $\mathbf{h}$ in Eq. (1) is a simple combination of polynomials, exponentials, sines, and cosines.

A very general procedure for handling nonhomogeneous systems will be presented in Section 26 under the heading "variation of parameters."

PROBLEMS

1. Use the general solution (7) of system (4) to find that solution which also satisfies the initial condition (8); that is, find the appropriate values of c_1 and c_2. Compare your answer with Eq. (9).
2. Verify that $\mathbf{y}_{(2)}$ defined in Eq. (20) is a solution of system (14) and is linearly independent of $\mathbf{y}_{(1)}$.
3. (a) Verify that the functions $\mathbf{y}_{(1)}$, $\mathbf{y}_{(2)}$, and $\mathbf{y}_{(3)}$ obtained in Example 3 are linearly independent solutions of system (21).
   (b) Same for Example 4.
4. (a) Solve system (21) by first solving its third equation, which is "uncoupled," for y_3. Then solve the second equation for y_2, and finally the first equation for y_1.
   (b) Same for system (26).
5. Find the general solutions of the following systems.

   (a) $\mathbf{x}' = \begin{pmatrix} 2 & -1 \\ -2 & 3 \end{pmatrix} \mathbf{x} + \begin{pmatrix} 4e^{4t} + 2t \\ e^{4t} + 2t \end{pmatrix}$.

   The coefficient matrix A used here occurred previously in Problem 22-4(a).

   (b) $\mathbf{y}' = \begin{pmatrix} 1 & 2 \\ -1 & -1 \end{pmatrix} \mathbf{y}$.   See Problem 22-4(b).

   (c) $\mathbf{x}' = \begin{pmatrix} -1 & -1 \\ 1 & -3 \end{pmatrix} \mathbf{x} + \begin{pmatrix} t - 2e^{-t} \\ 1 + e^{-t} \end{pmatrix}$.   Compare Example 2.

(d) $\mathbf{y}' = \begin{pmatrix} -1 & 0 & 3 \\ -8 & 1 & 12 \\ -2 & 0 & 4 \end{pmatrix} \mathbf{y}.$ (e) $\mathbf{y}' = \begin{pmatrix} -1 & 0 & 3 \\ -8 & 1 & 11 \\ -2 & 0 & 4 \end{pmatrix} \mathbf{y}.$

(f) $\mathbf{y}' = \begin{pmatrix} -1 & 1 & 0 \\ -1 & -4 & 1 \\ -1 & -2 & -1 \end{pmatrix} \mathbf{y}.$ (g) $\mathbf{y}' = \begin{pmatrix} -1 & 2 & -1 \\ -1 & -4 & 1 \\ -1 & -2 & -1 \end{pmatrix} \mathbf{y}.$

(h) $\mathbf{y}' = \begin{pmatrix} 0 & 1 & 2 \\ 1 & 0 & 3 \\ -1 & -1 & -3 \end{pmatrix} \mathbf{y}.$

6. Find the particular solutions of Problems 5(a)–(e) which satisfy respectively the initial conditions:

(a) $\mathbf{x}(0) = \mathrm{col}\,(2, -1).$ (b) $\mathbf{y}(\pi) = \mathrm{col}\,(-1, 1).$
(c) $\mathbf{x}(0) = \mathrm{col}\,(2, 0).$ (d) and (e) $\mathbf{y}(0) = \mathrm{col}\,(1, 2, 3).$

7. Solve each of the following scalar equations first by the method of Chapter Three and then by converting to an equivalent system through $y_1 = z$ and $y_2 = z'$. Compare the results.

(a) $z'' + 4z' + 3z = 0.$ (b) $z'' + 4z' + 4z = 0.$

8. The given 3×3 matrix in Problem 5(d) has only two (distinct) eigenvalues, and yet has three linearly independent eigenvectors. Prove that it is impossible for any nondiagonal $n \times n$ matrix to have n linearly independent eigenvectors if it has *only one* (distinct) eigenvalue.

*25. OSCILLATIONS AND DAMPING IN APPLICATIONS

Systems of linear equations with constant coefficients arise naturally in the analysis of mechanical systems and electrical circuits. However, the resulting differential equations may be rather cumbersome to solve exactly, even though such solution is actually possible. Thus it would be useful to have ways of gaining *some* information without carrying out the complete solution.

In the analysis of a system

$$\mathbf{x}' = A\mathbf{x} + \mathbf{h}(t) \tag{1}$$

with initial conditions

$$\mathbf{x}(t_0) = \mathbf{x}_0, \tag{2}$$

it is useful to first study the associated homogeneous system

$$\mathbf{y}' = A\mathbf{y}. \tag{3}$$

If A is a constant $n \times n$ matrix with real elements, various conclusions about the behaviour of solutions of (3) can be obtained by examining the eigenvalues of A. Again we take $J = \mathbf{R}$.

THEOREM A

Let ρ be any number such that

$$\text{Re } \lambda < \rho$$

for every eigenvalue λ of A. Then for each solution $\mathbf{y}$ of (3) there exists some constant $H > 0$ such that

$$\|\mathbf{y}(t)\| \leq He^{\rho t} \quad \text{for} \quad t \geq 0.$$

Proof. Choose $\delta > 0$ such that

$$\text{Re } \lambda < \rho - \delta$$

for every eigenvalue λ. Then, if p is any polynomial, we will have $|p(t)| \leq Be^{\delta t}$ for all $t \geq 0$ for some $B > 0$. Why? Thus, for any eigenvalue,

$$|e^{\lambda t}p(t)| \leq e^{\text{Re } \lambda t}Be^{\delta t} < Be^{\rho t} \quad \text{for} \quad t \geq 0.$$

The assertion of the theorem then follows from Theorem 24-B.

Of course, since we proved Theorem 24-B only for the case $n = 2$, the present proof is complete only for the case $n = 2$. ∎

COROLLARY B

Every solution of system (3) tends to zero as $t \to \infty$ if and only if every eigenvalue of A has negative real part, that is, $\text{Re } \lambda < 0$ for every eigenvalue λ. (In this case A is called a *stable* matrix.)

The proof is left to the reader as Problem 1.

If every solution of system (3) with constant coefficients tends to zero as $t \to \infty$ one says that the trivial solution ($\mathbf{y} \equiv \mathbf{0}$) of (3) is "asymptotically stable."

COROLLARY C

If A has n linearly independent eigenvectors and if $\text{Re } \lambda \leq 0$ for every eigenvalue λ of A, then every solution of (3) remains bounded as $t \to \infty$.

The proof is Problem 2.

In order to apply Corollaries B and C we require some way of determining the algebraic sign of the real parts of the eigenvalues. Since the eigenvalues of a matrix A are solutions of the polynomial equation $\det (A - \lambda I) = 0$ our problem is to study the zeros of a polynomial. If we are interested in systems of 2, 3, or 4 linear first-order equations the following Lemmas D, E, and F will be useful.

LEMMA D

Both roots of the quadratic equation

$$\lambda^2 + a_1\lambda + a_0 = 0 \tag{4}$$

with real coefficients (a_1 and a_0) have negative real parts if and only if

$$a_1 > 0 \quad \text{and} \quad a_0 > 0. \tag{5}$$

Proof. Let λ_1 and λ_2 be the roots of Eq. (4). Then $\lambda^2 + a_1\lambda + a_0 = (\lambda - \lambda_1)(\lambda - \lambda_2)$, and so

$$\lambda_1 + \lambda_2 = -a_1 \quad \text{and} \quad \lambda_1\lambda_2 = a_0.$$

If λ_1 and λ_2 are both real (possibly identical), then it is easily seen that they are both negative if and only if $a_1 > 0$ and $a_0 > 0$.

If λ_1 and λ_2 are not both real, then they must be complex conjugates, say $\lambda_2 = \bar{\lambda}_1$. Thus,

$$2 \operatorname{Re} \lambda_1 = -a_1 \quad \text{and} \quad |\lambda_1|^2 = a_0.$$

So it follows that $\operatorname{Re} \lambda_1 < 0$ if and only if $a_1 > 0$. And in this case $a_0 > 0$ also. ∎

LEMMA E

The roots of the cubic equation

$$\lambda^3 + a_2\lambda^2 + a_1\lambda + a_0 = 0 \tag{6}$$

with real coefficients all have negative real parts if and only if

$$a_2 > 0, \quad a_1 > 0, \quad a_0 > 0, \quad \text{and} \quad a_2 a_1 > a_0. \tag{7}$$

Proof. Since the coefficients in Eq. (6) are all real, it follows that any complex roots must occur in conjugate pairs. Thus, there must be at least one real root, call it λ_1. Hence, we can factor the polynomial in Eq. (6) and write

$$(\lambda^2 + b_1\lambda + b_0)(\lambda + c_0) = 0, \tag{8}$$

where b_1, b_0, and $c_0 = -\lambda_1$ are real. It follows that

$$a_2 = b_1 + c_0, \quad a_1 = b_0 + b_1 c_0, \quad \text{and} \quad a_0 = b_0 c_0. \tag{9}$$

From these we compute

$$\begin{aligned} a_2 a_1 - a_0 &= b_1 a_1 + c_0 a_1 - b_0 c_0 \\ &= b_1(a_1 + c_0^2). \end{aligned} \tag{10}$$

Now let conditions (7) hold. Then it follows from Eq. (10) that $b_1 > 0$. Also, from (9), the condition $a_0 = b_0 c_0$ shows that b_0 and c_0 have like signs. Now we could not have $b_0 < 0$ and $c_0 < 0$, or we would violate $b_0 + b_1 c_0 = a_1 > 0$. Thus, we conclude that $b_0 > 0$ and $c_0 > 0$.

The positivity of c_0 implies $\lambda_1 < 0$. And, by Lemma D, the positivity of b_1 and b_0 guarantees that the remaining two roots of (6), alias (8), have negative real parts.

The proof of the converse part of the lemma is Problem 5. ∎

LEMMA F

The roots of the quartic equation

$$\lambda^4 + a_3 \lambda^3 + a_2 \lambda^2 + a_1 \lambda + a_0 = 0 \tag{11}$$

with real coefficients all have negative real parts if and only if

$$a_3 > 0, \quad a_2 > 0, \quad a_1 > 0, \quad a_0 > 0, \quad \text{and} \quad a_3 a_2 a_1 > a_3^2 a_0 + a_1^2. \tag{12}$$

Proof. Since any complex roots of Eq. (11) must occur in conjugate pairs, we can factor the left-hand side of (11) and get

$$(\lambda^2 + b_1 \lambda + b_0)(\lambda^2 + c_1 \lambda + c_0) = 0, \tag{13}$$

where b_1, b_0, c_1, and c_0 are all real. It follows that

$$a_3 = b_1 + c_1, \quad a_2 = b_1 c_1 + b_0 + c_0,$$
$$a_1 = b_0 c_1 + b_1 c_0, \quad \text{and} \quad a_0 = b_0 c_0. \tag{14}$$

From these we can compute

$$(a_3 a_2 - a_1)a_1 - a_3^2 a_0$$

$$= (b_1^2 c_1 + b_1 b_0 + b_1 c_1^2 + c_1 c_0)(b_0 c_1 + b_1 c_0) - (b_1^2 + 2b_1 c_1 + c_1^2)b_0 c_0$$

$$= b_1 c_1 (b_0^2 - 2b_0 c_0 + c_0^2 + b_1 b_0 c_1 + b_0 c_1^2 + b_1^2 c_0 + b_1 c_1 c_0)$$

or

$$a_3 a_2 a_1 - a_1^2 - a_3^2 a_0 = b_1 c_1 [(b_0 - c_0)^2 + a_1 a_3]. \tag{15}$$

Now let conditions (12) hold. Then it follows from (15) that $b_1 c_1 > 0$ and from (14) that $b_1 + c_1 = a_3 > 0$. These can both hold only if $b_1 > 0$ and $c_1 > 0$. It also follows from (14) that $b_0 c_0 = a_0 > 0$ which means b_0 and c_0 must have the same sign. But b_0 and c_0 cannot both be negative since $b_0 c_1 + b_1 c_0 = a_1 > 0$. Thus we must have $b_0 > 0$ and $c_0 > 0$. It now follows from Lemma D that all the roots of (13), and hence of (11), have negative real parts.

Again the proof of the converse is left to the reader. (Problem 6.) ■

The unmotivated and mysterious looking criteria given in Lemmas D, E, and F are actually special cases of a systematic test which applies to polynomials of arbitrary order. If you ever need to know the nature of the zeros of a polynomial of degree $n \geq 5$, the general criteria, which includes Lemmas D, E, and F as special cases, can be found in books on matrix theory under the heading "Routh-Hurwitz" criteria.

Imagine that the homogeneous system (3) represents some real physical problem. Then we should expect all solutions to be "damped" out, that is, to approach zero as $t \to \infty$, because of loss of energy due to friction. Such damping was indeed observed in the equations for the simple mechanical and electrical problems of Example 14-1 and Section 18. Let us determine whether solutions also tend to zero in a more complicated case.

Example 1. Let two masses, m_1 and m_2, be suspended (and bouncing) on idealized weightless springs hanging from a fixed support as shown in Figure 1. Let $z_1(t)$ and $z_2(t)$ be their respective distances, at time t, below their equilibrium positions. Let k_1 and k_2 be the two spring constants and let b_1 and b_2 be coefficients of frictional resistance for the two masses. These are all positive numbers. Then, by an argument like that in Example 10-1, one finds

$$m_1 z_1'' = -k_1 z_1 - k_2 z_1 + k_2 z_2 - b_1 z_1',$$
$$m_2 z_2'' = k_2 z_1 - k_2 z_2 - b_2 z_2'. \tag{16}$$

To analyze this system using the methods of Section 24 we should change it into an equivalent system of first-order equations.
    Let

$$y_1 = z_1, \quad y_2 = z_1', \quad y_3 = z_2, \quad y_4 = z_2'.$$

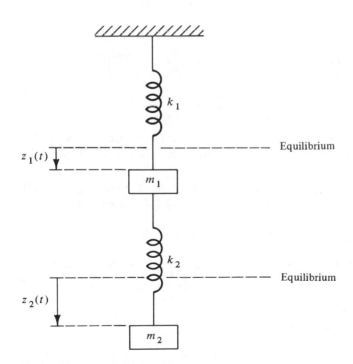

FIGURE 1

Then

$$y_1' = y_2,$$

$$y_2' = -\frac{k_1 + k_2}{m_1} y_1 - \frac{b_1}{m_1} y_2 + \frac{k_2}{m_1} y_3,$$

$$y_3' = y_4,$$

$$y_4' = \frac{k_2}{m_2} y_1 - \frac{k_2}{m_2} y_3 - \frac{b_2}{m_2} y_4.$$

(17)

Computation of the general solution of this system would be straightforward but tedious. Let us be satisfied with an examination of the eigenvalues.

The characteristic equation for system (17) is

$$\begin{vmatrix} -\lambda & 1 & 0 & 0 \\ -\dfrac{k_1 + k_2}{m_1} & -\dfrac{b_1}{m_1} - \lambda & \dfrac{k_2}{m_1} & 0 \\ 0 & 0 & -\lambda & 1 \\ \dfrac{k_2}{m_2} & 0 & -\dfrac{k_2}{m_2} & -\dfrac{b_2}{m_2} - \lambda \end{vmatrix} = 0,$$

(18)

which gives (Problem 7)

$$\lambda^4 + \left(\frac{b_1}{m_1} + \frac{b_2}{m_2}\right)\lambda^3 + \left(\frac{k_1 + k_2}{m_1} + \frac{k_2}{m_2} + \frac{b_1 b_2}{m_1 m_2}\right)\lambda^2$$

$$+ \left(\frac{b_1}{m_1}\frac{k_2}{m_2} + \frac{k_1 + k_2}{m_1}\frac{b_2}{m_2}\right)\lambda + \frac{k_1 k_2}{m_1 m_2} = 0.$$

(19)

Now if friction means what we want it to mean then, according to Corollary B, we would expect all solutions of Eq. (19) to have negative real parts. Let us apply Lemma F to check this.

The condition that a_3, a_2, a_1, and a_0 all be positive is clearly satisfied by Eq. (19). To meet the last requirement of (12) the reader should verify (Problem 8) that

$$(a_3 a_2 - a_1)a_1 - a_3^2 a_0$$

$$= \left(\frac{b_1 k_1}{m_1^2} + \frac{b_1 k_2}{m_1^2} + \frac{b_1^2 b_2}{m_1^2 m_2} + \frac{b_2 k_2}{m_2^2} + \frac{b_1 b_2^2}{m_1 m_2^2}\right)\left(\frac{b_1 k_2}{m_1 m_2} + \frac{k_1 + k_2}{m_1}\frac{b_2}{m_2}\right)$$

$$- \left(\frac{b_1^2}{m_1^2} + \frac{2b_1 b_2}{m_1 m_2} + \frac{b_2^2}{m_2^2}\right)\frac{k_1 k_2}{m_1 m_2}$$

$$= \frac{b_1 b_2}{m_1 m_2}\left(\frac{k_1}{m_1} - \frac{k_2}{m_2}\right)^2 + \frac{b_1 k_2}{m_1^3 m_2}(2b_2 k_1 + b_1 k_2 + b_2 k_2)$$

$$+ \frac{b_1 b_2}{m_1^3 m_2^2} (b_1^2 k_2 + b_1 b_2 k_1 + b_1 b_2 k_2) + \frac{b_2^2 k_2^2}{m_1 m_2^3} \tag{20}$$

$$+ \frac{b_1 b_2^2}{m_1^2 m_2^3} (b_1 k_2 + b_2 k_1 + b_2 k_2),$$

which is clearly positive.

Example 2. Sometimes one considers the idealized case of Example 1 without friction, that is, with $b_1 = 0$ and $b_2 = 0$. Then we might expect, on the basis of "physical intuition," that solutions would be undamped oscillations of some sort. Again this can be checked by considering the characteristic equation (19) which now reduces to

$$\lambda^4 + \left(\frac{k_1 + k_2}{m_1} + \frac{k_2}{m_2} \right) \lambda^2 + \frac{k_1 k_2}{m_1 m_2} = 0. \tag{21}$$

This is a quadratic equation in λ^2 and the discriminant is

$$\left(\frac{k_1 + k_2}{m_1} + \frac{k_2}{m_2} \right)^2 - 4 \frac{k_1 k_2}{m_1 m_2} = \left(\frac{k_1}{m_1} - \frac{k_2}{m_2} \right)^2 + 2 \frac{k_1 k_2}{m_1^2} + \frac{k_2^2}{m_1^2} + 2 \frac{k_2^2}{m_1 m_2} > 0.$$

Hence, the values of λ^2 satisfying Eq. (21) must be real and distinct. But, since the coefficients in Eq. (21) are all positive, λ^2 must be negative. We conclude that there are two negative values of λ^2 which satisfy Eq. (21). Thus, the roots λ are distinct pairs of conjugate pure imaginary numbers,

$$\lambda = \pm i\omega_1, \quad \pm i\omega_2.$$

Corollary C then asserts that the solutions of system (16) or (17) are bounded.

Example 3. The electrical currents $x_1(t)$ and $x_2(t)$ flowing in the network of Figure 2 satisfy the differential equations

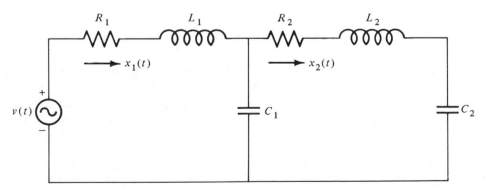

FIGURE 2

$$L_1 x_1'' + R_1 x_1' + \frac{1}{C_1} x_1 - \frac{1}{C_1} x_2 = v'(t),$$

$$L_2 x_2'' + R_2 x_2' + \left(\frac{1}{C_1} + \frac{1}{C_2} \right) x_2 - \frac{1}{C_1} x_1 = 0, \tag{22}$$

where L_1, L_2, R_1, R_2, C_1, and C_2 are given positive constants. But in the case $v(t) \equiv 0$ this system is equivalent to (16). To see this, one merely sets $x_1 = z_2$, $x_2 = z_1$, $L_1 = m_2$, $L_2 = m_1$, $R_1 = b_2$, $R_2 = b_1$, $1/C_1 = k_2$, and $1/C_2 = k_1$. Thus, without any further discussion, we conclude that all solutions of system (22) with $v(t) \equiv 0$ tend to zero as $t \to \infty$.

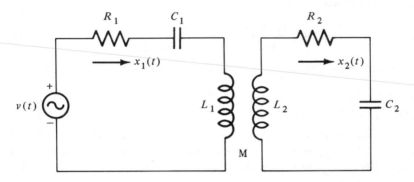

FIGURE 3

Example 4. Consider the two electrical circuits of Figure 3 which are coupled only by the "mutual inductance" M between the two coils of the transformer. The currents $x_1(t)$ and $x_2(t)$ flowing in these circuits satisfy

$$L_1 x_1'' + M x_2'' + R_1 x_1' + \frac{1}{C_1} x_1 = v'(t),$$

$$M x_1'' + L_2 x_2'' + R_2 x_2' + \frac{1}{C_2} x_2 = 0, \tag{23}$$

where M, L_1, L_2, R_1, R_2, C_1, and C_2 are given positive constants with $M^2 < L_1 L_2$. Note that the equations of system (23) are not "solved" for x_1'' and x_2''. It is left as Problem 9 for the reader to transform (23) into a system of four equations of first order, and then show that if $v(t) \equiv 0$, all solutions tend to zero as $t \to \infty$.

After discussing all these examples of the homogeneous system (3), we should show how the results of this section can also give valuable information about important special cases of the nonhomogeneous system (1). The basic observation is simply this:

The general solution of (1) has the form

$$\mathbf{x} = \tilde{\mathbf{x}} + \mathbf{y},$$

where $\tilde{\mathbf{x}}$ is any particular solution of (1) and $\mathbf{y}$ is the general solution of (3). If A is a stable matrix so that $\mathbf{y}(t) \to 0$ as $t \to \infty$, then every solution of system (1) will be close to $\tilde{\mathbf{x}}(t)$ for large t. In other words, if we know one particular solution of system (1), then we essentially know all solutions for large t— regardless of the initial conditions.

Example 5. Analyze the solutions of

$$\mathbf{x}' = A\mathbf{x} + \mathbf{a} \cos \omega t + \mathbf{b} \sin \omega t, \tag{24}$$

where A is a stable matrix (with real elements), $\omega > 0$, and $\mathbf{a}$ and $\mathbf{b}$ are constant (real) vectors.

It will be convenient to first consider the system

$$\mathbf{z}' = A\mathbf{z} + \mathbf{a}e^{i\omega t}. \tag{25}$$

If $\mathbf{z}$ is a solution of (25), then $\mathbf{x}_{(1)} = \operatorname{Re} \mathbf{z}$ will satisfy

$$\mathbf{x}'_{(1)} = A\mathbf{x}_{(1)} + \mathbf{a} \cos \omega t.$$

Why? Let us now seek a particular solution of (25) in the form

$$\mathbf{z}(t) = \mathbf{c}e^{i\omega t},$$

where $\mathbf{c}$ is a constant "undetermined coefficient" vector--probably complex. Substitution into (25) yields the condition

$$i\omega \mathbf{c}e^{i\omega t} = A\mathbf{c}e^{i\omega t} + \mathbf{a}e^{i\omega t},$$

or

$$(A - i\omega I)\mathbf{c} = -\mathbf{a}. \tag{26}$$

Now, since all eigenvalues of A have negative real parts, $i\omega$ cannot be an eigenvalue. Therefore

$$\det (A - i\omega I) \neq 0.$$

So system (26) has a unique solution,

$$\mathbf{c} = -(A - i\omega I)^{-1}\mathbf{a}.$$

Similarly, one can find a solution of the system

$$\mathbf{z}' = A\mathbf{z} + \mathbf{b}e^{i\omega t}$$

in the form $\mathbf{z}(t) = \mathbf{d}e^{i\omega t}$, and argue that $\mathbf{x}_{(2)} = \operatorname{Im} \mathbf{z}$ will then satisfy

$$\mathbf{x}'_{(2)} = A\mathbf{x}_{(2)} + \mathbf{b} \sin \omega t.$$

It follows by superposition that a particular solution of (24) is given by

$$\tilde{\mathbf{x}}(t) = \mathbf{x}_{(1)}(t) + \mathbf{x}_{(2)}(t) = \operatorname{Re} [\mathbf{c}e^{i\omega t}] + \operatorname{Im} [\mathbf{d}e^{i\omega t}]$$

$$= \operatorname{Re} [\mathbf{c}(\cos \omega t + i \sin \omega t)] + \operatorname{Im} [\mathbf{d}(\cos \omega t + i \sin \omega t)],$$

or

$$\tilde{\mathbf{x}}(t) = (\operatorname{Re} \mathbf{c} + \operatorname{Im} \mathbf{d}) \cos \omega t + (-\operatorname{Im} \mathbf{c} + \operatorname{Re} \mathbf{d}) \sin \omega t. \tag{27}$$

The general solution of (24) has the form $\mathbf{x} = \tilde{\mathbf{x}} + \mathbf{y}$, where $\tilde{\mathbf{x}}$ is given by (27) and $\mathbf{y}$ is the general solution of the homogeneous system (3). So, since $\mathbf{y}(t) \to \mathbf{0}$ as $t \to \infty$, $\mathbf{x}(t)$ looks more and more like the periodic function $\tilde{\mathbf{x}}(t)$ as $t \to \infty$ (regardless of the initial conditions). Borrowing some terminology introduced in Section 18, $\tilde{\mathbf{x}}$ is the "steady state" solution and $\mathbf{y}$ is a decaying "transient" solution—often of less interest.

In Problem 11, the reader is asked to apply the conclusions of Example 5 to system (21) or system (22) with $v(t) = 17 \cos \pi t$.

In conclusion let us add the following observation about the role of the imaginary parts of the eigenvalues of A.

THEOREM G

If Im $\lambda \neq 0$ for some eigenvalue λ of A, then system (3) has oscillatory solutions. This means that some component of some solution changes sign infinitely often as $t \to \infty$.

The proof is Problem 12.

The last theorem serves in particular to assure the existence of oscillatory solutions in Example 2. The ambitious reader may be able to prove a stronger result. It is not too difficult to show in that case that *every* component of *every* nontrivial solution changes sign infinitely often as $t \to \infty$,

PROBLEMS

1. Prove Corollary B.
2. Prove Corollary C.
3. Examine each of the following polynomial equations to determine whether or not all its roots have negative real parts.

   (a) $\lambda^2 + \lambda + 1 = 0$.   (b) $\lambda^3 + \lambda^2 + \lambda + 1 = 0$.
   (c) $3\lambda^3 + 2\lambda^2 + \lambda + 1 = 0$.   (d) $\lambda^3 + 2\lambda^2 + \lambda + 1 = 0$.
   (e) $2\lambda^4 + 3\lambda^3 + 4\lambda^2 + \lambda + 1 = 0$.

4. Prove that all solutions of the following systems tend to zero as $t \to \infty$, so that the trivial solution in each case is asymptotically stable.

   (a) $y' = \begin{pmatrix} 0 & 2 & 3 \\ 2 & 0 & 4 \\ -1 & -1 & -3 \end{pmatrix} y.$
   (b) $y' = \begin{pmatrix} -1 & 2 & -1 & -3 \\ -3 & -7 & 9 & -7 \\ -2 & -3 & 5 & -8 \\ 0 & 1 & 0 & -4 \end{pmatrix} y.$

5. Complete the proof of Lemma E by showing that conditions (7) are necessary conditions for the roots of Eq. (6) to have negative real parts.
6. Complete the proof of Lemma F by showing the necessity of conditions (12).
7. Verify that Eqs. (19) and (18) are equivalent.

8. Verify Eqs. (20).
9. Transform (23) into a system of four equations of first order. Then prove that if $v(t) \equiv 0$, all eigenvalues for this system have negative real parts.
10. Consider the nonhomogeneous system

$$\mathbf{x}' = \begin{pmatrix} 0 & 2 & 3 \\ 2 & 0 & 4 \\ -1 & -1 & -3 \end{pmatrix}\mathbf{x} + \begin{pmatrix} 3\cos t - \sin t \\ \cos t + 4\sin t \\ -2\cos t - \sin t \end{pmatrix}.$$

(a) Find a particular solution by the method of "undetermined coefficients" illustrated in Example 23-1.
(b) Invoke the result of Problem 4(a) to say something about the behavior of every solution as $t \to \infty$.
11. Apply the conclusion of Example 5 to system (22) or (23) with $v(t) = 17 \cos \pi t$ to show that for large t the solution in either case is approximately

$$x_1(t) = A_1 \cos \pi t + B_1 \sin \pi t, \quad x_2(t) = A_2 \cos \pi t + B_2 \sin \pi t,$$

where A_1, B_1, A_2, and B_2 are constants.
12. Prove Theorem G.

*26. VARIATION OF PARAMETERS

Let us now return to the general linear system

$$\mathbf{x}' = A(t)\mathbf{x} + \mathbf{h}(t), \tag{1}$$

where A is a given continuous $n \times n$-matrix-valued function and $\mathbf{h}$ is a given continuous n-vector-valued function on an interval $J = (\alpha, \beta)$. Let us assume an initial condition of the form

$$\mathbf{x}(t_0) = \mathbf{x}_0, \tag{2}$$

where $t_0 \in J$ and $\mathbf{x}_0$ is an n-vector.

In Section 19 we studied the corresponding problem for a scalar equation of order n. The important method introduced there was the method of variation of parameters (or variation of constants). This provided a systematic procedure for solving a nonhomogeneous equation, assuming we could first find the general solution of the associated homogeneous equation.

We will soon obtain an analogous method for the system (1) of n first-order equations. More precisely, we will express the solutions of (1) in terms of the solutions of the associated homogeneous system

$$\mathbf{y}' = A(t)\mathbf{y}, \tag{3}$$

provided we can find the *general solution* of (3). Thus we need n linearly independent solutions of (3).

If A is a constant $n \times n$ matrix, the methods of Section 24 will provide n linearly independent solutions of (3).

It will be proved in Section 28 that whenever A is continuous, then (3) does have n linearly independent solutions. It may, however, be difficult to find these solutions. For the purposes of this section, let us assume that A is continuous *and* that system (3) has n linearly independent solutions.

Let $\mathbf{y}_{(1)}, \ldots, \mathbf{y}_{(n)}$ be any n linearly independent solutions of (3) on J, and form a matrix-valued function Y by using these solutions as columns:

$$Y(t) = \begin{pmatrix} y_{(1)1}(t) & \cdots & y_{(n)1}(t) \\ \vdots & & \vdots \\ y_{(1)n}(t) & \cdots & y_{(n)n}(t) \end{pmatrix}. \tag{4}$$

Any such Y is called a *fundamental matrix* for system (3). (Because (3) has many different sets of n linearly independent solutions, there are many different fundamental matrices.) In terms of the n given linearly independent solutions, the general solution of (3) on J can be written as

$$\mathbf{y}(t) = c_1 \mathbf{y}_{(1)}(t) + \cdots + c_n \mathbf{y}_{(n)}(t) = Y(t)\mathbf{c}, \tag{5}$$

where $\mathbf{c} = \operatorname{col}(c_1, \ldots, c_n)$.

By analogy with Section 19, we are going to replace the constant (vector) parameter $\mathbf{c}$ in (5) by a vector-valued *function* $\mathbf{p}$ and seek a solution of the nonhomogeneous system (1) in the form

$$\mathbf{x}(t) = Y(t)\mathbf{p}(t). \tag{6}$$

Again, this method is naturally called "variation of parameters" (or "variation of constants").

Note that, since $\mathbf{y}_{(1)}, \ldots, \mathbf{y}_{(n)}$ are linearly independent solutions of (3), Corollary 23-C assures us that $\det Y(t) \neq 0$ and hence $Y(t)^{-1}$ exists for all t in J. From this it follows that *any* solution $\mathbf{x}$ of system (1) on J can be written in the form (6). One simply defines

$$\mathbf{p}(t) = Y(t)^{-1}\mathbf{x}(t).$$

When we substitute (6) into Eq. (1) the following results will be useful.

LEMMA A

If B is a differentiable matrix-valued function and $\mathbf{p}$ is a differentiable vector-valued function, then

$$\frac{d}{dt}[B(t)\mathbf{p}(t)] = B'(t)\mathbf{p}(t) + B(t)\mathbf{p}'(t). \tag{7}$$

If B is constant and $\mathbf{p}$ is continuous, then

$$\int_{t_0}^{t} B\mathbf{p}(s)\, ds = B \int_{t_0}^{t} \mathbf{p}(s)\, ds. \tag{8}$$

Proof. To prove (7), we consider the ith component of the left-hand side:

$$\frac{d}{dt} \sum_{j=1}^{n} b_{ij}(t) p_j(t) = \sum_{j=1}^{n} b'_{ij}(t) p_j(t) + \sum_{j=1}^{n} b_{ij}(t) p'_j(t).$$

The result is the ith component of the right-hand side of (7).

The proof of (8) is similar. ∎

Regarding $\mathbf{p}$ as an unknown but differentiable function, we now substitute (6) into Eq. (1) to find

$$Y'(t)\mathbf{p}(t) + Y(t)\mathbf{p}'(t) = A(t)Y(t)\mathbf{p}(t) + \mathbf{h}(t). \tag{9}$$

Since $\mathbf{y}'_{(j)}(t) = A(t)\mathbf{y}_{(j)}(t)$ for each j, it follows from the definition of matrix multiplication that

$$Y'(t) = A(t)Y(t) \quad \text{on} \quad J.$$

Using this in Eq. (9) we are left with

$$Y(t)\mathbf{p}'(t) = \mathbf{h}(t).$$

or, multiplying from the left by $Y(t)^{-1}$,

$$\mathbf{p}'(t) = Y(t)^{-1}\mathbf{h}(t).$$

So, for some constant vector $\mathbf{c}$,

$$\mathbf{p}(t) = \mathbf{c} + \int_{t_0}^{t} Y(s)^{-1}\mathbf{h}(s)\, ds$$

and

$$\mathbf{x}(t) = Y(t)\mathbf{p}(t) = Y(t)\mathbf{c} + Y(t) \int_{t_0}^{t} Y(s)^{-1}\mathbf{h}(s)\, ds. \tag{10}$$

If we also insist that $\mathbf{x}$ satisfy Eq. (2), we find $\mathbf{x}_0 = Y(t_0)\mathbf{c} + \mathbf{0}$ or $\mathbf{c} = Y(t_0)^{-1}\mathbf{x}_0$. Using this and the fact that $Y(t)$ can be regarded as a constant matrix as far as integration with respect to s is concerned, Eqs. (10) and (8) give

$$\mathbf{x}(t) = Y(t)Y(t_0)^{-1}\mathbf{x}_0 + \int_{t_0}^{t} Y(t)Y(s)^{-1}\mathbf{h}(s)\, ds. \tag{11}$$

The above analysis shows that *if* Eqs. (1) and (2) have a solution, it must be given by (11). To prove that (11) really *is* a solution of Eqs. (1) and (2) it suffices to verify that the above calculations can be reversed, leading to $\mathbf{x}' = A(t)\mathbf{x} + \mathbf{h}(t)$, and that substitution of $t = t_0$ in (11) gives $\mathbf{x}(t_0) = \mathbf{x}_0$.

We have thus proved the following.

THEOREM B (VARIATION OF PARAMETERS)

Equations (1) and (2) have a (unique) solution $\mathbf{x}$ on J, and $\mathbf{x}$ is given by Eq. (11), where $Y(t)$ is any fundamental matrix for (3).

In the application of Theorem B it is often convenient to use a special type of fundamental matrix described below.

For a given s in J, let $\mathbf{y}_{(j)}(t)$ be the unique solution of system (3) satisfying the initial condition $\mathbf{y}_{(j)}(s) = \mathbf{I}_{(j)}$. If we denote this special solution by $\mathbf{y}_{(j)}(t; s)$, then

$$\mathbf{y}_{(j)}(s; s) = \mathbf{I}_{(j)}.$$

It follows that, for each fixed s in J,

$$\mathbf{y}_{(1)}(t; s), \quad \ldots, \quad \mathbf{y}_{(n)}(t, s)$$

are linearly independent solutions of Eq. (3). Why? Using this special set of n linearly independent solutions, let us define the special fundamental matrix called the *transition matrix*, by

$$T(t; s) = \begin{pmatrix} y_{(1)1}(t; s) & \cdots & y_{(n)1}(t; s) \\ \vdots & & \vdots \\ y_{(1)n}(t; s) & \cdots & y_{(n)n}(t; s) \end{pmatrix}. \tag{12}$$

Note that $T(s; s) = I$. So, given any s in J and any n-vector ξ,

$$\mathbf{y}(t) = T(t; s)\xi \tag{13}$$

is the (unique) solution of Eq. (3) such that $\mathbf{y}(s) = \xi$.

Now consider the function of u defined (for a given t in J) by

$$T(u; t)\mathbf{y}(t) \quad \text{for all } u \text{ in } J.$$

This must be the solution of Eq. (3), considered as a function of u, which takes the value $y(t)$ when $u = t$. Thus, by uniqueness, $T(u; t)\mathbf{y}(t) = \mathbf{y}(u)$ on J. So if $\mathbf{y}$ is defined by (13)

$$T(u; s)\xi = \mathbf{y}(u) = T(u; t)\mathbf{y}(t) = T(u; t)T(t; s)\xi \quad \text{for all } s, t, \text{ and } u \text{ in } J. \tag{14}$$

Since (14) holds for every vector ξ, application of Problem 22-8 shows that

$$T(u; s) = T(u; t)T(t; s) \quad \text{for all } s, t, \text{ and } u \text{ in } J. \tag{15}$$

In particular $T(s; t)T(t; s) = I = T(t; s)T(s; t)$. So

$$T(s; t)^{-1} = T(t; s). \tag{16}$$

Now let $Y(t) = T(t; t_0)$, and Eq. (11) reduces to the equivalent form

$$\mathbf{x}(t) = T(t, t_0)\mathbf{x}_0 + \int_{t_0}^{t} T(t; s)\mathbf{h}(s) \, ds. \tag{11'}$$

Example 1. Use Eq. (11′) to re-solve the system of Example 24-1,

$$\mathbf{x}' = \begin{pmatrix} -2 & 2 \\ 2 & -5 \end{pmatrix} \mathbf{x} + \begin{pmatrix} 5 \sin 2t \\ 0 \end{pmatrix} \tag{17}$$

on **R**, with

$$\mathbf{x}(0) = \mathbf{x}_0 = \text{col } (2, -1). \tag{18}$$

Proceeding as in Example 23-1, we obtain the general solution of the associated homogeneous system, $\mathbf{y}' = A\mathbf{y}$, in the form

$$\mathbf{y}(t) = c_1 \begin{pmatrix} 2e^{-t} \\ e^{-t} \end{pmatrix} + c_2 \begin{pmatrix} e^{-6t} \\ -2e^{-6t} \end{pmatrix}.$$

To obtain $\mathbf{y}_{(1)}(t; s)$ we must choose c_1 and c_2 so that $\mathbf{y}(s) = \text{col } (1, 0)$, that is, so that

$$2e^{-s}c_1 + e^{-6s}c_2 = 1,$$

$$e^{-s}c_1 - 2e^{-6s}c_2 = 0.$$

This gives $c_1 = \frac{2}{5}e^s$ and $c_2 = \frac{1}{5}e^{6s}$. Thus,

$$\mathbf{y}_{(1)}(t; s) = \frac{1}{5} \begin{pmatrix} 4e^{-(t-s)} & | & e^{-6(t-s)} \\ 2e^{-(t-s)} & - & 2e^{-6(t-s)} \end{pmatrix}.$$

By a similar computation, the reader should find $\mathbf{y}_{(2)}(t; s)$ and hence,

$$T(t; s) = \frac{1}{5} \begin{pmatrix} 4e^{-(t-s)} + e^{-6(t-s)} & 2e^{-(t-s)} - 2e^{-6(t-s)} \\ 2e^{-(t-s)} - 2e^{-6(t-s)} & e^{-(t-s)} + 4e^{-6(t-s)} \end{pmatrix}.$$

Substituting this into Eq. (11′) we have

$$\mathbf{x}(t) = \frac{1}{5} \begin{pmatrix} 6e^{-t} + 4e^{-6t} \\ 3e^{-t} - 8e^{-6t} \end{pmatrix} + \begin{pmatrix} \int_0^t [4e^{-(t-s)} + e^{-6(t-s)}]\sin 2s \ ds \\ \int_0^t [2e^{-(t-s)} - 2e^{-6(t-s)}]\sin 2s \ ds \end{pmatrix} \tag{19}$$

as the solution of (17) and (18). The reader should perform the indicated integrations in (19) and compare this with the solution obtained previously, namely, Eq. 24-(9). (Problem 2.)

For a linear system with constant coefficients, such as Eq. (17), we make the following labor-saving observation:

If $A(t) = A$ is constant, then one can take $J = \mathbf{R}$. Now observe that if $\mathbf{y}$ is any solution of Eq. (3) and if s is a constant (real number), then

$$\frac{d}{dt} \mathbf{y}(t - s) = \mathbf{y}'(t - s) = A\mathbf{y}(t - s).$$

So, in particular, $\mathbf{y}_{(j)}(t - s; 0)$ is a solution of Eq. (3). Moreover, when $t = s$, we find $\mathbf{y}_{(j)}(t - s; 0) = \mathbf{y}_{(j)}(0; 0) = \mathbf{I}_{(j)}$. Thus, by uniqueness, $\mathbf{y}_{(j)}(t - s; 0)$ $= \mathbf{y}_{(j)}(t; s)$ for each $j = 1, \ldots, n$. So

$$T(t; s) = T(t - s; 0). \tag{20}$$

To find $T(t; s)$ when A is constant, it is simplest to first compute $T(t; 0)$ and then apply (20).

Example 2. Find the transition matrix $T(t; s)$ for the system of Example 24-2,

$$\mathbf{y}' = \begin{pmatrix} -1 & -1 \\ 1 & -3 \end{pmatrix} \mathbf{y}. \tag{21}$$

Recall that this matrix, A, has only one eigenvalue, $\lambda = -2$, and only one linearly independent eigenvector. Nevertheless, we found in Example 24-2 that the general solution is given by

$$\mathbf{y}(t) = c_1 \begin{pmatrix} e^{-2t} \\ e^{-2t} \end{pmatrix} + c_2 \begin{pmatrix} (1 + t)e^{-2t} \\ te^{-2t} \end{pmatrix}. \tag{22}$$

We shall proceed by first finding $T(t; 0)$. The condition $\mathbf{y}(0) = \mathbf{I}_{(1)}$ gives $c_1 = 0$ and $c_2 = 1$. Substitution of these values into Eq. (22) yields $\mathbf{y}_{(1)}(t; 0)$, the first column of $T(t, 0)$. Similarly, to get $\mathbf{y}_{(2)}(t; 0)$ we require $c_1 = 1$ and $c_2 = -1$ in Eq. (22). Thus,

$$T(t; 0) = \begin{pmatrix} (1 + t)e^{-2t} & -te^{-2t} \\ te^{-2t} & (1 - t)e^{-2t} \end{pmatrix}.$$

Finally we apply Eq. (20) to find

$$T(t; s) = \begin{pmatrix} (1 + t - s)e^{-2(t-s)} & -(t - s)e^{-2(t-s)} \\ (t - s)e^{-2(t-s)} & (1 - t + s)e^{-2(t-s)} \end{pmatrix}, \tag{23}$$

for all t and s in **R**.

Remark. This chapter has presented only the most basic results for linear systems. To appreciate the full power of matrix methods for differential systems, the reader should see the more thorough discussions in advanced books on differential equations (for example, Coddington and Levinson [3]) or various books on matrix theory.

PROBLEMS

1. Let Y be any fundamental matrix for system (3). Prove that for all t and s in J, $T(t; s) = Y(t)Y(s)^{-1}$. [*Hint:* Use Eq. (11) with $\mathbf{h} \equiv \mathbf{0}$, $t_0 = s$, and $\mathbf{x}_0 = \mathbf{I}_{(j)}$.]

2. Evaluate the integrals in (19) and compare with the solution found in Example 24-1. [*Hint:*

$$\int e^{as} \sin bs \ ds = \frac{1}{a^2 + b^2} \left(a e^{as} \sin bs - b e^{as} \cos bs \right) + C. \Bigg]$$

3. Find the transition matrices $T(t, s)$ for the systems in Problem 24-5(a), (b), (d), (e), and (f).

4. Use the transition matrices (found above) for Problem 24-5(b) and for Example 2 of this section to solve the following equations with $x(0) = x_0$.

(a) $\mathbf{x}' = \begin{pmatrix} 1 & 2 \\ -1 & -1 \end{pmatrix} \mathbf{x} + \begin{pmatrix} 1 \\ -2 \end{pmatrix}$, $J = \mathbf{R}$.

(b) $\mathbf{x}' = \begin{pmatrix} 1 & 2 \\ -1 & -1 \end{pmatrix} \mathbf{x} + \begin{pmatrix} 2 \cos t \\ 0 \end{pmatrix}$, $J = \mathbf{R}$.

(c) $\mathbf{x}' = \begin{pmatrix} 1 & 2 \\ -1 & -1 \end{pmatrix} \mathbf{x} + \begin{pmatrix} 0 \\ \tan t \end{pmatrix}$, $J = \left(-\frac{\pi}{2}, \frac{\pi}{2} \right)$.

(d) $\mathbf{x}' = \begin{pmatrix} -1 & 1 \\ 1 & -3 \end{pmatrix} \mathbf{x} + \begin{pmatrix} 0 \\ e^{-2t} \end{pmatrix}$, $J = \mathbf{R}$.

5. Verify that the transition matrix for

$$\mathbf{y}' = \begin{pmatrix} 0 & t \\ -t & 0 \end{pmatrix} \mathbf{y} \quad \text{is} \quad T(t, s) = \begin{pmatrix} \cos \dfrac{t^2 - s^2}{2} & \sin \dfrac{t^2 - s^2}{2} \\ -\sin \dfrac{t^2 - s^2}{2} & \cos \dfrac{t^2 - s^2}{2} \end{pmatrix}.$$

6. Solve on $J = \mathbf{R}$,

$$\mathbf{x}' = \begin{pmatrix} 0 & t \\ -t & 0 \end{pmatrix} \mathbf{x} + \begin{pmatrix} 0 \\ 2t \end{pmatrix}, \qquad \mathbf{x}(0) = \begin{pmatrix} x_{01} \\ x_{02} \end{pmatrix}.$$

7. Give a brief proof of Theorem B by directly substituting (10) or (11) into Eqs. (1) and (2), and then invoking the appropriate uniqueness theorem from Chapter Two.

*Chapter Five

EXISTENCE AND COMPUTATIONS

Up to this point we have postponed (with no loss of rigor) any worries about the existence of solutions of differential equations. In all the examples and problems it has been possible to explicitly exhibit solutions. And in certain theorems on linear equations with nonconstant coefficients, we have simply hypothesized the existence of enough linearly independent solutions of the associated homogeneous equations.

At last, in Section 28, we shall show that linear systems with continuous coefficients and continuous forcing functions always do have solutions. An existence theorem for nonlinear systems will be stated without proof.

The tools (definitions and lemmas) needed for this make it appropriate to also include a section on the numerical computation of solutions.

The entire chapter is "starred" because several proofs depend on Section 12. However, it *is* possible to understand and use the existence theorems of Section 28 without reading the proofs. Also one can read Section 29 on numerical computations and work most of the problems without having covered any previous starred material if one passes over the discussion of error estimates.

27. A MATRIX NORM

In Section 12 the norm of a vector was introduced and found to be a useful tool in the study of systems of equations.

In the case of a linear system

$$\mathbf{x}' = A(t)\mathbf{x} + \mathbf{h}(t) \tag{1}$$

on $J = (\alpha, \beta)$, the matrix $A(t)$ plays a central role; and it will be useful to have a "norm" for matrices. We now define such a norm, and illustrate its usefulness by re-proving uniqueness and continuous dependence of the solutions of (1) with the usual initial conditions,

$$\mathbf{x}(t_0) = \mathbf{x}_0, \tag{2}$$

where t_0 in J and the vector $\mathbf{x}_0$ are given.

DEFINITION

The *norm* of a matrix A with elements a_{ij} will be defined as

$$\|A\| = \max_{j=1,\ldots,n} \sum_{i=1}^{n} |a_{ij}|, \tag{3}$$

that is, the maximum of the sums of the absolute values of the elements in each column of A.

For Example, $\|I\| = 1$. (Problem 2.)

Many other definitions of norm are also available. One is given in Problem 7.

Recall that the norm of a vector ξ with components $\xi_1, \ldots, \xi_n$ was defined (in Section 12) by

$$\|\xi\| = \sum_{j=1}^{n} |\xi_j|. \tag{4}$$

We shall use the same symbol $\|\cdot\|$ for both the matrix norm and the vector norm.

Note that for an $n \times n$ matrix A

$$\max_{i,j} |a_{ij}| \leq \|A\| \leq n \cdot \max_{i,j} |a_{ij}|. \tag{5}$$

(Problem 3.)

Other important properties of the matrix norm and its relation to the vector norm are given in the following theorem. Note that (i), (ii), (iii), and (iv) below correspond to the properties found in Section 12 for the norm of a vector.

THEOREM A (PROPERTIES OF THE MATRIX NORM)

Let A and B be any $n \times n$ matrices and let ξ be any column n-vector. Then

(i) $\|A\| \geq 0$,
(ii) $\|A\| = 0$ if and only if $A = 0$ (the zero matrix),
(iii) $\|cA\| = |c| \cdot \|A\|$ for every scalar c,
(iv) $\|A + B\| \leq \|A\| + \|B\|$ (the triangle inequality),
(v) $\|A\xi\| \leq \|A\| \cdot \|\xi\|$,
(vi) $\|AB\| \leq \|A\| \cdot \|B\|$.

Proof. The proofs of (i), (ii), (iii), and (iv) are straightforward and will be left as Problem 4.

The proof of (v) is as follows:

$$\|A\xi\| = \sum_{i=1}^{n} \left| \sum_{j=1}^{n} a_{ij}\xi_j \right| \leq \sum_{i=1}^{n} \sum_{j=1}^{n} |a_{ij}| \cdot |\xi_j| = \sum_{j=1}^{n} \sum_{i=1}^{n} |a_{ij}| \cdot |\xi_j|$$

$$\leq \sum_{j=1}^{n} \|A\| \cdot |\xi_j| = \|A\| \sum_{j=1}^{n} |\xi_j| = \|A\| \cdot \|\xi\|.$$

Similarly for (vi),

$$\|AB\| = \max_j \sum_{i=1}^{n} \left| \sum_{k=1}^{n} a_{ik} b_{kj} \right|$$

$$\leq \max_j \sum_{i=1}^{n} \sum_{k=1}^{n} |a_{ik}| \cdot |b_{kj}| = \max_j \sum_{k=1}^{n} |b_{kj}| \sum_{i=1}^{n} |a_{ik}|$$

$$\leq \max_j \sum_{k=1}^{n} |b_{kj}| \cdot \|A\| = \|B\| \cdot \|A\|. \quad \blacksquare$$

COROLLARY B

(Compare Problem 12-5.) If A and B are any two $n \times n$ matrices, then

$$\big| \|A\| - \|B\| \big| \leq \|A - B\|. \tag{6}$$

Proof. From Theorem A-(iv) it follows that

$$\|A\| = \|A - B + B\| \leq \|A - B\| + \|B\|$$

or

$$\|A\| - \|B\| \leq \|A - B\|.$$

Similarly one finds $\|B\| - \|A\| \leq \|B - A\|$. These two inequalities combined give (6). $\blacksquare$

COROLLARY C

A matrix-valued function A is continuous at a point t_1 if and only if

$$\lim_{t \to t_1} \|A(t) - A(t_1)\| = 0. \tag{7}$$

Proof. If $A(t)$ is continuous at t_1, then each $a_{ij}(t)$ is continuous at t_1, and (5) applied to the matrix $A(t) - A(t_1)$ gives

$$\|A(t) - A(t_1)\| \leq n \max_{i,j} |a_{ij}(t) - a_{ij}(t_1)| \to 0$$

as $t \to t_1$. This is (7).

Conversely, if (7) holds, then (5) implies that for each i and j

$$|a_{ij}(t) - a_{ij}(t_1)| \leq \|A(t) - A(t_1)\| \to 0,$$

which says that $A(t)$ is continuous at t_1. $\blacksquare$

COROLLARY D

If $A(t)$ is continuous at t_1, then so is $\|A(t)\|$.

Proof. From Corollary B,

$$\big| \|A(t)\| - \|A(t_1)\| \big| \leq \|A(t) - A(t_1)\|.$$

But the latter tends to zero as $t \to t_1$ by (7). $\blacksquare$

To illustrate the use of the norm notation for matrices, we will prove a minor generalization of Corollary 12-D. Assuming the continuity of A, the following theorem asserts that the solution of (1) and (2) is unique and "depends continuously" on the initial conditions.

THEOREM E

Let A be a continuous matrix-valued function and let $\mathbf{h}$ be any vector-valued function on J. Let $t_0 \in J$ and let $\mathbf{x}_0$ and $\tilde{\mathbf{x}}_0$ be any two (constant) n-vectors. Then if $\mathbf{x}$ and $\tilde{\mathbf{x}}$ are solutions of Eq. (1) on J which satisfy $\mathbf{x}(t_0) = \mathbf{x}_0$ and $\tilde{\mathbf{x}}(t_0) = \tilde{\mathbf{x}}_0$, it follows that for all t in J

$$\|\mathbf{x}(t) - \tilde{\mathbf{x}}(t)\| \le \|\mathbf{x}_0 - \tilde{\mathbf{x}}_0\| e^{|\int_{t_0}^t \|A(s)\| \, ds|}. \tag{8}$$

Proof. It is given that

$$\mathbf{x}'(t) = A(t)\mathbf{x}(t) + \mathbf{h}(t) \quad \text{and} \quad \tilde{\mathbf{x}}'(t) = A(t)\tilde{\mathbf{x}}(t) + \mathbf{h}(t).$$

Thus, for all s in J,

$$\frac{d}{ds}[\mathbf{x}(s) - \tilde{\mathbf{x}}(s)] = A(s)[\mathbf{x}(s) - \tilde{\mathbf{x}}(s)].$$

Now integrate from t_0 to t, using the initial condition $\mathbf{x}(t_0) - \tilde{\mathbf{x}}(t_0) = \mathbf{x}_0 - \tilde{\mathbf{x}}_0$, to find

$$\mathbf{x}(t) - \tilde{\mathbf{x}}(t) = \mathbf{x}_0 - \tilde{\mathbf{x}}_0 + \int_{t_0}^t A(s)[\mathbf{x}(s) - \tilde{\mathbf{x}}(s)] \, ds.$$

Applying 12-(7) and Theorem A-(v), one finds

$$\|\mathbf{x}(t) - \tilde{\mathbf{x}}(t)\| \le \|\mathbf{x}_0 - \tilde{\mathbf{x}}_0\| + \left\| \int_{t_0}^t A(s)[\mathbf{x}(s) - \tilde{\mathbf{x}}(s)] \, ds \right\|$$

$$\le \|\mathbf{x}_0 - \tilde{\mathbf{x}}_0\| + \left| \int_{t_0}^t \|A(s)[\mathbf{x}(s) - \tilde{\mathbf{x}}(s)]\| \, ds \right|$$

$$\le \|\mathbf{x}_0 - \tilde{\mathbf{x}}_0\| + \left| \int_{t_0}^t \|A(s)\| \cdot \|\mathbf{x}(s) - \tilde{\mathbf{x}}(s)\| \, ds \right|.$$

Inequality (8) now follows from Lemma 12-A. ∎

To see the relationship between Theorem E and Corollary 12-D, note that if $\mathbf{f}(t, \xi) = A(t)\xi + \mathbf{h}(t)$ and $\|A(t)\| \le K$, then

$$\|f(t,\xi) - f(t, \tilde{\xi})\| = \|A(t)(\xi - \tilde{\xi})\| \le K\|\xi - \tilde{\xi}\|.$$

This Lipschitz condition can be applied in Theorem 12-C to get the estimate for the difference of two solutions of Eq. (1),

$$\|x(t) - \tilde{\mathbf{x}}(t)\| \le \|\mathbf{x}_0 - \tilde{\mathbf{x}}_0\| e^{K|t - t_0|} \tag{9}$$

as in Corollary 12-D. But clearly this is also a consequence of Theorem E
when $\|A(t)\| \leq K$. Hence Theorem E is the stronger result.

Example 1. Consider the differential system

$$x'_1 = -(\sin t)x_2 + 4,$$

$$x'_2 = -x_1 + 2tx_2 - x_3 + e^t,$$

$$x'_3 = 3(\cos t)x_1 + x_2 + \frac{1}{t}x_3 - 5t^2.$$

Assuming vector solutions $\mathbf{x}$ and $\tilde{\mathbf{x}}$ exist on the interval $(1, 3)$ with

$$\mathbf{x}(2) = \text{col}\,(7, 3, -2) \quad \text{and} \quad \tilde{\mathbf{x}}(2) = \text{col}\,(6.7, 3.2, -1.9)$$

let us estimate the "error," $\|\mathbf{x}(t) - \tilde{\mathbf{x}}(t)\|$, for $1 < t < 3$.

In order to apply Theorem E we must first compute the norm of the
matrix

$$A(t) = \begin{pmatrix} 0 & -\sin t & 0 \\ -1 & 2t & -1 \\ 3\cos t & 1 & 1/t \end{pmatrix},$$

namely,

$$\|A(t)\| = \max\left\{1 + 3|\cos t|, \quad |\sin t| + 2t + 1, \quad 1 + \frac{1}{t}\right\}.$$

From this we can easily obtain an upper bound for $\|A(t)\|$ when $1 < t < 3$,
namely,

$$\|A(t)\| \leq \max\,\{4, \quad 2 + 2t, \quad 2\} = 2 + 2t \leq 8.$$

It therefore follows from either (8) or (9) that

$$\|\mathbf{x}(t) - \tilde{\mathbf{x}}(t)\| \leq \|\mathbf{x}(2) - \tilde{\mathbf{x}}(2)\| e^{|\int_2^t \|A(s)\|\,ds|}$$

$$\leq (0.3 + 0.2 + 0.1)e^{|\int_2^t 8\,ds|}$$

$$= 0.6e^{8|t-2|} < 0.6e^8 \approx 1790$$

for all t in $(1, 3)$.

A sharper estimate results if we use $\|A(t)\| \leq 2 + 2t$ in (8):

$$\|\mathbf{x}(t) - \tilde{\mathbf{x}}(t)\| \leq 0.6e^{|\int_2^t (2 + 2s)\,ds|} \leq 0.6e^7 \approx 660.$$

It is not surprising that error bounds obtained this way are generally quite
pessimistic. For our use of absolute values in all calculations prevents any
possible beneficial "cancellation" which might actually occur between
positive and negative quantities. (Compare Problem 12-7.)

In addition to the estimate of the growth rate of errors, the following estimate for the growth of a solution itself is often useful.

THEOREM F

Let A be a continuous matrix-valued function on J with $\|A(t)\| \le K$ for some constant $K > 0$, and let $\mathbf{h}$ be a continuous vector-valued function on J. Then, if $\mathbf{x}$ is a solution of (1) and (2),

$$\|\mathbf{x}(t)\| \le \left[\|\mathbf{x}_0\| + \left| \int_{t_0}^t \|\mathbf{h}(s)\| \, ds \right| \right] e^{K|t - t_0|}$$

for t in J as far as the solution exists.

Proof. The solution of (1) and (2) satisfies

$$\mathbf{x}(t) = \mathbf{x}_0 + \int_{t_0}^t [\mathbf{h}(s) + A(s)\mathbf{x}(s)] \, ds.$$

Thus,

$$\|\mathbf{x}(t)\| \le \|\mathbf{x}_0\| + \left| \int_{t_0}^t \|\mathbf{h}(s)\| \, ds \right| + \left| \int_{t_0}^t K \|\mathbf{x}(s)\| \, ds \right|.$$

The assertion of the theorem now follows from Corollary 12-E. ∎

PROBLEMS

1. Let $A = \begin{pmatrix} 0 & 2 \\ -4 & 1 \end{pmatrix}$ and $B = \begin{pmatrix} -1 & 3 \\ 1 & 2 \end{pmatrix}$. Then verify that, as asserted by Theorem A, $\|A + B\| \le \|A\| + \|B\|$, $\|AB\| \le \|A\| \cdot \|B\|$, and $\|BA\| \le \|B\| \cdot \|A\|$.
2. Show that $\|I\| = 1$.
3. Prove inequalities (5).
4. Prove parts (i), (ii), (iii), and (iv) of Theorem A.
5. If A is an $n \times n$ invertible matrix, prove that

$$\|A\| > 0 \quad \text{and} \quad \|A^{-1}\| \ge \frac{1}{\|A\|}.$$

6. Let A be an $n \times n$ matrix with $\|A\| < 1$. For each integer $k = 1, 2, \ldots$ let $a_{ij}^{(k)}$ be the ijth element of the matrix A^k. Then prove that $\lim_{k \to \infty} a_{ij}^{(k)} = 0$ for each $i, j = 1, \ldots, n$. (We then say $\lim_{k \to \infty} A^k = 0$, the zero matrix.)
7. Another possible way of defining the norm of an $n \times n$ matrix A is

$$\|A\|_s = \sum_{j=1}^n \sum_{i=1}^n |a_{ij}|.$$

Prove that with this definition Theorem A remains true.
8. Compute $\|I\|_s$ if the norm is defined as in Problem 7.

9. Assume the hypotheses of Theorem F. Then:

(a) Show that $\|T(t, s)\| \le e^{K|t-s|}$, where $T(t, s)$ is the transition matrix of Section 26.

(b) Using variation of parameters Eq. 26-(11′) show that if $\|h(t)\| \le M$, then

$$\|x(t)\| \le \|x_0\| e^{K|t-t_0|} + \frac{M}{K} (e^{K|t-t_0|} - 1).$$

(c) Show that the estimate in (b) is sharper than the estimate

$$\|x(t)\| \le [\|x_0\| + M|t-t_0|] e^{K|t-t_0|}$$

given by Theorem F when $\|h(t)\| \le M$.

10. Let A be a constant matrix such that, for some $\delta > 0$, Re $\lambda < -\delta$ for every eigenvalue of A (i.e., A is a stable matrix.)

(a) Show that, for some constant H, $\|T(t, s)\| \le He^{-\delta(t-s)}$ for all $t \ge s$. (See Theorem 25-A.)

(b) Use variation of parameters to show that if $h(t)$ is bounded, say $\|h(t)\| \le M$, then the solution of Eqs. (1) and (2) is bounded.

(c) Observe how easily this would have given *some* of the information obtained by other means in Example 25-5 and Problems 25-10 and 25-11.

28. EXISTENCE OF SOLUTIONS

This section gives a proof of the existence of solutions of general linear differential systems with variable (but continuous) coefficients, plus a statement without proof of an existence theorem for nonlinear systems.

Consider first the linear system

$$x' = A(t)x + h(t), \tag{1}$$

where the given $n \times n$ matrix function A and vector function h are continuous on $J = (\alpha, \beta)$. We shall then prove that this equation together with an initial condition

$$x(t_0) = x_0 \tag{2}$$

has a (unique) solution valid on the entire interval J. As a corollary it will follow that the homogeneous equation

$$y' = A(t)y \tag{3}$$

has n linearly independent solutions on J. We shall also prove the existence of solutions of the general nth-order linear equations discussed in Chapter Three.

The proof of existence for (1) and (2) is based on the equivalent system of "integral equations"

$$x(t) = x_0 + \int_{t_0}^{t} [A(s)x(s) + h(s)] \, ds, \tag{4}$$

for t in J. [Recall Eq. 9-(7).] We shall use a simple and intuitively appealing method of "successive approximations." This method was used by J. Liouville and others in the early 1800s. But the method is often referred to as "Picard iteration," after E. Picard who further developed it in 1893.

Before proceeding with a general proof let us introduce the method of successive approximations by treating a simple and familiar example.

Example 1. Consider the scalar equation

$$x' = cx \quad \text{with} \quad x(0) = 1, \tag{5}$$

where c is a given constant and $J = \mathbf{R}$. (Of course we already know how to solve this elementary problem.) The problem represented by Eqs. (5) is equivalent to

$$x(t) = 1 + \int_0^t cx(s)\, ds. \tag{6}$$

Let us now pick some continuous function $x_{(0)}$ to be considered as a "first approximation" or "first guess" for the solution. For convenience in computation, let

$$x_{(0)}(t) \equiv 1 \quad \text{for all} \quad t.$$

Now, to generate a "next approximation," let us substitute $x_{(0)}$ into the right-hand side of Eq. (6) and define

$$x_{(1)}(t) = 1 + \int_0^t cx_{(0)}(s)\, ds = 1 + ct.$$

Similarly, put $x_{(1)}$ into the right-hand side of (6) and define

$$x_{(2)}(t) = 1 + \int_0^t cx_{(1)}(s)\, ds$$

$$= 1 + c \int_0^t (1 + cs)\, ds = 1 + ct + \frac{(ct)^2}{2}.$$

Continuing in this manner, we find

$$x_{(3)}(t) = 1 + \int_0^t cx_{(2)}(s)\, ds = 1 + ct + \frac{(ct)^2}{2!} + \frac{(ct)^3}{3!},$$

and in general the "successive approximations" are

$$x_{(l)}(t) = 1 + ct + \frac{(ct)^2}{2!} + \cdots + \frac{(ct)^l}{l!},$$

for $l = 1, 2, \ldots$. Thus, for each t,

$$\lim_{l \to \infty} x_{(l)}(t) = e^{ct}.$$

And as we well know, $x(t) = e^{ct}$ *is the unique solution of* (5).

This nice result is much more than a coincidence. We shall now show that this same method provides a solution of Eqs. (1) and (2). In other words, successive approximations can be defined as in the example; they form a convergent sequence; and the limit of that sequence is a solution of (1) and (2).

THEOREM A (LINEAR SYSTEM)

Let A and $\mathbf{h}$ be continuous on $J = (\alpha, \beta)$ and let $(t_0, \mathbf{x}_0) \in J \times \mathbf{R}^n$. Then Eqs. (1) and (2) have a (unique) solution on the entire interval J.

Proof. As in Example 1, we are going to try to use the right-hand side of Eq. (4) to generate a sequence of "successive approximations" for a solution of Eqs. (1) and (2). Thus we will begin with an arbitrary first guess for the continuous function $\mathbf{x}_{(0)}: J \to \mathbf{R}^n$. Then we hope to define successively better approximations $\mathbf{x}_{(1)}, \mathbf{x}_{(2)}, \ldots,$ by

$$\mathbf{x}_{(1)}(t) = \mathbf{x}_0 + \int_{t_0}^{t} [A(s)\mathbf{x}_{(0)}(s) + \mathbf{h}(s)] \, ds, \tag{7}$$

$$\mathbf{x}_{(2)}(t) = \mathbf{x}_0 + \int_{t_0}^{t} [A(s)\mathbf{x}_{(1)}(s) + \mathbf{h}(s)] \, ds, \tag{8}$$

$$\vdots$$

$$\mathbf{x}_{(l+1)}(t) = \mathbf{x}_0 + \int_{t_0}^{t} [A(s)\mathbf{x}_{(l)}(s) + \mathbf{h}(s)] \, ds. \tag{9}$$

But can we do this? Since the functions A and $\mathbf{h}$ are unspecified, we cannot actually evaluate the integrals in Eqs. (7), (8), and (9) as we did in Example 1. So we must ask: Will the integrands in the right-hand sides of these equations always be well defined and continuous? And if so will the resulting sequence of functions $\{\mathbf{x}_{(l)}\}$ converge? And if it does converge, say to $\mathbf{x}$, will $\mathbf{x}$ be a solution of Eqs. (1) and (2)?

Since A, $\mathbf{x}_{(0)}$, and $\mathbf{h}$ are continuous on J it follows that the integrand in Eq. (7) is continuous. Thus $\mathbf{x}_{(1)}$ exists and is continuous on J. This in turn shows that $\mathbf{x}_{(2)}$, defined by Eq. (8), exists and is continuous on J. The existence of $\mathbf{x}_{(l)}$ for all $l = 1, 2, \ldots$ follows by induction.

Our next objective is to show that the sequence $\{\mathbf{x}_{(l)}\}$ of "successive approximations" does indeed converge.

Let α_0 and β_0 be any two numbers such that

$$\alpha < \alpha_0 < t_0 < \beta_0 < \beta.$$

Then the continuity of A on the closed bounded interval $[\alpha_0, \beta_0]$ assures the continuity of $\|A(t)\|$ (by Corollary 27-D). Hence there exists a constant $K > 0$ such that

$$\|A(t)\| \leq K \quad \text{for} \quad \alpha_0 \leq t \leq \beta_0.$$

Now consider the magnitude of the difference between two consecutive functions of the sequence $\{x_{(l)}\}$, namely,

$$\|x_{(l+1)}(t) - x_{(l)}(t)\| \quad \text{for} \quad \alpha_0 \le t \le \beta_0.$$

We shall obtain a recursive relation for these differences. For $\alpha_0 \le t \le \beta_0$ and $l = 0, 1, 2, \ldots$ one easily finds

$$\|x_{(l+2)}(t) - x_{(l+1)}(t)\| = \left\| \int_{t_0}^{t} A(s)[x_{(l+1)}(s) - x_{(l)}(s)]\, ds \right\|.$$

Thus

$$\|x_{(l+2)}(t) - x_{(l+1)}(t)\| \le K \left| \int_{t_0}^{t} \|x_{(l+1)}(s) - x_{(l)}(s)\|\, ds \right|. \tag{10}$$

By the continuity of $x_{(1)}$ and $x_{(0)}$ it follows that $\|x_{(1)}(t) - x_{(0)}(t)\|$ is continuous. Why? (Compare with Corollary 27-D.) Thus, since $[\alpha_0, \beta_0]$ is closed and bounded, there exists a constant M such that

$$\|x_{(1)}(t) - x_{(0)}(t)\| \le M \quad \text{for} \quad \alpha_0 \le t \le \beta_0.$$

Putting this into (10) with $l = 0$ gives, for $\alpha_0 \le t \le \beta_0$,

$$\|x_{(2)}(t) - x_{(1)}(t)\| \le K \left| \int_{t_0}^{t} M\, ds \right| = MK|t - t_0|.$$

Now put this last estimate into (10) with $l = 1$ to find

$$\|x_{(3)}(t) - x_{(2)}(t)\| \le K \left| \int_{t_0}^{t} MK|s - t_0|\, ds \right| = M \frac{K^2|t - t_0|^2}{2}.$$

You should verify the evaluation of this integral by considering two cases, $t \ge t_0$ and $t \le t_0$. Then carry out the next similar computation to show that

$$\|x_{(4)}(t) - x_{(3)}(t)\| \le M \frac{K^3|t - t_0|^3}{3!}.$$

(Problem 2.)

At about this point one might suspect that a pattern is emerging. It appears likely that

$$\|x_{(l+1)}(t) - x_{(l)}(t)\| \le M \frac{K^l|t - t_0|^l}{l!} \quad \text{for} \quad \alpha_0 \le t \le \beta_0, \tag{11}$$

for $l = 0, 1, 2, \ldots$. This conjecture is verified by mathematical induction as follows. Assuming (11) holds for some $l = 0$, or 1, or 2, $\ldots$ we compute from (10)

$$\|x_{(l+2)}(t) - x_{(l+1)}(t)\| \le K \left| \int_{t_0}^{t} M \frac{K^l|s - t_0|^l}{l!}\, ds \right| = M \frac{K^{l+1}|t - t_0|^{l+1}}{(l+1)!}.$$

(Again one evaluates the integral for the two cases $t \geq t_0$ and $t \leq t_0$.) Since this estimate is just (11) with l replaced by $l + 1$ the verification of (11) is complete.

Now, for each $l = 1, 2, \ldots$ and each t in $[\alpha_0, \beta_0]$, we can write

$$\mathbf{x}_{(l)}(t) = \mathbf{x}_{(0)}(t) + \sum_{p=0}^{l-1} [\mathbf{x}_{(p+1)}(t) - \mathbf{x}_{(p)}(t)]. \tag{12}$$

Why?

Inequality (11) gives

$$\|\mathbf{x}_{(p+1)}(t) - \mathbf{x}_{(p)}(t)\| \leq M \frac{(Ka)^p}{p!} \quad \text{for} \quad \alpha_0 \leq t \leq \beta_0,$$

where $a = \max \{\beta_0 - t_0, t_0 - \alpha_0\}$. Hence, for each $k = 1, \ldots, n$,

$$|x_{(p+1)k}(t) - x_{(p)k}(t)| \leq M \frac{(Ka)^p}{p!} \quad \text{for} \quad \alpha_0 \leq t \leq \beta_0.$$

Applying this and the comparison test, Theorem A2-H(ii), we conclude that each scalar series

$$\sum_{p=0}^{\infty} [x_{(p+1)k}(t) - x_{(p)k}(t)] \qquad (k = 1, \ldots, n) \tag{13}$$

converges for all t in $[\alpha_0, \beta_0]$ because of the convergence of $\sum_{p=0}^{\infty} M(Ka)^p/p!$, the well-known series for Me^{Ka}. This means that the sequence of "partial sums" of (13) involved in (12) must converge on $[\alpha_0, \beta_0]$. But, by the arbitrariness of α_0 and β_0, this sequence must converge for all t in J. Let

$$\mathbf{x}(t) = \lim_{l \to \infty} \mathbf{x}_{(l)}(t) \quad \text{for} \quad t \quad \text{in} \quad J.$$

We note further that for all t in $[\alpha_0, \beta_0]$ and for any integers $l \geq 0$ and $m > l$

$$\|\mathbf{x}_{(m)}(t) - \mathbf{x}_{(l)}(t)\| = \left\| \sum_{p=l}^{m-1} [\mathbf{x}_{(p+1)}(t) - \mathbf{x}_{(p)}(t)] \right\| \leq M \sum_{p=l}^{\infty} \frac{(Ka)^p}{p!}.$$

Hence, letting $m \to \infty$, we find (Problem 3)

$$\|\mathbf{x}(t) - \mathbf{x}_{(l)}(t)\| \leq M \sum_{p=l}^{\infty} \frac{(Ka)^p}{p!} \quad \text{for all} \quad t \quad \text{in} \quad [\alpha_0, \beta_0]. \tag{14}$$

This proves that $x_{(l)k}$ converges to x_k *uniformly* on $[\alpha_0, \beta_0]$ as $l \to \infty$ for each $k = 1, \ldots, n$. (See Appendix 2 for a discussion of uniform convergence.) Thus, since each $x_{(l)k}$ is continuous it follows that the limit x_k is continuous on $[\alpha_0, \beta_0]$ by Theorem A2-I. Hence the vector-valued function $\mathbf{x}$ is also continuous on $[\alpha_0, \beta_0]$.

Now we should like to take limits as $l \to \infty$ in Eq. (9) and conclude that

x satisfies Eq. (4). To justify this we note that, for $\alpha_0 \leq t \leq \beta_0$,

$$\left\| \int_{t_0}^{t} A(s)\mathbf{x}(s)\, ds - \int_{t_0}^{t} A(s)\mathbf{x}_{(l)}(s)\, ds \right\| \leq K \left| \int_{t_0}^{t} \|\mathbf{x}(s) - \mathbf{x}_{(l)}(s)\|\, ds \right|. \quad (15)$$

But, by (14), the scalar sequence $\{\|\mathbf{x}(s) - \mathbf{x}_{(l)}(s)\|\}$ converges uniformly to zero for s in $[\alpha_0, \beta_0]$. So, by part (ii) of Theorem A2-1, the integral on the right-hand side of (15) tends to zero as $l \to \infty$. It follows that **x** satisfies Eq. (4) on $[\alpha_0, \beta_0]$. This means **x** is a solution of (1) and (2) on (α_0, β_0). Hence, invoking the arbitrariness of α_0 and β_0 once more, **x** is a solution on J.

And, of course, we have known since Chapter Two that there can be at most one solution. ∎

The proof of Theorem A shows that, *starting with any continuous function* $\mathbf{x}_{(0)}: J \to \mathbf{R}^n$, the successive approximations defined by (9) will always converge to the one and only solution of (1) and (2) on J.

Note further that inequality (14) gives a specific error estimate for the successive approximations. In other words, even though we may not know the solution **x**, we can say that $\mathbf{x}_{(l)}$ lies within a certain small distance of **x** for all sufficiently large l.

In Section 26, when we solved Eqs. (1) and (2) by variation of parameters, we assumed that the associated homogeneous equation (3) had n linearly independent solutions. The following important corollary of Theorem A asserts that such an assumption was superfluous.

COROLLARY B

Let A be a continuous $n \times n$-matrix-valued function on J. Then system (3) has n linearly independent solutions on J.

Proof. Choose any t_0 in J and any n linearly independent vectors $\xi_{(1)}, \ldots, \xi_{(n)}$. For example, the latter could be $\mathbf{I}_{(1)}, \ldots, \mathbf{I}_{(n)}$. Then for $j = 1, \ldots, n$, let $\mathbf{y}_{(j)}$ be the unique solution of (3) on J which satisfies $\mathbf{y}_{(j)}(t_0) = \xi_{(j)}$. The solutions $\mathbf{y}_{(j)}$ exist by Theorem A, and they are linearly independent by Theorem 23-B. ∎

Chapter Three treated the linear scalar differential equation of order n

$$x^{n} + a_{n-1}(t)x^{(n-1)} + \cdots + a_1(t)x' + a_0(t)x = h(t). \quad (16)$$

But most of the results were restricted to the case of constant coefficients. Now, at last, we are ready to prove a general existence theorem for Eq. (16) in case of variable continuous coefficients and with arbitrary initial conditions of the form

$$x(t_0) = b_0, \quad x'(t_0) = b_1, \quad \ldots, \quad x^{(n-1)}(t_0) = b_{n-1}. \quad (17)$$

THEOREM C

Let $a_{n-1}, \ldots, a_0$, and h be continuous (real-valued) functions on J and let $(t_0, b_0, b_1, \ldots, b_{n-1})$ be any point in $J \times \mathbf{R}^n$. Then Eqs. (16) and (17) have a (unique) solution on J.

Moreover, the associated homogeneous equation [Eq. (16) with $h(t) \equiv 0$] has n linearly independent solutions $y_1, \ldots, y_n$ on J.

Remark. This theorem shows that it was unnecessary to hypothesize, in Theorem 16-D, the existence of a particular solution of Eq. (16) and n linearly independent solutions of the associated homogeneous equation.

Proof of Theorem C. Defining $z_1 = x$, $z_2 = x'$, $\ldots$, $z_n = x^{(n-1)}$, we can convert Eq. (16) into the equivalent system

$$z_1' = z_2,$$
$$z_2' = z_3, \tag{18}$$
$$\vdots$$
$$z_n' = -a_0(t)z_1 - a_1(t)z_2 - \cdots - a_{n-1}(t)z_n + h(t),$$

with initial condition

$$\mathbf{z}(t_0) = \mathrm{col}\,(b_0, b_1, \ldots, b_{n-1}). \tag{19}$$

By Theorem A, Eqs. (18) and (19) have a unique solution $\mathbf{z}$ on J. Thus, taking $x = z_1$ we have a (unique) solution of (16) and (17) on J.

To prove the existence of n linearly independent solutions of the homogeneous equation (16) with $h(t) \equiv 0$ it suffices to apply the above result using, in turn, $\mathrm{col}\,(b_0, b_1, \ldots, b_{n-1}) = \mathbf{I}_{(1)}, \mathbf{I}_{(2)}, \ldots$, and $\mathbf{I}_{(n)}$. ∎

When it comes to the more general system

$$\mathbf{x}' = \mathbf{f}(t, \mathbf{x}), \tag{20}$$

not assumed linear, we must not expect such a simple and complete result as was given by Theorem A.

For example, the scalar-valued function $f(\xi) = \xi^2$ is as nice a function (defined everywhere) as most people would normally ask for. Yet the solution of the scalar differential equation

$$x' = x^2 \quad \text{with} \quad x(0) = 1 \tag{21}$$

does *not* exist on all of $\mathbf{R}$. It exists only on $(-\infty, 1)$. (Problem 4.)

In general, let $\mathbf{f}$ be continuous on $J \times D$, where $J = (\alpha, \beta)$ and D is some open set in $\mathbf{R}^n$—perhaps an open rectangle. (Note that the set D used in Sections 9 and 12 actually corresponds to $J \times D$ here.) Recall that $\mathbf{f}$ is said to be Lipschitzian on a set $A \subset J \times D$ if, for some constant K,

$$\|\mathbf{f}(t, \xi) - \mathbf{f}(t, \tilde{\xi})\| \le K\|\xi - \tilde{\xi}\| \tag{22}$$

whenever (t, ξ) and $(t, \tilde{\xi}) \in A$.

If it should happen that $D = \mathbf{R}^n$ and that $\mathbf{f}$ satisfies a Lipschitz condition (22) on each set of the form $[\alpha_0, \beta_0] \times \mathbf{R}^n$ where $[\alpha_0, \beta_0] \subset J$, then one does get the same conclusion for system (20) as Theorem A gave for the linear case (Problem 14). But, in general, one cannot expect such a strong Lipschitz condition for a nonlinear system. The following theorem, stated without proof, uses more natural hypotheses, and gives a weaker conclusion.

THEOREM D (NONLINEAR SYSTEM)

Let $\mathbf{f}: J \times D \to \mathbf{R}^n$ be continuous and Lipschitzian on each closed bounded $(n + 1)$-dimensional rectangle, say $[a_0, b_0] \times [a_1, b_1] \times \cdots \times [a_n, b_n]$ in $J \times D$. Let $(t_0, \mathbf{x}_0)$ in $J \times D$ be given. Then Eqs. (20) and (2) have a unique solution $\mathbf{x}$ on $[t_0, \beta_1)$ where $\beta_1 \leq \beta$; and if $\beta_1 < \beta$, then $\mathbf{x}(t)$ becomes arbitrarily close to the boundary of D (or else $\|\mathbf{x}(t)\|$ becomes arbitrarily large) as $t \to \beta_1$. (An analogous statement holds for solutions on $(\alpha_1, t_0]$.)

The uniqueness follows from Theorem 12-B. The existence of a solution as described above is proved in intermediate and advanced books on differential equations (see, for example, Theorem 24-C of [4]). The first part of such a proof is indicated in Problem 12.

Recall, from Section 12, that $\mathbf{f}$ is Lipschitzian on each closed bounded rectangle if $D_{1+j} f_i = \partial f_i / \partial \zeta_j$ is continuous on $J \times D$ for each $i, j = 1, \ldots, n$.

The following two examples should clarify the meaning and use of Theorem D.

Example 2. Let us try to apply Theorem D to Eqs. (21) (assuming we did not already know how to solve them exactly).

In this example we can take $J = \mathbf{R}$, and we *could* take $D = \mathbf{R}$ since f and $\partial f / \partial \xi$ are continuous everywhere. (See Problem 5.) However, there is some advantage in using a bounded set for D.

We must have $x_0 \in D$. Let us try $D = (0, 10)$. Then Theorem D asserts the existence of a unique solution on $[0, \beta_1)$, where either $\beta_1 = \infty$ or $x(t)$ gets arbitrarily close to 0 or 10 as $t \to \beta_1$. Now, as long as $x(t)$ remains in D, (21) shows that $0 < x'(t) < 100$ and so $1 \leq x(t) < 1 + 100t$ for $0 \leq t < \beta_1$. Hence $x(t)$ cannot approach the boundary of D as long as t remains less than $9/100$. Thus β_1 must be at least as big as $9/100$.

Had we chosen $D = (0, 2)$ we would have obtained the stronger conclusion that $\beta_1 \geq \frac{1}{4}$. (Problem 6.) Of course we know that, in reality, $\beta_1 = 1$.

Example 3. The equation of motion for a simple pendulum, as in Figure 1, acting under the sole external influence of gravity can be taken to be

$$ml\theta'' = -mg \sin \theta - bl\theta'. \tag{23}$$

Here m is the mass of the bob, l the length of the massless rod, g the accelera-

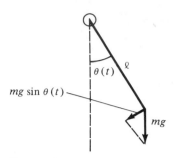

FIGURE 1

tion of gravity, and b a coefficient of friction—all positive constants. Let initial data

$$\theta(t_0) = \theta_0, \qquad \theta'(t_0) = \omega_0 \tag{24}$$

be specified. In order to apply Theorem D, we define $x_1(t) = \theta(t)$ and $x_2(t) = \theta'(t)$ to transform Eq. (23) into the ordinary differential system

$$x_1' = x_2,$$

$$x_2' = -\frac{g}{l} \sin x_1 - \frac{b}{m} x_2, \tag{25}$$

with

$$x_1(t_0) = \theta_0, \qquad x_2(t_0) = \omega_0. \tag{26}$$

We can now take $J = \mathbf{R}$ and $D = \mathbf{R}^2$, and verify that

$$\mathbf{f}(t, \xi) = \mathbf{f}(\xi) = \begin{pmatrix} \xi_2 \\ -\dfrac{g}{l} \sin \xi_1 - \dfrac{b}{m} \xi_2 \end{pmatrix} \tag{27}$$

satisfies a Lipschitz condition everywhere. (Problem 7.) Then the existence of a unique solution on $[t_0, \beta_1)$ for some $\beta_1 > t_0$ follows from Theorem D. We shall show that $\beta_1 = \infty$.

Consider the function $v: [t_0, \beta_1) \to \mathbf{R}$ defined by

$$v(t) = \frac{1}{2} x_2^2(t) + \frac{g}{l} [1 - \cos x_1(t)]. \tag{28}$$

(This essentially represents the energy—kinetic plus potential—of the bob at time t.) Since x_1 and x_2 satisfy (25),

$$v' = x_2 x_2' + \frac{g}{l} (\sin x_1) x_1' = -\frac{b}{m} x_2^2 \le 0.$$

Thus v is nonincreasing. It follows that

$$|x_2(t)| \le \sqrt{2v(t)} \le \sqrt{2v(t_0)} \equiv M.$$

Using this and the first of Eqs. (25) we find

$$|x_1(t)| \leq |\theta_0| + M(t - t_0) \quad \text{for} \quad t_0 \leq t < \beta_1.$$

Now suppose (for contradiction) that $\beta_1 < \infty$. Then on $[t_0, \beta_1)$,

$$|x_1(t)| \leq |\theta_0| + M(\beta_1 - t_0) \quad \text{and} \quad |x_2(t)| \leq M,$$

so that $\mathbf{x}(t) = \text{col}\,(x_1(t), x_2(t))$ remains bounded. By Theorem D, this contradicts the assumption that $\beta_1 < \infty$. Hence $\beta_1 = \infty$.

The natural question regarding this example is, How was the appropriate function v obtained? In the present example, and in many others, the function v represents the "energy" in the system.
    More generally, for the second-order equation of the form

$$x'' + f_1(x)x' + f_2(x) = 0, \tag{29}$$

one often uses, as in Example 1, the function

$$v(t) = \frac{1}{2}(x'(t))^2 + \int_0^{x(t)} f_2(\xi)\, d\xi. \tag{30}$$

Here $(x'(t))^2/2$ corresponds to the "kinetic energy" and $\int_0^{x(t)} f_2(\xi)\, d\xi$, computed from the "restoring force" f_2, represents "potential energy."
    The method illustrated in Example 3 is often useful. But it should be noted that for the special case of system (25) the following corollary provides a quicker proof that solutions exist on all of $J = \mathbf{R}$.

COROLLARY E (GLOBAL EXISTENCE)

Let $D = \mathbf{R}^n$. Assume the hypotheses of Theorem D and assume further that

$$\|\mathbf{f}(t, \xi)\| \leq M(t) + N(t)\|\xi\| \quad \text{on} \quad J \times \mathbf{R}^n, \tag{31}$$

where M and N are continuous nonnegative functions on J. Then Eqs. (20) and (2) have a unique solution on the entire interval J.

Proof. Let $\mathbf{x}$ on (α_1, β_1) be the unique solution of (20) and (2) described in Theorem D. Suppose (for contradiction) that $\beta_1 < \beta$. Then M and N are bounded on $[t_0, \beta_1]$. Say $M(s) \leq M_1$ and $N(s) \leq N_1$ for $t_0 \leq s \leq \beta_1$. So, for $t_0 \leq t < \beta_1$,

$$\|\mathbf{x}(t)\| = \left\| \mathbf{x}_0 + \int_{t_0}^t \mathbf{f}(s, \mathbf{x}(s))\, ds \right\|$$

$$\leq \|\mathbf{x}_0\| + \int_{t_0}^t (M_1 + N_1\|\mathbf{x}(s)\|)\, ds$$

$$\leq \|\mathbf{x}_0\| + M_1(\beta_1 - t_0) + \int_{t_0}^t N_1\|\mathbf{x}(s)\|\, ds$$

Now one simply applies Lemma 12-A to conclude that

$$\|x(t)\| \leq [\|x_0\| + M_1(\beta_1 - t_0)]e^{N_1(\beta_1 - t_0)}.$$

This shows that $\|x(t)\|$ is bounded on $[t_0, \beta_1)$, contradicting the supposition that $\beta_1 < \beta$. Hence $\beta_1 = \beta$.

Similarly, one shows that $\alpha_1 = \alpha$. ∎

PROBLEMS

1. Obtain the first few successive approximations for the following problems and compare with the known solutions obtained by other methods.

   (a) $x' = cx$ with $x(0) = 1$ using $x_{(0)}(t) = 1 - t$.
   (b) $x' = x + t$ with $x(1) = 2$ using $x_{(0)}(t) = 2$.
   (c) $x' = \begin{pmatrix} -1 & -2 \\ 1 & -4 \end{pmatrix} x$ with $x(0) = \begin{pmatrix} 1 \\ 0 \end{pmatrix}$ using $x_{(0)}(t) = x(0)$.

2. Verify the estimates given in the proof of Theorem A for $\|x_{(3)}(t) - x_{(2)}(t)\|$ and $\|x_{(4)}(t) - x_{(3)}(t)\|$.

3. Complete the deduction of inequality (14) with the aid of the triangle inequality

$$\|x(t) - x_{(l)}(t)\| \leq \|x(t) - x_{(m)}(t)\| + \|x_{(m)}(t) - x_{(l)}(t)\|.$$

4. Find the exact solution of Eqs. (21) and verify that this solution exists only on $(-\infty, 1)$.

5. Taking $f(\xi) = \xi^2$ as in Eqs. (21), show that no Lipschitz condition holds if one takes $D = \mathbf{R}$, but a Lipschitz condition does hold if $D \subset (-b \ b)$ for any $b > 0$.

6. Show that if one takes $D = (0, 2)$ one finds $\beta_1 \geq \frac{1}{4}$ for Eqs. (21).

7. Prove that the function $f: \mathbf{R} \times \mathbf{R}^2 \to \mathbf{R}$ defined by (27) satisfies Lipschitz condition (22) with $K = \max \{g/l, 1 + b/m\}$ on $\mathbf{R} \times \mathbf{R}^2$. Show that f also satisfies (31) with $M(t) = 0$ and $N(t) = K$.

8. Show that the scalar equation

$$x' = \sin x - \sin t \quad \text{with} \quad x(1) = 1$$

   has a unique solution on the entire interval $J = \mathbf{R}$.

9. Show that, given any initial data $x(t_0) = x_0 \in \mathbf{R}^3$, the system

$$x_1' = tx_1 - \cos x_2 + 17 \sin t,$$

$$x_2' = 3x_1 + 2e^t x_2 + t^2 x_3,$$

$$x_3' = (\cos t)(\sin x_1)(\sin x_2)x_3,$$

   has a unique solution on the entire interval $\mathbf{R}$.

10. Prove that every solution of

$$x'' - x' + t^2 x = t^3$$

   exists on all of $\mathbf{R}$.

11. Show that Theorem A follows from Corollary E.

12. Let $\mathbf{f}: J \times D \to \mathbf{R}^n$ satisfy the hypotheses of Theorem D. Choose $a > 0$ and $b > 0$ such that the closed bounded $(n + 1)$-dimensional rectangle

$$A \equiv [t_0, t_0 + a] \times [x_{01} - b, x_{01} + b] \times \cdots \times [x_{0n} - b, x_{0n} + b]$$

is contained in $J \times D$, and let $|f_i(t, \xi)| \leq M$ on A for each $i = 1, \ldots, n$. Then show (by adapting the proof of Theorem A) that Eqs. (20) and (2) have a unique solution on $[t_0, \beta_1)$ where $\beta_1 \geq t_0 + \min \{a, b/M\}$.

*13. The equation

$$x'' + \mu(x^2 - 1)x' + x = 0 \qquad (\mu > 0)$$

was studied by van der Pol as a model for the operation of an electronic vacuum tube oscillator. Prove that every solution, with initial data at t_0, exists on $[t_0, \infty)$. Why does Corollary E not apply here?

14. Let $D = \mathbf{R}^n$, let $\mathbf{f}$ be continuous on $J \times \mathbf{R}^n$, and assume that for each interval $[\alpha_0, \beta_0] \subset J$ there exists a constant K such that $\mathbf{f}$ satisfies Lipschitz condition (22) on the entire set $[\alpha_0, \beta_0] \times \mathbf{R}^n$. Show, using a successive-approximations argument like that in the proof of Theorem A, that given any t_0 in J and $\mathbf{x}_0$ in $\mathbf{R}^n$, Eqs. (20) and (2) have a unique solution on the entire interval J. [Note that this alternative to Corollary E would handle Example 3 and Problems 9 and 10. But, like Corollary E, it would be quite useless in Problem 13.]

29. THE EULER METHOD FOR NUMERICAL COMPUTATION

Section 28 provided theorems asserting the existence of a unique solution for a fairly general differential system

$$\mathbf{x}' = \mathbf{f}(t, \mathbf{x}) \tag{1}$$

with initial condition

$$\mathbf{x}(t_0) = \mathbf{x}_0. \tag{2}$$

The present section describes the most elementary methods (associated with the names of Cauchy, Euler, Lipschitz, and Peano) for the numerical computation of solutions.

We shall treat the problem of approximating the solution of Eqs. (1) and (2) for $t \geq t_0$. Similar procedures work for $t \leq t_0$.

We begin by constructing a *polygonal curve* (a "curve" made up of straight line segments) by which we hope to approximate a solution.

If the unknown solution is known to exist on $[t_0, \beta_1)$ we will consider an interval $[t_0, \gamma] \subset [t_0, \beta_1)$. Partition the interval $[t_0, \gamma]$ into p equal subintervals by introducing the points

$$t_i \equiv t_0 + i \frac{\gamma - t_0}{p} \quad \text{for} \quad i = 0, 1, \ldots, p. \tag{3}$$

Then define a function $z : [t_0, \gamma] \rightarrow \mathbf{R}^n$ (if possible) by

$$
\mathbf{z}(t) = \begin{cases}
\mathbf{x}_0 & \text{for } t = t_0, \\
\mathbf{x}_0 + \mathbf{f}(t_0, \mathbf{x}_0)(t - t_0) & \text{for } t_0 < t \le t_1, \\
\mathbf{z}(t_1) + \mathbf{f}(t_1, \mathbf{z}(t_1))(t - t_1) & \text{for } t_1 < t \le t_2, \\
\mathbf{z}(t_2) + \mathbf{f}(t_2, \mathbf{z}(t_2))(t - t_2) & \text{for } t_2 < t \le t_3, \\
\quad\vdots & \\
\mathbf{z}(t_{p-1}) + \mathbf{f}(t_{p-1}, \mathbf{z}(t_{p-1}))(t - t_{p-1}) & \text{for } t_{p-1} < t \le \gamma.
\end{cases}
\tag{4}
$$

On the subinterval $(t_i, t_{i+1}]$ for $i = 0, 1, \ldots, p - 1$ we extend each component of the vector-valued function z by means of a straight line segment with slope computed from the differential equation at $(t_i, \mathbf{z}(t_i))$. This requires that $\mathbf{f}$ be defined at $(t_i, \mathbf{z}(t_i))$—a matter to be verified.

Assuming it exists, the function z defined by (4) is called a polygonal curve or a Cauchy-Euler polygon. It is reasonable to expect that if $\mathbf{f}$ is sufficiently "nice" and if the partition (3) is fine enough, that is, if $(\gamma - t_0)/p$ is sufficiently small, then z ought to approximate the solution x, as suggested by Figure 1 (for the case $n = 1$).

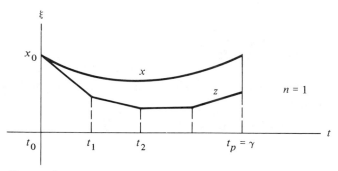

FIGURE 1

Example 1. Let us introduce the method by applying it to the simple and familiar scalar equation,

$$x' = cx \quad \text{for} \quad t \ge 0 \quad \text{with} \quad x(0) = 1.$$

Since the solution exists on $[0, \infty)$ we can take any interval $[0, \gamma]$. Let us partition $[0, \gamma]$ into p equal subintervals of length $\delta = \gamma/p$. Thus $t_i = i\delta$ for $i = 0, 1, \ldots, p$. From (4) we then find

$$
z(t) = \begin{cases}
1 + ct & \text{for } 0 \le t \le \delta \\
(1 + c\delta)[1 + c(t - \delta)] & \text{for } \delta < t \le 2\delta \\
(1 + c\delta)^2[1 + c(t - 2\delta)] & \text{for } 2\delta < t \le 3\delta \\
\quad\vdots & \\
(1 + c\delta)^{p-1}[1 + c(t - p\delta + \delta)] & \text{for } (p - 1)\delta < t \le p\delta = \gamma.
\end{cases}
$$

In particular we have

$$z(\gamma) = (1 + c\delta)^p = \left(1 + \frac{c\gamma}{p}\right)^p;$$

and perhaps the reader will recognize that

$$\lim_{p \to \infty} z(\gamma) = \lim_{p \to \infty} \left(1 + \frac{c\gamma}{p}\right)^p = e^{c\gamma}.$$

Thus, in this case, by taking a sufficiently fine uniform partition of the interval $[0, \gamma]$ one obtains an approximate solution which is arbitrarily close to the exact solution at γ.

But how shall we justify the polygonal-curve-approximation method when we do not have a known exact solution to compare with? Such, of course, are the interesting cases.

Our theorem answering this question is going to assume that **f** is *bounded* and *Lipschitzian* with respect to *all* its arguments on $J \times D$ where $J = (\alpha, \beta)$ and D is an open set in $\mathbf{R}^n$. These sound like severe hypotheses, but Examples 2 and 3 will show how the conditions can sometimes be met by appropriate choices of J and D. Under these hypotheses we shall prove that the unique solution of Eqs. (1) and (2) can be approximated to any desired accuracy by the polygonal curve **z** defined in (4). One need only use a sufficiently fine partition.

THEOREM A

Let $\mathbf{f}: J \times D \to \mathbf{R}^n$ satisfy

$$\|\mathbf{f}(t, \xi)\| \le M \quad \text{and} \quad \|\mathbf{f}(t, \xi) - \mathbf{f}(\tilde{t}, \tilde{\xi})\| \le K\|\xi - \tilde{\xi}\| + L|t - \tilde{t}| \qquad (5)$$

for all (t, ξ), $(\tilde{t}, \tilde{\xi}) \in J \times D$, where M, K, and L are positive constants. Let $(t_0, \mathbf{x}_0) \in J \times D$ and let **x** be the unique solution of Eqs. (1) and (2) on $[t_0, \beta_1)$. Choose any γ in (t_0, β_1) and partition the interval $[t_0, \gamma]$ into p equal subintervals each of length $\delta = (\gamma - t_0)/p$.

Then, if **z** is defined by (4),

$$\|\mathbf{z}(t) - \mathbf{x}(t)\| \le \frac{1}{2}\left(M + \frac{L}{K}\right)\delta[e^{K(t - t_0)} - 1] \qquad (6)$$

for $t_0 \le t \le \gamma$ as long as $\mathbf{z}(t)$ remains in D.

Proof. Assuming $\mathbf{z}(t)$ remains in D, we shall prove, by induction on i, that (6) holds for $t_0 \le t \le t_i$.

For $i = 0$ the assertion is trivial.

Now let i be an integer in $[0, p)$ such that (6) holds for $t_0 \le t \le t_i$. Then for $t_i < t \le t_{i+1}$,

$$\mathbf{z}(t) = \mathbf{z}(t_i) + \mathbf{f}(t_i, \mathbf{z}(t_i))(t - t_i)$$

and

$$\|\mathbf{z}(t) - \mathbf{x}(t)\| = \left\| \mathbf{z}(t_i) + \mathbf{f}(t_i, \mathbf{z}(t_i))(t - t_i) - \mathbf{x}(t_i) - \int_{t_i}^{t} \mathbf{f}(s, \mathbf{x}(s)) \, ds \right\|$$

$$\leq \|\mathbf{z}(t_i) - \mathbf{x}(t_i)\| + \int_{t_i}^{t} \|\mathbf{f}(t_i, \mathbf{z}(t_i)) - \mathbf{f}(t_i, \mathbf{x}(t_i))\| \, ds$$

$$+ \int_{t_i}^{t} \|\mathbf{f}(t_i, \mathbf{x}(t_i)) - \mathbf{f}(s, \mathbf{x}(s))\| \, ds.$$

Apply (5) to the integrands and note, by the mean value theorem, that $\|\mathbf{x}(t_i) - \mathbf{x}(s)\| \leq M(s - t_i)$. Thus

$$\|\mathbf{z}(t) - \mathbf{x}(t)\| \leq \|\mathbf{z}(t_i) - \mathbf{x}(t_i)\| [1 + K(t - t_i)] + \int_{t_i}^{t} (KM + L)(s - t_i) \, ds.$$

Now use the induction hypothesis and evaluate the integral to obtain

$$\|\mathbf{z}(t) - \mathbf{x}(t)\| \leq \frac{1}{2} \left(M + \frac{L}{K} \right) \{\delta[e^{K(t_i - t_0)} - 1][1 + K(t - t_i)] + K(t - t_i)^2\}$$

$$\leq \frac{1}{2} \left(M + \frac{L}{K} \right) \{\delta e^{K(t_i - t_0)}[1 + K(t - t_i)] - \delta\}$$

for $t_i \leq t \leq t_{i+1}$. It remains only to use the simple inequality $1 + a \leq e^a$ to see that (6) holds for $t_i \leq t \leq t_{i+1}$. This completes the verification of (6) for $t_0 \leq t \leq t_{i+1}$. ∎

Examples 2 and 3 will illustrate the use of Theorem A to determine an upper bound for the error in an approximate solution.

Example 2. Approximate the solution of

$$x' = 1 + x^2 \quad \text{with} \quad x(0) = 0. \tag{7}$$

This again is a familiar example. You may recall, and can easily verify, that the unique solution is given by

$$x(t) = \tan t \quad \text{for} \quad -\frac{\pi}{2} < t < \frac{\pi}{2}.$$

But even in cases when the solution of a differential equation can be expressed in terms of "elementary functions"—trigonometric functions, exponentials, and polynomials—it may sometimes be preferable to have a table of numerical values for the solution instead.

In the present example numerical values for the solution of (7) can, of course, be obtained from a table for the tangent function. On the other hand, the direct construction of an approximate numerical solution of (7) provides a method of actually *making* a table for tan t.

Using a step size of $\delta = 0.1$ (radians) the reader can easily verify the following results of computation by the Euler method for (7) on $[0, 0.5]$. Also tabulated for comparison are the corresponding values of tan t from a four-place table.

i	t_i	$z(t_i) = z(t_{i-1}) + f_{i-1}\delta$	$f_i = 1 + z^2(t_i)$	$x(t_i) = \tan t_i$ (for comparison)
0	0	0	1	0
1	0.1	0.100	1.01	0.1003
2	0.2	0.201	1.04	0.2027
3	0.3	0.305	1.09	0.3093
4	0.4	0.414	1.17	0.4228
5	0.5	0.531	—	0.5463

Now if we were really trying to construct a table for tan t, say on a desert island, we would not be able to compare our results with a "real" table to see how we were doing. Here is a way of estimating the error without using any tables:

Note first that the solution of (7) must be an increasing function. Moreover $0.5 < \pi/6$. Thus without access to any trigonometric tables we can conclude that the solution x of (7) satisfies

$$0 \leq x(t) = \tan t \leq \tan \frac{\pi}{6} = \frac{\sqrt{3}}{3} \leq 0.6$$

for $0 \leq t \leq 0.5$. Let us take

$$J = \mathbf{R} \quad \text{and} \quad D = (-0.1, 0.6).$$

Then inequalities (5) will be satisfied with

$$M = 1.36, \quad K = 1.2, \quad L = 0.$$

With $\delta = 0.1$, inequality (6) now becomes

$$|z(t) - \tan t| \leq 0.068(e^{1.2t} - 1) \quad \text{for} \quad 0 \leq t \leq 0.5,$$

valid since $z(t)$ *does* remain in D. For example, we conclude that

$$|z(0.2) - \tan 0.2| \leq 0.02$$

and

$$|z(0.5) - \tan 0.5| \leq 0.06.$$

It is easily seen from the comparisons in the table above that our approximate solution is really much better than these error estimates show. In fact our error estimates are more than three times the actual error.

As Theorem A suggests, we could improve the approximate solution by reducing the step size δ. (Problem 1.)

Example 3. Show how to approximate the unique solution of

$$x' = t^2 + x^2 \quad \text{with} \quad x(0) = 0 \tag{8}$$

on some appropriate interval. This simple looking problem is one which we cannot solve exactly. In fact it is not even clear how far the solution will exist.

Suppose we use $J = \mathbf{R}$ and $D = \mathbf{R}$ in Theorem 28-D. Then, taking $A = [-a, a] \times [-b, b]$ for some $a > 0$ and $b > 0$, one finds

$$|f(t, \xi)| \le a^2 + b^2 \quad \text{on} \quad A,$$

and Problem 28-12 asserts the existence of a unique solution on $[0, \beta_1)$ where

$$\beta_1 \ge \min \left\{ a, \frac{b}{a^2 + b^2} \right\}. \tag{9}$$

To maximize this expression one must set $b = a = 1/\sqrt{2}$. This yields $\beta_1 \ge 0.7$. (Problem 2.)

But, in fact, we can do better—showing that $\beta_1 \ge 1.2$ at least. For let $J = (-0.1, 1.2)$ and $D = \mathbf{R}$. Then, if x satisfies (8) and if $0 \le t < 1.2$,

$$0 \le x' \le (1.2)^2 + x^2 \quad \text{with} \quad x(0) = 0. \tag{10}$$

From this

$$\int_0^t \frac{x'(s)}{(1.2)^2 + x^2(s)} \, ds \le t.$$

Thus

$$0 \le \frac{1}{1.2} \text{Arctan} \frac{x(t)}{1.2} \le t$$

for $0 \le t < 1.2$, or

$$0 \le x(t) \le 1.2 \tan 1.2t < 1.2 \tan 1.44 < 9.2.$$

The existence of the solution on $[0, 1.2)$ now follows from Theorem 28-D. Why? Thus $\beta_1 \ge 1.2$.

(It is still unclear just how big β_1 really is; but it is not difficult to see that *the above method would fail* to prove existence on $[0, 1.3)$. Problem 3.)

Let $\gamma = 1$. Then on $[0, 1]$, $x' \le 1 + x^2$. So, calculating as above, one finds $0 \le x(t) \le \tan t \le \tan 1 < 1.56$. It also follows that, if z is constructed according to (4), then

$$0 \le z(t) \le x(t) \quad \text{for} \quad 0 \le t \le 1.$$

Why? We can now take, say,

$$J = (-0.1, 1.01) \quad \text{and} \quad D = (-0.1, 1.56).$$

Then in (5) we can use

$$M = 3.46, \quad K = 3.12, \quad L = 2.02,$$

so that inequality (6) becomes

$$|z(t) - x(t)| \le \frac{4.11}{2} \delta(e^{3.12t} - 1) < 45\delta$$

for $0 \le t \le 1$.

If, for example, we want to be sure that $|z(t) - x(t)| < 0.1$, we can partition the interval $[0, 1]$ into 450 subintervals, each of length $\delta = 1/450$. However, an error of 0.1 in this example is nothing to be very proud of and, furthermore, 450 steps in the computation is a very large number.

With some effort we could sharpen the above error estimate of 45δ and thus show that the approximate solution is really better than indicated so far. [Problem 7(b).] But the effort is better spent on an improved method of computing the approximate solution. [Problem 6(b).]

On the subject of errors, it should be remarked that the cumulative error estimated by Theorem A is what is called the "truncation error." In any actual computation, whether by hand or computer, a complete analysis would also require examination of the "round-off error."

An Improved Euler Method

More sophisticated and more efficient methods of computing approximate solutions are available; but these are matters for a course in numerical analysis. We shall settle for a brief and intuitive presentation of one particularly simple improvement on Euler's method.

In Eq. (4) one computes $z(t_i)$ from $z(t_{i-1})$ by using as the "slope" the value of f at $(t_{i-1}, z(t_{i-1}))$. Now it seems reasonable that we could improve the accuracy if we then replaced $z(t_i)$ by a new value computed from $z(t_{i-1})$ by using as the slope the average of $f(t_{i-1}, z(t_{i-1}))$ and $f(t_i, z(t_i))$.

Let y_{i-1} be the approximate value of $x(t_{i-1})$ found by this method. To get the next value y_i in the approximate solution, we first compute

$$z_i \equiv y_{i-1} + f(t_{i-1}, y_{i-1})\delta. \tag{11}$$

Then we improve this calculation by taking

$$y_i \equiv y_{i-1} + \frac{f(t_{i-1}, y_{i-1}) + f(t_i, z_i)}{2} \delta. \tag{12}$$

We will not give a *proof* that this actually gives better results than the simple Euler method. But the following example strongly suggests that it does.

Example 4. Let us use the improved Euler method to compute an approximate solution for the equation of Example 2,

$$x' = 1 + x^2 \quad \text{with} \quad x(0) = 0. \tag{7}$$

In order to be able to compare with the results in Example 2, we again take $\delta = 0.1$ and consider (7) on $[0, 0.5]$.

The computations, based on Eqs. (11) and (12), are presented in the following self-explanatory table.

i	t_i	$z_i = y_{i-1} + f_{i-1}\delta$	$g_i = \frac{1}{2}[f_{i-1} + 1 + z_i^2]$	$y_i = y_{i-1} + g_i\delta$	$f_i = 1 + y_i^2$
0	0	—	—	0	1.000
1	0.1	0.1000	1.005	0.1005	1.010
2	0.2	0.2015	1.025	0.2030	1.041
3	0.3	0.3071	1.068	0.3098	1.096
4	0.4	0.4194	1.136	0.4234	1.179
5	0.5	0.5413	1.236	0.5470	—

Clearly the values obtained here for y_i represent a marked improvement over the values of $z(t_i)$ found in Example 2.

PROBLEMS

1. Apply the Euler method to Eq. (7) as in Example 2, but now use $\delta = 0.05$, half the step size, and go only as far as $t_4 = 0.2$. Your results should be better than those of Example 2. But comparison with the results of Example 4 shows that halving the step size was less beneficial than switching to the "improved Euler method."

2. Show that, in order to maximize the lower bound for β_1 given by (9), one should first set $b = a$ and then set $a = 1/\sqrt{2}$.

3. Show that calculations analogous to those associated with (10) would *fail* to prove existence of a solution of Eqs. (8) on $[0, 1.3)$.

4. For the equation

$$x' = x + t + 1 \quad \text{with} \quad x(1) = 1$$

   find an approximate solution on $[1, 2]$ using
   (a) the Euler method with $\delta = 0.2$,
   (b) the improved Euler method with $\delta = 0.2$.
   Compare the results with the exact solution.

5. Apply the Euler method to the system

$$x' = \begin{pmatrix} -1 & -1 \\ 1 & -3 \end{pmatrix} x \quad \text{with} \quad x(0) = \begin{pmatrix} 0 \\ 1 \end{pmatrix}$$

   on $[0, 0.5]$ using $\delta = 0.1$. Compare your results with the exact solution. (Use Example 24-2.)

6. Find approximate solutions for Eqs. (8) on $[0, 1]$ using (a) the Euler method with $\delta = 0.2$, (b) the improved Euler method with $\delta = 0.2$.

7. Prove that the exact solution of Eqs. (8) on [0, 1] satisfies $t^3/3 \le x(t) \le t^3/2.8$.
   [*Hints*: One inequality follows from the observation that $x' \ge t^2$. To prove the other, show that if $0 \le x_{(0)}(t) \le t^3/2.8$ on [0, 1], then

$$x_{(1)}(t) \equiv \int_0^t [s^2 + x_{(0)}^2(s)] \, ds \le \frac{t^3}{2.8}.$$

Now adapt the *method of proof* used for Theorem 28-A.]
   (a) Compare this with the results of your calculations for Problem 6.
   (b) Use the a priori estimate $0 \le x(t) \le 1/2.8$ to obtain (as in Example 3) the improved error estimate $|z(t) - x(t)| < 2.1\delta$ for $0 \le t \le 1$.

Chapter Six

DELAY DIFFERENTIAL EQUATIONS

The previous five chapters have considered only those ordinary differential equations in which the unknown function and its derivatives are all evaluated at the *same instant*, t.

A more general type of differential equation, often called a functional differential equation, is one in which the unknown function occurs with various different arguments. For example, we might have

$$x'(t) = -2x(t-1),$$

$$x'(t) = x(t) - x\left(\frac{t}{2}\right) + x'(t-1),$$

$$x'(t) = x(t)x(t-1) + t^2 x(t+2), \text{ or}$$

$$x''(t) = -x'(t) - x'(t-1) - 3\sin x(t) + \cos t.$$

In the Russian literature these are called "differential equations with deviating argument."

The simplest and perhaps most natural type of functional differential equation is a "delay differential equation" (or "differential equation with retarded argument"). This means an equation expressing some derivative of x at time t in terms of x (and its lower-order derivatives if any) at t and at *earlier* instants. The first and fourth examples above are of this type. The other two are not, for the second example involves the highest-order derivative at two different instants and the third contains an advanced argument.

This book treats only delay differential equations.

Clearly, when working with delay differential equations, one must write out the arguments of the unknown function. We cannot omit them as we have been doing in ordinary differential equations.

30. EXAMPLES AND THE METHOD OF STEPS

In this section we give some examples of physical and biological systems in which the present rate of change of some unknown function(s) depends on past values of the same function(s).

Mixing of Liquids

Consider a tank containing B liters of salt water brine. Fresh water flows in at the top of the tank at a rate of q liters per minute. The brine in the tank is continually stirred, and the mixed solution flows out through a hole at the bottom, also at the rate of q liters per minute. See Figure 2-1.

Let $x(t)$ be the amount (in grams) of salt in the brine in the tank at time t. If we assume continual instantaneous perfect mixing throughout the tank, then the brine leaving the tank contains $x(t)/B$ gm of salt per liter and hence

$$x'(t) = \frac{-qx(t)}{B}.$$

But, more realistically, let us agree that mixing cannot occur instantaneously throughout the tank. Thus the concentration of the brine leaving the tank at time t will equal the average concentration at some earlier instant, say $t - r$. We shall assume that r is a positive constant (although this assumption may also be subject to improvement, by allowing r to depend on x). The differential equation for x then becomes a delay differential equation, $x'(t) = -qx(t - r)/B$ or, setting $c = q/B$,

$$x'(t) = -cx(t - r), \tag{1}$$

where r is the "delay" or "time lag."

The first mathematical question now is, what kind of initial conditions should one use, if any, in order to obtain a unique solution of Eq. (1)? Various possibilities will be considered in the problems and in the next section.

From these considerations we will find that the most natural answer appears to be that one should specify an *initial function* on some interval of length r, say $[t_0 - r, t_0]$, and then try to satisfy Eq. (1) for $t \geq t_0$. Thus we set

$$x(t) = \theta(t) \quad \text{for} \quad t_0 - r \leq t \leq t_0, \tag{2}$$

where θ is some given function. Then we seek a continuous extension of θ into the future, to a function x which satisfies Eq. (1) for $t \geq t_0$. See Figure 1.

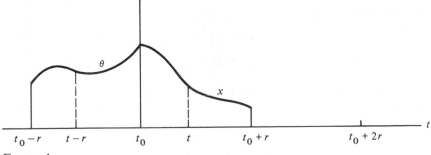

FIGURE 1

[At t_0, $x'(t_0)$ should be interpreted as a right-hand derivative.] We are assuming that the past history of the salt content of the tank is known, and represented by θ, *without regard to whether or not that past history satisfies the delay differential equation.*

For example, let us take $\theta(t) \equiv \theta_0$, a positive constant. In other words, assume that the tank contained θ_0 grams of salt thoroughly mixed in B liters of brine prior to time t_0. Then at t_0 valves were opened allowing fresh water to flow in at the top and mixed brine to flow out at the bottom, each at the rate of q liters per minute.

With this constant initial function, it is easy to solve Eqs. (1) and (2) on $[t_0, t_0 + r]$. For there Eq. (1) becomes simply

$$x'(t) = -c\theta_0$$

with initial condition $x(t_0) = \theta_0$. The solution, of course, is

$$x(t) = \theta_0 - c\theta_0(t - t_0) \quad \text{for} \quad t_0 \le t \le t_0 + r. \tag{3}$$

Now that x is known up to $t_0 + r$, we consider the interval $[t_0 + r, t_0 + 2r]$. There Eq. (1) becomes

$$x'(t) = -c\theta_0 + c^2\theta_0(t - r - t_0)$$

with initial condition, obtained from (3), $x(t_0 + r) = \theta_0 - cr\theta_0$. This can easily be solved, and one can then consider the interval $[t_0 + 2r, t_0 + 3r]$, and so on. This procedure can, in principle, be continued as far as desired. It is called, quite naturally, *the method of steps.*

But the calculations quickly become unwieldy, and it is difficult to determine even the most essential properties of the solution. For example, since $x(t)$ represents the amount of salt in the tank we must have $x(t) \ge 0$ for all t. Hence Eq. (3) shows that we should have $cr < 1$ if the model is to be meaningful. But $cr < 1$ is not a sufficient smallness condition on r. (Problem 1.) How do we know that $x(t)$ does not eventually become negative regardless of how small r is?

Population Growth

If $N(t)$ is the population at time t of an isolated colony of animals, the most naive model for the growth of the population is

$$N'(t) = kN(t), \tag{4}$$

where k is a positive constant. This implies exponential growth, $N(t) = N_0 e^{kt}$ where $N_0 = N(0)$.

A more realistic model is obtained if we admit that the growth rate coefficient k will not be constant but will diminish as $N(t)$ grows, because of overcrowding and shortage of food. This leads to the differential equation

$$N'(t) = k\left[1 - \frac{N(t)}{P}\right]N(t), \tag{5}$$

where k and P are both positive constants. This equation with $N(0) = N_0$ was solved and several solution curves were sketched in Example 6-1. We found that, regardless of the size of $N_0 > 0$, $N(t)$ approaches the equilibrium value P as $t \to \infty$.

Now suppose that the biological self-regulatory reaction represented by the factor $[1 - N(t)/P]$ in (5) is not instantaneous, but responds only after some time lag $r > 0$. Then instead of (5) we have the delay differential (or difference differential) equation

$$N'(t) = k\left[1 - \frac{N(t-r)}{P}\right]N(t). \tag{6}$$

This equation (studied by Wright [16] and others) is often rewritten, by introducing $x(s) = N(rs)/P - 1$ and $c = kr$, in the form

$$x'(s) = -cx(s-1)[1 + x(s)].$$

Having made this change of variables (Problem 2), let us now relabel s as t and write

$$x'(t) = -cx(t-1)[1 + x(t)]. \tag{7}$$

It is customary to specify an initial function θ on $[-1, 0]$ and then seek a function x such that

$$x(t) = \theta(t) \quad \text{for} \quad -1 < t \le 0, \tag{8}$$

and x satisfies Eq. (7) for $t \ge 0$—as we did for Eq. (1).

If we assume θ to be continuous, then it is quite easy to establish the existence and uniqueness of a solution of Eq. (7) subject to initial condition (8) by the method of steps. On $[0, 1]$, Eq. (7) becomes a linear first-order *ordinary* differential equation

$$x'(t) + c\theta(t-1)x(t) = -c\theta(t-1)$$

with initial condition $x(0) = \theta(0)$. As in Section 2, the solution of this problem is found with the aid of an integrating factor. We get

$$\frac{d}{dt}\left[x(t)e^{\int_0^t c\theta(s-1)\,ds}\right] = -c\theta(t-1)e^{\int_0^t c\theta(s-1)\,ds}$$

which leads to the unique solution on $[0, 1]$,

$$x(t) = [\theta(0) + 1]e^{-\int_0^t c\theta(s-1)\,ds} - 1. \tag{9}$$

Having found the exact solution on $[0, 1]$, we can now, in a similar manner, reduce Eq. (7) on $[1, 2]$ to a linear first-order ordinary differential equation and solve on that interval. And so on.

Here too, the procedure can apparently be continued to obtain the exact solution of Eqs. (7) and (8) as far as desired. But, once again, the integrals quickly become very cumbersome, and it is difficult to draw any general conclusions about the solution from this exact procedure. It would be of interest to determine whether a solution of Eq. (7) is bounded, whether it oscillates, and whether like the solutions of (5), it approaches a limit as $t \to \infty$. Some questions of this sort are formulated in Problem 10.

Prey-Predator Population Models

Let $x(t)$ be the population at time t of some species of animal called prey and let $y(t)$ be the population of a predator species which lives off these prey. We assume that $x(t)$ would increase at a rate proportional to $x(t)$ if the prey were left alone; that is, we would have $x'(t) = a_1 x(t)$, where $a_1 > 0$. However, the predators are hungry, and the rate at which each of them eats prey is limited only by his ability to find prey. (This seems like a reasonable assumption as long as there are not too many prey available.) Thus we shall assume that the activities of the predators reduce the growth rate of $x(t)$ by an amount proportional to the product $x(t)y(t)$, that is,

$$x'(t) = a_1 x(t) - b_1 x(t)y(t),$$

where b_1 is another positive constant.

Now let us also assume that the predators are completely dependent on the prey as their food supply. If there were no prey, we assume $y'(t) = -a_2 y(t)$, where $a_2 > 0$; that is, the predator species would die out exponentially. However, given food the predators breed at a rate proportional to their number and to the amount of food available to them. Thus we consider the pair of equations

$$x'(t) = a_1 x(t) - b_1 x(t)y(t),$$
$$y'(t) = -a_2 y(t) + b_2 x(t)y(t),$$

(10)

where a_1, a_2, b_1, and b_2 are positive constants. This well-known model was invented and studied by A. J. Lotka [11] and Vito Volterra [15] in the 1920s.

Volterra was trying to understand the observed fluctuations in the sizes of populations $x(t)$ of commercially desirable fish and $y(t)$ of larger fish which fed on the smaller ones in the Adriatic Sea in the decade from 1914 to 1923. He succeeded in showing that system (10), naive as this model is, does lead to the prediction of "population cycles."

If we now assume that the birth rate of prey has a further limitation as in Eq. (5) and that the birth rate of predators responds to changes in the magnitudes of x and y only after a delay $r > 0$, then system (10) is replaced by

$$x'(t) = a_1\left[1 - \frac{x(t)}{P}\right]x(t) - b_1 y(t)x(t)$$
$$y'(t) = -a_2 y(t) + b_2 x(t-r)y(t-r).$$

(11)

System (11) can be solved by the method of steps. Indeed, the reader will find that the solution of (11) is, in principle, more elementary than the solution of

(10). (Problem 8.) But the discovery of useful information about the solutions turns out to be much more difficult.

Volterra studied several generalizations of (10). One of these was the system

$$x'(t) = \left[a_1 - b_1 y(t) - \int_{-r}^{0} h_1(\sigma) y(t + \sigma) \, d\sigma \right] x(t)$$

$$y'(t) = \left[-a_2 + b_2 x(t) + \int_{-r}^{0} h_2(\sigma) x(t + \sigma) \, d\sigma \right] y(t),$$

(12)

where h_1 and h_2 are given continuous nonnegative functions. Here the past histories of x and y over an entire interval of length r enter the equations through the integrals. Such equations may be said to have infinitely many delays.

The method of steps cannot be applied to system (12). Why not? Try it. The trouble is that (12) contains arbitrarily small delays.

Control Systems

Any system involving a feedback control will almost certainly involve time delays. These arise because a finite time is required to sense information and then react to it.

Consider, as a simple example, a system whose motion is governed by a second-order linear homogeneous differential equation with *positive constant* coefficients

$$mx''(t) + bx'(t) + kx(t) = 0. \tag{13}$$

The solution of this equation with arbitrarily specified initial conditions, $x(t_0)$ and $x'(t_0)$, is a function which decays exponentially toward zero. Recall that we classified the equation (in Example 14-1) in terms of the magnitude of the damping coefficient b.

The case $b^2 < 4mk$ is called "underdamped," and the solution oscillates as it decays.

The case $b^2 > 4mk$ is called "overdamped." The solutions do not oscillate, but die out exponentially at a slower rate the larger b becomes.

The case $b^2 = 4mk$ is called "critically damped" because, in a sense, solutions approach zero most rapidly in this case.

Let us assume that the system is underdamped ($b^2 < 4mk$) and we wish to somehow increase the damping coefficient to bring the system closer to critical damping, thereby diminishing the oscillations more rapidly. Perhaps we would even prefer a slight overdamping so as to eliminate oscillations.

If our physical system is a simple mass hanging from a spring in the laboratory, then it is quite simple to increase b. We might simply immerse the whole system in molasses. Or if that makes b *too* big we could try No. 10 motor oil instead.

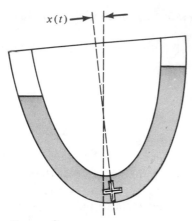

$x(t)$

FIGURE 2

However, if our system is a ship rolling in the waves and x is the angle of tilt from the normal upright position (Figure 2), we must be more ingenious in trying to increase b. We might, for example, introduce ballast tanks, partially filled with water, in each side of the ship. We would also have a servo-mechanism designed to pump water from one tank to the other in an attempt to counteract the roll of the ship. Hopefully, this would introduce another term proportional to $x'(t)$ in the equation, say $qx'(t)$. Thus we consider

$$mx''(t) + bx'(t) + qx'(t) + kx(t) = 0. \qquad (14)$$

But now one must recognize that the servomechanism cannot respond instantaneously. Thus, instead of Eq. (14), we should consider

$$mx''(t) + bx'(t) + qx'(t - r) + kx(t) = 0. \qquad (15)$$

The control takes time $r > 0$ to respond and thus the control term is proportional to the velocity at the earlier instant, $t - r$. It seems possible that such a time lag could result in the force represented by $qx'(t - r)$ being in the opposite direction to that which is desired. Thus it is conceivable that rather than helping to stabilize the system, such a control might make matters worse, and even *cause* undesired oscillations. (Problem 9.)

This is a simplified version of a problem which actually arose during tests of systems for antirolling stabilization of a ship before World War II. (Minorsky [12].) The delay in that case was found to be due to "cavitation" at the pump itself.

Others

Hopefully one or more of the examples cited above has convinced the reader that delay differential equations are worth studying. But, if not, we could cite others which have arisen in studies of: nuclear reactors, neutron shielding, electron energy distribution in a gas discharge, liquid fuel rocket engines, transistor circuits, electrodynamics, photoemulsions, transmission lines, elasticity theory, the spread of infectious diseases, neurology, the respiratory system, business cycles and economic growth, inventory maintenance, the production and death of red blood cells, metal rolling control systems, and number theory. Specific references to the literature on these sources of delay differential equations can be found in [4].

PROBLEMS

1. (a) How small must cr be in order that the solution of Eqs. (1) and (2) with $\theta(t) \equiv \theta_0 > 0$ remain positive for $t_0 \leq t \leq t_0 + 2r$?
   (b) Can you find a smallness condition on r which would guarantee $x(t) > 0$ for all $t \geq t_0$?
   (c) In the "mixing-of-liquids" example, instead of fresh water, let the incoming liquid be brine containing a grams of salt per liter. Then show that

   $$x'(t) = qa - cx(t - r).$$

   Find a change of variable which transforms this into Eq. (1).
2. Justify the transformation of Eq. (6) into Eq. (7).
3. Apply the method of steps to solve on $[0, 2]$ the equation

   $$x'(t) = ax(t) + bx(t - 1), \tag{16}$$

   with $x(t) = \theta(t) = 1 + t$ on $[-1, 0]$. Assume that a and b are constants with $b \neq 0$. [*Caution:* Most students get this problem wrong. Check your answer carefully on $[0, 1]$ before proceeding to $[1, 2]$.]
4. (a) As another approach to Eq. (16), we might seek exponential solutions of the form $x(t) = e^{\lambda t}$, where λ is a constant. Find appropriate values of the coefficients a and b in Eq. (16) so that $x(t) = e^{\lambda t}$ will be a solution both for $\lambda = 0$ and $\lambda = 1$.
   (b) Using these values of a and b, show that Eq. (16) with $x(0) = x_0$ can have infinitely many solutions *valid for all t* in **R**.
   (c) Do you suppose the solution of the equation in part (b) would be unique if we specified both $x(0) = x_0$ and $x'(0) = x_1$?
5. Example 3-5 showed that the equation $x'(t) = [x(t)]^{2/3}$ with $x(0) = 0$ has many solutions. What can you say about existence and uniqueness of solutions of the equation

   $$x'(t) = [x(t - r)]^{2/3} \quad \text{for} \quad t \geq 0,$$

   with $x(t) = \theta(t)$ for $-r \leq t \leq 0$, where r is a positive constant and θ is a given continuous initial function?
6. We found in Example 3-2 that the equation $x'(t) = 1 + x^2(t)$ with $x(0) = 1$ has a unique solution, but the solution cannot be continued beyond $\pi/4$. What can you say about existence and uniqueness of solutions of the equation

   $$x'(t) = 1 + x^2(t - r) \quad \text{for} \quad t \geq 0,$$

with $x(t) = \theta(t)$, for $-r \leq t \leq 0$, where r is a positive constant and θ is a given continuous initial function?

7. Prove that the equation $x'(t) = x(t)x(t - r)$ with continuous $\theta: [-r, 0] \to \mathbf{R}$ has a unique solution for all $t \geq 0$. Compare this result, which holds for any $r > 0$ (no matter how small), with the situation when $r = 0$.

8. (a) Show that, in principle, it is always possible to solve system (11) exactly for $t \geq 0$ when $x(t) = \theta_1(t)$ and $y(t) = \theta_2(t)$ for $-r \leq t \leq 0$, where θ_1 and θ_2 are given continuous functions. [*Hint:* You will encounter a Bernoulli equation.]

   (b) Can you say the same for system (10)?

   (c) Show that the solution of system (10) will be unique if $x(0) = x_0$ and $y(0) = y_0$ are given.

9. Show that it is possible for Eq. (15) to have *undamped* periodic solutions when m, k, b, q, and r are positive constants. [*Hint:* Try to find a solution of the form $x(t) = e^{\lambda t}$ where $\lambda = i\omega$ with ω real. Then verify that $\cos \omega t$ and $\sin \omega t$ are solutions. For simplicity set $m = 1$, $b = 1$, and $r = 1$.] What initial functions on $[-1, 0]$ would produce these solutions?

*10. Consider Eq. (7) for $t > 0$ with $x(t) = \theta(t)$ for t in $[-1, 0]$ where θ is continuous Prove the following results (from a paper of Wright [16]).

   (a) A unique solution exists for all $t \geq 0$, and

$$x(t) > -1 \quad \text{if} \quad \theta(0) > -1,$$

$$x(t) = -1 \quad \text{if} \quad \theta(0) = -1,$$

$$x(t) < -1 \quad \text{if} \quad \theta(0) < -1.$$

[*Hint:* First show that on $[0, 1]$

$$x(t) + 1 = [\theta(0) + 1]e^{\int_0^t c\theta(s-1)\,ds}.]$$

   (b) If $c > 0$ and $\theta(0) > -1$, then either (i) $x(t) \to 0$ as $t \to \infty$ or (ii) $x(t)$ oscillates about $x = 0$, that is, assumes both positive and negative values for arbitrarily large t. Interpret these results for Eq. (6). [*Hint:* If (ii) does not hold, show that for large t, $x(t)$ and $x'(t)$ must have constant but opposite signs. Thus $L = \lim_{t \to \infty} x(t)$ exists. Now argue from Eq. (7) that L must be either 0 or -1, and we know $L \neq -1$.]

   (c) If $c > 0$ and $\theta(0) < -1$, then $x(t) \to -\infty$ as $t \to \infty$. [*Hint:* Since $x(t) < -1$ for $t \geq 0$, $x'(t) \leq 0$ for $t \geq 1$. Now $x(t) \not\to L$ since L would have to be either 0 or -1.]

31. WHY NOT SOME OTHER FORMULATION OF THE PROBLEM?

In Section 30 we established the existence of unique solutions of some delay differential equations in the future (that is, for $t \geq t_0$) provided that a suitable initial function was specified (for $t \leq t_0$).

Let us pause now to consider what would happen if we were to seek solutions into the past instead of the future, or if we were to specify initial conditions only at a point t_0, instead of over an interval. We shall see that such approaches can lead to nonuniqueness or nonexistence—in marked contrast to the situation for ordinary differential equations.

But, after getting this out of the way, much of the remaining discussion (in Sections 32 and 33) will be aimed at showing that delay differential equations are, in many respects, similar to ordinary differential equations when the " basic initial problem " is properly formulated.

In discussing delay differential equations we adopt some terminology from ordinary differential equations. For example, the *order* of a delay differential equation will mean the order of the highest derivative involved in the equation. Thus

$$x'(t) = ax(t) + bx(t - r) \tag{1}$$

and

$$x'(t) = -cx(t - 1)[1 + x(t)] \tag{2}$$

are of first order, while

$$mx''(t) + bx'(t) + qx'(t - r) + kx(t) = h(t) \tag{3}$$

is of second order. Similarly, we will refer to a system like 30-(11) or 30-(12) as a *system* of first-order equations.

We shall say Eqs. (1) and (3) are linear, while Eq. (2) is nonlinear. Equation (1) is homogeneous (linear), but Eq. (3) is nonhomogeneous (linear).

This section contains five observations about delay differential equations —illustrated with examples. Each of the examples will be a special case of the scalar linear homogeneous delay differential equation

$$x'(t) = a(t)x(t) + b(t)x(t - r), \tag{4}$$

where the coefficients a and b are continuous real-valued functions on $\mathbf{R}$ and $r > 0$ is a constant.

OBSERVATION 1

A first-order (scalar), linear, homogeneous delay differential equation with real coefficients can have a nontrivial oscillating solution (that is, a solution taking both positive and negative values for arbitrarily large values of t). Show that this is impossible if $r = 0$. (Problem 1.)

Example 1. The equation

$$x'(t) = -x\left(t - \frac{\pi}{2}\right) \tag{5}$$

has solutions of the form $x(t) = c_1 \cos t + c_2 \sin t$ for arbitrary constants c_1 and c_2. Verify it.

OBSERVATION 2

If one specifies

$$x(t) = \theta(t) \text{ (continuous)} \quad \text{for} \quad t_0 - r \le t \le t_0, \tag{6}$$

and then tries to satisfy Eq. (4) for $t \le t_0$, there will, in general, be no (continuous) solution. This remains true even if θ is infinitely differentiable.

Example 2. Consider Eq. (1) with $b \ne 0$, $r > 0$, and $a \ne -b$. Let $\theta(t) = k$, a nonzero constant, on $[-r, 0]$. Then, if x were to be a solution, we would require

$$x(t) = -\frac{ak}{b} \quad \text{on} \quad [-2r, -r].$$

Why? But, since $-ak/b \ne k$, we would have a discontinuity at $-r$ and so x would not be a solution.

OBSERVATION 3

If θ is chosen so that Eqs. (4) and (6) do have a (backwards) solution for $t \le t_0$, this solution will not, in general, be unique.

Example 3. Consider the scalar equation

$$x'(t) = b(t)x(t - 1), \tag{7}$$

where

$$b(t) = \begin{cases} 0 & \text{for} \quad t \le 0 \\ \cos 2\pi t - 1 & \text{for} \quad 0 < t \le 1 \\ 0 & \text{for} \quad t > 1. \end{cases}$$

(Note that b is continuous.) Let $x(t) = k$ for $t \le 0$, where k is *any* constant. Then Eq. (7) is clearly satisfied on $(-\infty, 0]$. Thus, on $[0, 1]$,

$$x'(t) = (\cos 2\pi t - 1)k, \quad \text{with} \quad x(0) = k.$$

Hence

$$x(1) = k + k \int_0^1 (\cos 2\pi t - 1)\, dt = 0.$$

But then the fact that $x'(t) = 0$ for $t \ge 1$ implies $x(t) = 0$ for all $t \ge 1$.

This shows that, given $x(t) = \theta(t) \equiv 0$ for $1 \le t \le 2$, Eq. (7) has infinitely many continuous solutions valid for $t \le 2$.

Note that this example also shows nonexistence for $t \le t_0 = 2$ (as asserted in Observation 2) if $x(t) = \theta(t) \not\equiv 0$ on $[1, 2]$.

OBSERVATION 4

If one merely specifies

$$x(t_0) = x_0, \tag{8}$$

there may still be no solution of (4) for $t \leq t_0$.

Example 4. Consider Eq. (7) with $x(1) = x_0 \neq 0$.

(Examples 3 and 4 are due to E. Winston and J. A. Yorke. Related examples have also been given by A. M. Zverkin, A. B. Nersesjan, V. M. Popov, and others.)

OBSERVATION 5

If a solution of (4) and (8) does exist on $(-\infty, t_0]$ it will, in general, be nonunique. In fact, even if one specifies x plus *all* its derivatives at t_0,

$$x(t_0) = x_0, \quad x'(t_0) = b_1, \quad x''(t_0) = b_2, \quad \dots, \tag{9}$$

there may be infinitely many different solutions of (4) on $(-\infty, t_0]$ which satisfy (9).

Nonuniqueness on $(-\infty, t_0]$ for Eqs. (4) and (8) was encountered in Problem 30-4(b). It is also illustrated by Examples 1 and 3 above. How?

To show that even (9), the specification of x and all its derivatives at t_0, does not assure uniqueness on $(-\infty, t_0]$ one could use Eq. (7) with $x(t_0) = 0$, $x'(t_0) = 0$, $x''(t_0) = 0$, ... for any $t_0 > 1$. But let us also give an example using the scalar Eq. (1) with *constant* coefficients. This will then answer Problem 30-4(c).

*****Example 5.** We shall consider Eq. (1) with $b \neq 0$ and $r > 0$. For simplicity, let $t_0 = 0$. Let

$$\theta(t) \equiv \begin{cases} 0 & \text{for} \quad t = -r \\ e^{-t^{-2}} e^{-(t+r)^{-2}} & \text{for} \quad -r < t < 0 \\ 0 & \text{for} \quad t = 0. \end{cases} \tag{10}$$

It is left as a nontrivial Problem 2 for the reader to show that:

(i) θ has continuous derivatives of *all* orders on $[-r, 0]$, and
(ii) θ and *all* its derivatives are zero at $t = -r$ and $t = 0$. (We interpret the derivatives as one-sided at the ends of the interval.)

Now let us try to use θ as an "initial function" to obtain a solution y of Eq. (1) for *all* t. We first consider the past. If Eq. (1) is to be satisfied for $-r \leq t \leq 0$, we must have

$$y(t - r) = \frac{\theta'(t) - a\theta(t)}{b} \quad \text{for} \quad -r \leq t \leq 0$$

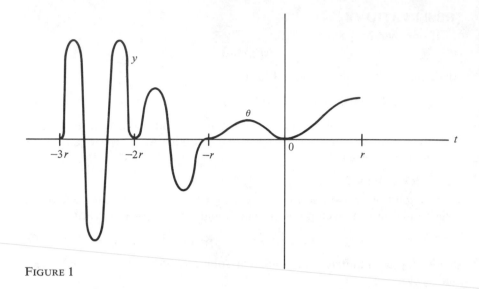

FIGURE 1

or

$$y(t) = \frac{\theta'(t + r) - a\theta(t + r)}{b} \quad \text{for} \quad -2r \le t \le -r. \tag{11}$$

The important thing to note is that (11) defines a function y on $[-2r, -r]$ having properties like (i) and (ii) listed above for θ. Thus we can perform a similar computation to extend y to $[-3r, -2r]$. This construction can be continued backwards by steps indefinitely. One can verify that the graph of y on each succeeding interval to the left has more oscillations and greater amplitudes, as suggested by Figure 1. (Clearly y could also be continued into the future by the method of steps.) Thus y satisfies Eq. (1) for all t, and

$$y(0) = 0, \quad y'(0) = 0, \quad y''(0) = 0, \quad \ldots . \tag{12}$$

But this solution of Eqs. (1) and (12) is not unique. For any multiple of y is also a solution (including the identically zero function).

Furthermore it is easily verified (Problem 3) that if x on $(-\infty, \infty)$ is a solution of Eqs. (1) and (9) with $t_0 = 0$ and if y is a solution on $(-\infty, \infty)$ of Eqs. (1) and (12), then, for any constant c, $x + cy$ is another solution on $(-\infty, \infty)$ of (1) and (9). Thus, if we have a solution of Eq. (1) on $(-\infty, \infty)$ satisfying *any* given conditions of the type (9), there must also be infinitely many other such solutions. For instance, Eq. (5) with $x^{(2k)}(0) = (-1)^k$ and $x^{(2k+1)}(0) = 0$ $(k = 0, 1, \ldots)$ is satisfied by $x(t) = \cos t$ and by infinitely many other functions.

Remarks. The phenomena of nonexistence and nonuniqueness described in Observations 4 and 5 could not occur for a linear ordinary differential equation or system. For we know from Sections 9 and 28 that if A is a continuous matrix-valued function and h (not necessarily zero) is a continuous vector-valued function on **R**, then the ordinary differential system

$$x'(t) = A(t)x(t) + h(t)$$

always has a unique solution on **R** satisfying any given initial condition of the form $x(t_0) = x_0$.

It should be mentioned that, despite Observations 4 and 5, there are some special delay differential equations arising in applications for which it *does* make sense to talk about solutions on $(-\infty, t_0]$ subject to a given initial condition $x(t_0) = x_0$.

PROBLEMS

1. If a is a continuous real-valued function on **R**, use the method of Section 2 to prove that the equation $x'(t) = a(t)x(t)$ cannot have an oscillating solution.

*2. Verify properties (i) and (ii) for the function θ defined by Eq. (10).

3. Verify that if x and y are two solutions of Eq. (1) which satisfy conditions (9) with $t_0 = 0$ and condition (12), respectively, then $x + cy$ is another solution which satisfies (9).

32. UNIQUENESS AND EXISTENCE FOR THE BASIC INITIAL PROBLEM

The difficulties encountered in Section 31 for various "other problems" associated with delay differential equations lead us to return now to the initial-function problem described in Section 30. Thus we specify $x(t) = \theta(t)$ for $t \le t_0$ and seek a continuation to a function x which satisfies a given delay differential equation for $t \ge t_0$. We shall refer to this as the *basic initial problem* for the delay differential equation.

Our first observation with respect to the basic initial problem is that it is sometimes easier to prove uniqueness or existence for a differential equation with delay than for one without.

Example 1. If $r > 0$, the equation

$$x'(t) = f(t, x(t - r)) \quad \text{for} \quad t \ge 0$$

with $x(t) = \theta(t)$ given for $-r \le t \le 0$ can be treated unambiguously by the method of steps. If f and θ are continuous functions on $\mathbf{R}^2$ and $\mathbf{R}$, respectively, then a unique solution, defined by

$$x(t) = \theta(0) + \int_0^t f(s, \theta(s - r))\, ds,$$

exists for $0 \leq t \leq r$. Arguing by the method of steps, we then conclude that a unique solution exists for *all* $t \geq 0$. This is valid no matter how small the value of $r > 0$.

But no such assertion can be made if $r = 0$—the case of the ordinary differential equation $x'(t) = f(t, x(t))$. For if

$$x'(t) = x^2(t) \quad \text{with} \quad x(0) = 1,$$

then the solution exists *only* for $t < 1$. While, on the other hand, the solution of

$$x'(t) = [x(t)]^{2/3} \quad \text{with} \quad x(0) = 0$$

is not unique.

Such examples, in which the presence of a delay *causes* uniqueness or "global" existence, were previously encountered in Problems 30-5, 6, and 7.

However, in the case of a delay which can become zero, as it does at $t = 0$ in

$$x'(t) = f\left(t, \, x\!\left(\frac{t}{2}\right)\right) \quad \text{for} \quad t \geq 0 \quad \text{with} \quad x(0) = \theta(0),$$

no such simplification results from the presence of the delay. In this equation we can think of $t/2 = t - t/2$ and consider the delay to be the variable quantity $r(t) = t/2$ for $t \geq 0$. (Note that, if $t_0 = 0$, there is no purpose in specifying initial data at any more than the single point $t = 0$.)

Example 2. The delay differential equation

$$x'(t) = \left[x\!\left(\frac{t}{2}\right)\right]^{2/3} \quad \text{for} \quad t \geq 0 \quad \text{with} \quad x(0) = 0$$

has a nonunique solution. (Problem 1).

It is possible to prove a uniqueness theorem (involving a Lipschitz condition) which covers very general delay differential equations with variable delays which may indeed become zero at certain points. (For example, see [4].) However, many practical applications involve only constant delays, and these are easier to handle.

So let us consider, as our most general problem, a system of n delay differential equations involving several different constant delays,

$$\mathbf{x}'(t) = \mathbf{f}(t, \mathbf{x}(t - r_1), \mathbf{x}(t - r_2), \ldots, \mathbf{x}(t - r_m)). \tag{1}$$

Here $\mathbf{x}$ and $\mathbf{f}$ are n-vector-valued functions, and the r_j's are nonnegative constants. Without loss of generality, let us assume that the r_j's are all distinct and are labeled in increasing order so that

$$0 \leq r_1 < r_2 < \cdots < r_m \equiv r.$$

(Often $r_1 = 0$.) We shall consider Eq. (1) for $t_0 \leq t < \beta$ with initial condition

$$\mathbf{x}(t) = \mathbf{\theta}(t) \quad \text{for} \quad t_0 - r \leq t \leq t_0, \tag{2}$$

where $\mathbf{\theta}$ is a given initial function.

Let D be an open set in $\mathbf{R}^n$, and let $\mathbf{f}$ be defined on $[t_0, \beta) \times D^m$. Then $\mathbf{\theta}$ should be a function mapping $[t_0 - r, t_0] \to D$. We shall always assume $\mathbf{\theta}$ to be continuous.

DEFINITIONS

A *solution* of Eqs. (1) and (2) is a continuous function $\mathbf{x} : [t_0 - r, \beta_1] \to D$, where $t_0 < \beta_1 \leq \beta$, such that

(i) $\mathbf{x}(t) = \mathbf{\theta}(t)$ for $t_0 - r \leq t \leq t_0$, and
(ii) Eq. (1) is satisfied for $t_0 \leq t < \beta_1$.

[Understand $\mathbf{x}'(t_0)$ to mean the right-hand derivative.]

The solution of Eqs. (1) and (2) is said to be *unique* if every two solutions agree with each other as far as both are defined.

For simplicity of notation, let us now assume that $t_0 = 0$. (Otherwise, one could always introduce a new "time" variable $\tilde{t} = t - t_0$ so that the initial point would become $\tilde{t} = 0$ instead of $t = t_0$.)

With this notation and terminology we can now generalize Example 1 nicely to the case of Eq. (1) with $r_1 > 0$.

THEOREM A (UNIQUENESS AND EXISTENCE WHEN $r_1 > 0$)

Let $r_1 > 0$, let $\mathbf{f}$ be continuous on $[0, \beta) \times D^m$, and let $\mathbf{\theta}$ be continuous on $[-r, 0] \to D$. Then Eqs. (1) and (2) have a unique solution $\mathbf{x}$ on $[-r, \beta_1)$ where $0 < \beta_1 \leq \beta$; and if $\beta_1 < \beta$, then $\mathbf{x}(t)$ approaches the boundary of D as $t \to \beta_1$.

Remarks. The hypothesis that $r_1 > 0$ has eliminated the need for the additional assumption on $\mathbf{f}$—continuity of partial derivatives or the existence of a Lipschitz condition—used in our theorems on uniqueness (Chapter Two) and existence (Theorem 28-D) for ordinary differential systems. The conclusion of Theorem A above is actually stronger than that of Theorem 28-D since now there is no concern that $\|\mathbf{x}(t)\|$ may become arbitrarily large when $t < \beta$. And finally, the proof of Theorem A is quite elementary and is independent of our previous theorems for ordinary differential equations.

Proof of Theorem A. For $0 \leq t \leq r_1$,

$$\mathbf{x}'(t) = \mathbf{f}(t, \mathbf{\theta}(t - r_1), \ldots, \mathbf{\theta}(t - r_m))$$

and $\mathbf{x}(0) = \mathbf{\theta}(0)$. This uniquely determines

$$\mathbf{x}(t) = \mathbf{\theta}(0) + \int_0^t \mathbf{f}(s, \mathbf{\theta}(s - r_1), \ldots, \mathbf{\theta}(s - r_m)) \, ds$$

for $0 \le t \le r_1$ [provided $r_1 < \beta$ and $\mathbf{x}(t)$ remains in D].

Since we now know $\mathbf{x}(t)$ for $-r \le t \le r_1$, we can find $\mathbf{x}(t)$ for $r_1 \le t \le 2r_1$ [provided $2r_1 < \beta$ and $\mathbf{x}(t)$ remains in D].

This method-of-steps procedure continues until we reach β or until $\mathbf{x}(t)$ reaches the boundary of D. ∎

If $r_1 = 0$, then Eq. (1) includes ordinary differential systems as a special case. Why? Thus we cannot expect the uniqueness and existence theorems to be any better than they were for ordinary differential equations.

When $r_1 = 0$ we can still apply the method of steps, but now with r_2 playing the role which r_1 played in the proof of Theorem A. Each step will now involve an ordinary differential equation for which we need additional conditions on $\mathbf{f}$—either continuity of $D_{1+j}f_i$ for $i, j = 1, \ldots, n$ or a Lipschitz condition of the form

$$\|\mathbf{f}(t, \mathbf{\xi}_{(1)}, \mathbf{\xi}_{(2)}, \ldots, \mathbf{\xi}_{(m)}) - \mathbf{f}(t, \bar{\mathbf{\xi}}_{(1)}, \mathbf{\xi}_{(2)}, \ldots, \mathbf{\xi}_{(m)})\| \le K\|\mathbf{\xi}_{(1)} - \bar{\mathbf{\xi}}_{(1)}\|. \tag{3}$$

THEOREM B (UNIQUENESS WHEN $r_1 \ge 0$)

Let $\mathbf{f}$ be continuous on $[0, \beta) \times D^m$ and either assume each $D_{1+j}f_i$ for $i, j = 1, \ldots, n$ continuous there also or assume Lipschitz condition (3). Let $\mathbf{\theta}$ be continuous on $[-r, 0] \to D$. Then Eqs. (1) and (2) have at most one solution on any interval $[-r, \beta_1)$ where $0 < \beta_1 \le \beta$.

Proof. The case $r_1 > 0$ is covered by Theorem A. So let $r_1 = 0$. Then for $0 \le t \le r_2$,

$$\mathbf{x}'(t) = \mathbf{f}(t, \mathbf{x}(t), \mathbf{\theta}(t - r_2), \ldots, \mathbf{\theta}(t - r_m))$$

with $\mathbf{x}(0) = \mathbf{\theta}(0)$. By the theorems of Chapter Two, this has at most one solution.

If $r_2 < \beta$, we now consider $r_2 \le t \le 2r_2$. Etc. ∎

One could state a companion existence theorem covering the case $r_1 = 0$ by referring to Theorem 28-D. But let us not bother since the individual examples and problems we shall encounter can be handled without difficulty on their own merits.

Example 3. Consider the second-order equation 30-(15),

$$mz''(t) + bz'(t) + qz'(t - r) + kz(t) = 0 \tag{4}$$

for $t \ge 0$. By introducing $x_1(t) = z(t)$ and $x_2(t) = z'(t)$, we convert Eq. (4)

into the system

$$x_1'(t) = x_2(t),$$

$$x_2'(t) = -\frac{k}{m} x_1(t) - \frac{b}{m} x_2(t) - \frac{q}{m} x_2(t - r). \tag{5}$$

This is system (1) with $n = 2$, $m = 2$, $r_1 = 0$, $r_2 = r > 0$, and

$$\mathbf{f}(t, \xi_{(1)}, \xi_{(2)}) = \begin{pmatrix} \xi_{(1)2} \\ -\dfrac{k}{m}\xi_{(1)1} - \dfrac{b}{m}\xi_{(1)2} - \dfrac{q}{m}\xi_{(2)2} \end{pmatrix}.$$

Certainly $\mathbf{f}$ is continuous, and so too are the partial derivatives

$$\frac{\partial f_1}{\partial \xi_{(1)1}} = 0, \qquad \frac{\partial f_1}{\partial \xi_{(1)2}} = 1,$$

$$\frac{\partial f_2}{\partial \xi_{(1)1}} = -\frac{k}{m}, \qquad \frac{\partial f_2}{\partial \xi_{(1)2}} = -\frac{b}{m}.$$

Thus, by Theorem B, given any continuous function $\boldsymbol{\theta} : [-r, 0] \to \mathbf{R}^2$, Eqs. (5) and (2) have at most one solution on $[-r, \beta_1)$ for any $\beta_1 > 0$.

In this example we could just as well have applied the method of steps directly, to conclude that Eqs. (5) and (2) do have a unique solution on $[-r, \infty)$. (Problem 5.)

Note that if φ is the initial function for Eq. (4), then φ should be assumed continuously differentiable on $[-r, 0] \to \mathbf{R}$. And when we convert to system (5) we should define

$$\boldsymbol{\theta}(t) = \mathrm{col}\,(\varphi(t), \varphi'(t)) \quad \text{for} \quad -r \leq t \leq 0.$$

Often one has a stronger Lipschitz condition than (3), covering the effect of changing *all* the arguments $\xi_{(1)}, \ldots, \xi_{(m)}$ in $\mathbf{f}$. Then the following estimate for the growth of errors is available.

THEOREM C (GROWTH OF ERRORS)

Let $\mathbf{f}$ be continuous on $[0, \beta) \times D^m$ and let it satisfy the strong Lipschitz condition

$$\|\mathbf{f}(t, \xi_{(1)}, \ldots, \xi_{(m)}) - \mathbf{f}(t, \tilde{\xi}_{(1)}, \ldots, \tilde{\xi}_{(m)})\| \leq K \max_{1 \leq j \leq m} \|\xi_{(j)} - \tilde{\xi}_{(j)}\|. \tag{6}$$

Let $\mathbf{x}$ and $\tilde{\mathbf{x}}$ be solutions on $[-r, \beta)$ of Eqs. (1) and

$$\mathbf{x}(t) = \boldsymbol{\theta}(t), \quad \tilde{\mathbf{x}}(t) = \tilde{\boldsymbol{\theta}}(t) \quad \text{on } [-r, 0], \tag{7}$$

where $\boldsymbol{\theta}$ and $\tilde{\boldsymbol{\theta}}$ are continuous. Then

$$\|\mathbf{x}(t) - \tilde{\mathbf{x}}(t)\| \leq \sup_{-r \leq s \leq 0} \|\boldsymbol{\theta}(s) - \tilde{\boldsymbol{\theta}}(s)\| e^{Kt} \quad \text{for} \quad 0 \leq t < \beta. \tag{8}$$

The proof, using Lemma 12-A, is outlined in Problem 6.

We conclude this section with a theorem on solutions of the scalar linear homogeneous equation

$$y'(t) = a_1(t)y(t) + a_2(t)y(t - r_2) + \cdots + a_m(t)y(t - r_m), \tag{9}$$

with negative coefficients. Example 4 will use this theorem to answer a question which arose in Section 30.

For a linear delay differential equation we take $D = \mathbf{R}$, just as in the case of a linear ordinary differential equation.

THEOREM D (A COMPARISON THEOREM)

Let $a_1, \ldots, a_m$ ($m \geq 2$) be continuous functions with

$$a_2(t) < 0, \ldots, \quad \text{and} \quad a_m(t) < 0.$$

Let y and $\tilde{y}: [-r, \beta) \to \mathbf{R}$ be solutions of Eq. (9) with continuous initial functions θ and $\tilde{\theta}$, respectively, where

$$\theta(t) < \tilde{\theta}(t) \quad \text{on} \quad [-r, 0) \quad \text{and} \quad \theta(0) = \tilde{\theta}(0) > 0. \tag{10}$$

If $y(t)$ and $\tilde{y}(t)$ remain positive for $0 \leq t < \beta_1 \leq \beta$, then

$$y(t) > \tilde{y}(t) \quad \text{on} \quad (0, \beta_1). \tag{11}$$

It follows that if $y(\beta_1) = 0$, then $\tilde{y}(t) = 0$ for some t in $(0, \beta_1]$.

Remarks. This theorem is a special case of a result found in Chapter VI of Myškis [13]. The following rather simple proof was suggested by J. A. Yorke.

Proof of Theorem D. At $t = 0$ the right-hand derivatives of y and $\tilde{y}$ satisfy

$$y'(0) - \tilde{y}'(0) = \sum_{j=2}^{m} a_j(0)[\theta(-r_j) - \tilde{\theta}(-r_j)] > 0.$$

Since $y(0) = \tilde{y}(0) > 0$, it follows that

$$\frac{y'(t)}{y(t)} > \frac{\tilde{y}'(t)}{\tilde{y}(t)} \tag{12}$$

at $t = 0$. Then, by the continuity of y, y', $\tilde{y}$, and $\tilde{y}'$, inequality (12) must hold for all sufficiently small $t \geq 0$. If (12) is valid for $0 \leq t < t_1$, the mean value theorem gives

$$\ln y(t_1) - \ln y(t) > \ln \tilde{y}(t_1) - \ln \tilde{y}(t)$$

or

$$\frac{y(t_1)}{y(t)} > \frac{\tilde{y}(t_1)}{\tilde{y}(t)} \quad \text{for} \quad 0 \leq t < t_1. \tag{13}$$

Setting $t = 0$ in (13) yields

$$y(t_1) > \tilde{y}(t_1). \tag{14}$$

So it follows from (10) that

$$\frac{\theta(t)}{y(t_1)} < \frac{\tilde{\theta}(t)}{\tilde{y}(t_1)} \quad \text{for} \quad -r \le t \le 0.$$

Combining this with (13), we have

$$\frac{y(t)}{y(t_1)} < \frac{\tilde{y}(t)}{\tilde{y}(t_1)} \quad \text{for} \quad -r \le t < t_1. \tag{15}$$

Now suppose (for contradiction) that inequality (12) does not hold for all t in $[0, \beta_1)$, and let t_1 be the first point at which (12) becomes an equality. Then

$$\frac{y'(t_1)}{y(t_1)} - \frac{\tilde{y}'(t_1)}{\tilde{y}(t_1)} = \sum_{j=2}^{m} a_j(t) \left[\frac{y(t_1 - r_j)}{y(t_1)} - \frac{\tilde{y}(t_1 - r_j)}{\tilde{y}(t_1)} \right] > 0$$

—a contradiction. Thus (12) holds for all t in $[0, \beta_1)$, and so (14) holds for all t_1 in $(0, \beta_1)$. ■

Example 4. Consider Eq. 30-(1),

$$y'(t) = -cy(t - r) \quad \text{for} \quad t \ge 0, \tag{16}$$

where c and r are positive constants. We shall show that if

$$cr \le \frac{1}{e}, \tag{17}$$

then for any positive constant initial function

$$y(t) = \theta(t) \equiv k > 0 \quad \text{on} \quad [-r, 0] \tag{18}$$

the solution $y(t)$ is positive for all $t \ge 0$. This finally provides an answer for Problem 30-1(b).

Of course we already know (by the method of steps) that $y(t)$ is uniquely determined for all $t \ge -r$ by Eqs. (16) and (18), regardless of the size of cr.

Assuming (17), the idea is to compare y with another solution of (16) of the form

$$\tilde{y}(t) = ke^{\lambda t} \quad \text{for} \quad t \ge -r. \tag{19}$$

In order that $\tilde{y}$ be a solution for a constant λ, it is necessary and sufficient that

$$\lambda = -ce^{-\lambda r}. \tag{20}$$

Why? [Equation (20) is the "characteristic equation" for (16).]

Now consider the function defined by $\Delta(u) = u + ce^{-ur}$. Then

$$\Delta(0) = c > 0 \quad \text{and} \quad \Delta\left(-\frac{1}{r}\right) = -\frac{1}{r} + ce \le 0.$$

So there is at least one $\lambda < 0$ satisfying $\Delta(\lambda) = 0$, that is, satisfying Eq. (20).

Using such a value of λ, the function $\tilde{y}$ defined by (19) becomes a positive solution of Eq. (16) with

$$\theta(t) < \tilde{\theta}(t) = ke^{\lambda t} \quad \text{on} \quad [-r, 0) \quad \text{and} \quad \theta(0) = \tilde{\theta}(0) = k > 0.$$

So, by Theorem D,

$$y(t) > \tilde{y}(t) = ke^{\lambda t} \quad \text{for} \quad t > 0$$

as long as $y(t)$ *remains positive.* But $y(t)$ could not become zero without first reaching $\tilde{y}(t)$. Hence

$$y(t) > ke^{\lambda t} > 0 \quad \text{for all} \quad t > 0. \tag{21}$$

PROBLEMS

(In each of the following assume $\beta = \infty$ and $D = \mathbf{R}$.)

1. Show that the equation $x'(t) = [x(t/2)]^{2/3}$ has more than one solution for $t \geq t_0 = 0$ with $x(0) = 0$. [*Hint:* Seek solutions of the form $x(t) = at^p$, where a and p are constants.]

2. Solve $x'(t) = [1 + x^2(t)]x(t/2)$ for $t \geq 1$ if $x(t) = 1$ on $[\frac{1}{2}, 1]$. How far can the solution be continued?

3. Consider the equation $x'(t) = x^2(t - r)$ for $t \geq 0$ with $x(t) = 1$ for $-r \leq t \leq 0$ On one graph display the solutions for $-r \leq t < 1$ when $r = 1$, $r = \frac{1}{2}$, and $r = 0$. What do you suppose the solution would look like if r were a very small positive number?

*4. Consider the equation $x'(t) = [1 + x(t)][x(t - 10^{-6}) + 2x(t - 3)]$ for $t \geq 0$ with $x(t) = 1$ on $[-3, 0]$. How far can the solution be continued?

5. Show, by the method of steps, that Eqs. (5) and (2) do have a unique solution on the entire interval $[-r, \infty)$.

*6. Prove Theorem C as follows: Show that for $0 \leq t < \beta$,

$$\|\mathbf{x}(t) - \tilde{\mathbf{x}}(t)\| \leq \|\theta(0) - \tilde{\theta}(0)\| + \int_0^t K \max_{1 \leq j \leq m} \|\mathbf{x}(s - r_j) - \tilde{\mathbf{x}}(s - r_j)\| \, ds$$

$$\leq v(0) + \int_0^t Kv(s) \, ds,$$

where $v(s) \equiv \sup_{-r \leq \sigma \leq s} \|\mathbf{x}(\sigma) - \tilde{\mathbf{x}}(\sigma)\|$. From this show that

$$v(t) \leq v(0) + \int_0^t Kv(s) \, ds \quad \text{for} \quad 0 \leq t < \beta.$$

Apply Lemma 12-A to complete the proof.

7. By examining $\Delta'(u)$, show that Eq. (20) in Example 4 has *exactly* one solution $\lambda \geq -1/r$.

*8. By treating the three cases listed below, prove that if $r > 0$ and $br \geq -e^{-1+ar}$, then the solution of $y'(t) = ay(t) + by(t-r)$ for $t \geq 0$ with $y(t) = k > 0$ on $[-r, 0]$ remains positive for all $t \geq 0$—generalizing Example 4.
   (a) $b \geq 0$. [*Hint*: Find the delay differential equation for $z(t) = e^{-at}y(t)$.]
   (b) $a + b \geq 0$ with $b < 0$. [*Hint*: Apply Theorem D.]
   (c) $a + b \leq 0$ with $-e^{-1+ar} \leq br < 0$. [*Hint*: Apply Theorem D.]

33. LINEAR EQUATIONS WITH CONSTANT COEFFICIENTS

This section contains three elementary theorems plus some examples illustrating methods of analysis for two of the equations from Section 30. These items suggest methods which work in *some* practical examples.

Here again we shall assume $t_0 = 0$.

Consider the linear system

$$\mathbf{x}'(t) = \sum_{j=1}^{m} A_j \mathbf{x}(t - r_j) + \mathbf{h}(t) \quad \text{for} \quad t \geq 0, \tag{1}$$

where each A_j is a constant $n \times n$ matrix, the r_j's are constants with

$$0 = r_1 < r_2 < \cdots < r_m \equiv r,$$

and $\mathbf{h}: [0, \infty) \to \mathbf{R}^n$ is continuous. The initial condition for Eq. (1) will be

$$\mathbf{x}(t) = \mathbf{\theta}(t) \quad \text{for} \quad -r \leq t \leq 0, \tag{2}$$

where $\mathbf{\theta}: [-r, 0] \to \mathbf{R}^n$ is continuous.

THEOREM A (GLOBAL EXISTENCE)

Equations (1) and (2) have a unique solution on the entire interval $[-r, \infty)$.

Proof. For $0 \leq t \leq r_2$, Eqs. (1) and (2) give

$$\mathbf{x}'(t) = A_1 \mathbf{x}(t) + \sum_{j=2}^{m} A_j \mathbf{\theta}(t - r_j) + \mathbf{h}(t)$$

with $\mathbf{x}(0) = \mathbf{\theta}(0)$. This implies a unique solution on $[0, r_2]$. Why? (Problem 1.)

Now that $\mathbf{x}$ is known on $[-r, r_2]$ we consider the interval $[r_2, 2r_2]$ and again find a unique solution.

Continuation of this method-of-steps argument shows the existence of a unique solution on $[-r, \infty)$. ∎

Actually, Theorem A plus Theorems B and C below are true (and the proofs are unchanged) if each matrix A_j is a continuous function on $[0, \infty)$.

Just as in the study of linear ordinary differential equations, it is helpful to also consider the homogeneous equation associated with (1),

$$\mathbf{y}'(t) = \sum_{j=1}^{m} A_j \mathbf{y}(t - r_j) \quad \text{for} \quad t \geq 0. \tag{3}$$

We shall sometimes talk about a *solution* of system (1) or of system (3) on $[-r, \infty)$ without mentioning the initial function. Then we mean any continuous function mapping $[-r, \infty) \to \mathbf{R}^n$ which satisfies (1) or (3) respectively on $[0, \infty)$.

In these terms, as in the case of ordinary differential systems, we get the following theorem on linear combinations of solutions.

THEOREM B (SUPERPOSITION)

Let $\mathbf{h}(t) = c_1 \mathbf{h}_{(1)}(t) + \cdots + c_k \mathbf{h}_{(k)}(t)$, where $c_1, \ldots, c_k$ are constants and $\mathbf{h}_{(1)}, \ldots, \mathbf{h}_{(k)}$ are given vector-valued functions on $[0, \beta)$. Let $\mathbf{x}_{(p)}$ on $[-r, \beta)$ be any (particular) solution of

$$\mathbf{x}'(t) = \sum_{j=1}^{m} A_j \mathbf{x}(t - r_j) + \mathbf{h}_{(p)}(t)$$

(for each $p = 1, \ldots, k$). Then

$$\mathbf{x} = c_1 \mathbf{x}_{(1)} + \cdots + c_k \mathbf{x}_{(k)}$$

is a solution of Eq. (1) on $[-r, \beta)$. In case some $\mathbf{h}_{(p)}$'s are identically zero, the corresponding $\mathbf{x}_{(p)}$'s can be any solutions of system (3).

Proof. For $0 \leq t < \beta$,

$$\mathbf{x}'(t) = \sum_{p=1}^{k} c_p \mathbf{x}'_{(p)}(t)$$

$$= \sum_{p=1}^{k} \sum_{j=1}^{m} c_p A_j \mathbf{x}_{(p)}(t - r_j) + \sum_{p=1}^{k} c_p \mathbf{h}_{(p)}(t)$$

$$= \sum_{j=1}^{m} A_j \sum_{p=1}^{k} c_p \mathbf{x}_{(p)}(t - r_j) + \mathbf{h}(t)$$

$$= \sum_{j=1}^{m} A_j \mathbf{x}(t - r_j) + \mathbf{h}(t). \quad \blacksquare$$

In the case of an *ordinary* differential system we have a theorem (Corollary 23-E) asserting that the associated homogeneous system has no more than n linearly independent solutions. (In fact, by Corollary 28-B it has exactly n.) Unfortunately, this is not true for system (3). For example, one easily verifies that the scalar equation ($n = 1$)

$$y'(t) = -y\left(t - \frac{\pi}{2}\right)$$

has linearly independent solutions $\cos t$ and $\sin t$. As a matter of fact it can be shown that this equation, like most delay differential equations with constant coefficients, has *infinitely many* independent solutions valid on **R**. So we cannot have a theorem stating that the general solution of Eq. (1) is simply a linear combination of some particular solution of (1) plus n linearly independent solutions of (3).

But how serious a deficiency is this? In many cases we do not need a detailed description of the unique solution in terms of the initial conditions.

For example, in the application of resonance to radio tuning (Section 18) the thing that mattered was the behavior of the solution (the electrical current in the receiver) for large values of t. And this turned out to be independent of the initial conditions because all solutions of the homogeneous equation associated with Eq. 18-(2) approach zero as $t \to \infty$.

If we were studying an equation for the growth of a population we would again be interested primarily in the behavior of solutions for large values of t.

Now for Eqs. (1) and (3) the following is easily proved. (Problem 2.)

THEOREM C

Let $\tilde{x}$ be a solution of (1). Then any other solution of (1) has the form

$$\mathbf{x} = \tilde{\mathbf{x}} + \mathbf{y},$$

where $\mathbf{y}$ is some solution of Eq. (3).

So if, in some example, we can prove that all solutions of Eq. (3) tend to zero as $t \to \infty$, then knowledge of just *one solution* of Eq. (1) says a lot about the behavior of *all solutions* as $t \to \infty$.

TERMINOLOGY

Equation (3) is certainly satisfied by the function $\mathbf{y}(t) = \mathbf{0}$, called the *trivial solution*. If all solutions of Eq. (3) tend to zero (approach the trivial solution) as $t \to \infty$ it is customary to say the trivial solution is *asymptotically stable*.

The first two examples below concern a simple scalar special case of Eq. (3). We first illustrate a method for showing that all solutions are bounded. Then, in Example 2, we show that they actually tend to zero.

The method used in Example 1 (and again in Example 4) was introduced for ordinary differential systems by A. M. Lyapunov in 1892 and was extended to delay differential equations in the 1950s by N. N. Krasovskiĭ [9]. In this method one tries to show that for every solution $\mathbf{y}$ of Eq. (3) there is a non-negative function v on $[0, \infty)$ such that

$$y_1^2(t) + \cdots + y_n^2(t) \leq cv(t), \tag{4}$$

where c is a constant, and

$$v'(t) \leq 0 \tag{5}$$

for $t \geq 0$. Then it follows that $v(t)$ is nonincreasing and hence bounded. Thus $\mathbf{y}(t)$ is bounded.

The whole trick is to find a function v for which one can easily verify conditions (4) and (5).

Actually there is much more to the Lyapunov-Krasovskiĭ theory than this. But the special case described above will suffice for our examples.

In the case of ordinary differential equations, $v(t)$ is often related to "energy." (Problem 3.) For delay differential equations the choice of v requires considerable ingenuity. The reader is referred to Krasovskiĭ's book [9] for guidance.

Example 1. (Krasovskiĭ). If y defined on $[-r, \infty)$ is a solution of the scalar equation

$$y'(t) = ay(t) + by(t - r), \tag{6}$$

let us define a new function v on $[0, \infty)$ by

$$v(t) = y^2(t) + |a| \int_{t-r}^{t} y^2(s)\, ds.$$

Then condition (4) is clearly satisfied with $c = 1$ (and $n = 1$). Moreover,

$$v'(t) = 2y(t)y'(t) + |a|y^2(t) - |a|y^2(t - r)$$
$$= (2a + |a|)y^2(t) + 2by(t)y(t - r) - |a|y^2(t - r).$$

Assume now that

$$a < 0 \quad \text{and} \quad |b| \leq |a|. \tag{7}$$

So $a = -|a|$.

At this point, and in the future, we need the inequality

$$2yz \leq y^2 + z^2, \tag{8}$$

which follows for arbitrary real numbers y and z by multiplying out the obvious inequality $(y - z)^2 \geq 0$.

Applying (7) and (8), we now find for $t \geq 0$

$$v'(t) \leq -|a|y^2(t) + 2|b| \cdot |y(t)| \cdot |y(t - r)| - |a|y^2(t - r)$$
$$\leq (-|a| + |b|)[y^2(t) + y^2(t - r)] \leq 0,$$

verifying condition (5). Thus, under the hypotheses (7), every solution of Eq. (6) is bounded. More specifically, if $y(t) = \theta(t)$ on $[-r, 0]$, then for all $t \geq 0$

$$y(t) \leq [v(t)]^{1/2} \leq \left[\theta^2(0) + |a| \int_{-r}^{0} \theta^2(s)\, ds\right]^{1/2}.$$

Example 2. If conditions (7) are strengthened to

$$a < 0 \quad \text{and} \quad |b| < |a|, \tag{9}$$

the trivial solution of Eq. (6) is asymptotically stable. Here is a simple trick for proving this from the result of Example 1.

If y is a solution of Eq. (6), define $w(t) = e^{\delta t} y(t)$ for some small number $\delta > 0$. Then

$$y(t) = e^{-\delta t} w(t),$$

$$y'(t) = e^{-\delta t} w'(t) - \delta e^{-\delta t} w(t),$$

and

$$y(t - r) = e^{-\delta t} e^{\delta r} w(t - r).$$

Substituting into Eq. (6) we find

$$e^{-\delta t} w'(t) - \delta e^{-\delta t} w(t) = a e^{-\delta t} w(t) + b e^{-\delta t} e^{\delta r} w(t - r)$$

or, dividing out $e^{-\delta t}$,

$$w'(t) = (a + \delta) w(t) + b e^{\delta r} w(t - r). \tag{10}$$

Now let δ be any positive number sufficiently small so that

$$|b| e^{\delta r} < |a| - \delta = |a + \delta|. \tag{11}$$

Then it follows also that $a + \delta = -|a| + \delta < 0$. So one can apply the result of Example 1 to Eq. (10) and conclude that $w(t)$ is bounded, say $|w(t)| \leq c$ for some constant c. Thus

$$|y(t)| \leq c e^{-\delta t} \quad \text{for} \quad t \geq 0. \tag{12}$$

Not only have we proved that $y(t) \to 0$ as $t \to \infty$, but (12) gives an exponential estimate for the rate of decay.

We are now ready to exploit these examples still further by considering a nonhomogeneous equation.

Example 3. Consider the scalar equation

$$x'(t) = -2x(t) + x\left(t - \frac{\pi}{2}\right) + \sin t \quad \text{for} \quad t \geq 0 \tag{13}$$

with

$$x(t) = \theta(t) \quad \text{for} \quad -\frac{\pi}{2} \le t \le 0, \tag{14}$$

where θ is a given continuous function.

Following the idea used for linear ordinary differential equations with constant coefficients, let us seek a particular solution of Eq. (13) of the form

$$\tilde{x}(t) = A \cos t + B \sin t.$$

Substitution into (13) gives

$$-A \sin t + B \cos t = -2A \cos t - 2B \sin t$$

$$+ A \sin t - B \cos t + \sin t.$$

This requires

$$-2A + 2B = 1 \quad \text{and} \quad 2A + 2B = 0,$$

or $A = -\frac{1}{4}$, $B = \frac{1}{4}$. So a solution of (13) is defined by

$$\tilde{x}(t) = -\tfrac{1}{4} \cos t + \tfrac{1}{4} \sin t. \tag{15}$$

Letting x be the unique solution of (13) and (14) on $[-\pi/2, \infty)$, let us subtract off $\tilde{x}$ and consider $y = x - \tilde{x}$. Then y satisfies the homogeneous equation

$$y'(t) = -2y(t) + y\left(t - \frac{\pi}{2}\right) \quad \text{for} \quad t \ge 0$$

with

$$y(t) = \theta(t) + \tfrac{1}{4} \cos t - \tfrac{1}{4} \sin t \quad \text{for} \quad -\frac{\pi}{2} \le t \le 0,$$

and x is given by

$$x(t) = \tilde{x}(t) + y(t) \quad \text{for} \quad t \ge -\frac{\pi}{2}.$$

But Example 2 asserts that $y(t) \to 0$ as $t \to \infty$. So we can regard $\tilde{x}$ as the "steady-state" part of the solution of (13) and (14), and y as the "transient" part which dies out as $t \to \infty$. Note that only the transient part depends on the initial function θ. In other words, Eq. (15) describes the behavior of all solutions for large t, regardless of the initial conditions.

We now present a series of three examples illustrating an analogous procedure for a second-order scalar equation.

Example 4. Consider the equation

$$z''(t) + bz'(t) + qz'(t - r) + kz(t) + pz(t - \tau) = 0 \quad \text{for} \quad t \ge 0, \tag{16}$$

where b, k, q, r, p, and τ are constants with $b > 0$, $k > 0$, $r \geq 0$, and $\tau \geq 0$. We shall show that if

$$p > -k \quad \text{and} \quad |q| + \tau|p| \leq b, \tag{17}$$

then every solution of (16) is bounded for $t \geq 0$. [Note that Eq. (16) includes Eq. 30-(15) if we consider b, k, and q to be playing the roles formerly played by b/m, k/m, and q/m.]

Letting $y_1(t) = z(t)$ and $y_2(t) = z'(t)$, we convert (16) into the system

$$y_1'(t) = y_2(t),$$
$$y_2'(t) = -ky_1(t) - by_2(t) - py_1(t - \tau) - qy_2(t - r). \tag{18}$$

Now assume inequalities (17), and for any solution of (18) define

$$v(t) = (k + p)y_1^2(t) + y_2^2(t) + (b - \tau|p|)\int_{t-r}^{t} y_2^2(s)\, ds$$

$$+ |p|\int_{-\tau}^{0}\int_{t+\sigma}^{t} y_2^2(u)\, du\, d\sigma.$$

(This function was composed from two examples in Krasovskiĭ's book.) Clearly

$$v(t) \geq \min\{k + p, 1\}[y_1^2(t) + y_2^2(t)].$$

So condition (4) is satisfied. We must also compute

$$v'(t) = 2py_1(t)y_2(t) - 2by_2^2(t) - 2py_2(t)y_1(t - \tau)$$
$$- 2qy_2(t)y_2(t - r) + (b - \tau|p|)[y_2^2(t) - y_2^2(t - r)]$$
$$+ |p|\int_{-\tau}^{0} [y_2^2(t) - y_2^2(t + \sigma)]\, d\sigma.$$

Now for $t \geq \tau$ [or for $t \geq 0$ if we demand that $y_1'(t) = y_2(t)$ on $[-\tau, \infty)$]

$$y_1(t) - y_1(t - \tau) = \int_{t-\tau}^{t} y_2(s)\, ds.$$

So

$$v'(t) = 2p\int_{t-\tau}^{t} y_2(t)y_2(s)\, ds - (b + \tau|p|)y_2^2(t)$$

$$- 2qy_2(t)y_2(t - r) - (b - \tau|p|)y_2^2(t - r)$$

$$+ |p|\tau y_2^2(t) - |p|\int_{t-\tau}^{t} y_2^2(s)\, ds.$$

Next apply inequality (8) to $2y_2(t)y_2(s)$ and $2y_2(t)y_2(t - r)$ to conclude that

$$v'(t) \leq [-b + |q| + \tau|p|][y_2^2(t) + y_2^2(t - r)] \leq 0$$

for $t \geq \tau$. The fact that this inequality was proved only for $t \geq \tau$, and not for $t \geq 0$ as assumed in (5), clearly does not alter the fact that $v(t)$ is bounded, and hence so is $z(t) = y_1(t)$.

Example 5. Returning now to the simpler Eq. 30-(15),

$$z''(t) + bz'(t) + qz'(t - r) + kz(t) = 0 \quad \text{for} \quad t \geq 0, \tag{19}$$

where b, k, q, and r are constants with $b > 0$, $k > 0$ and $r \geq 0$, we shall show that the trivial solution is asymptotically stable if

$$|q| < b. \tag{20}$$

As in Example 2, if z is a solution of Eq. (19), let $w(t) = e^{\delta t}z(t)$ where $\delta > 0$. Then $z(t) = e^{-\delta t}w(t)$ and the reader should show by substitution that

$$w''(t) + (b - 2\delta)w'(t) + qe^{\delta r}w'(t - r) + (k - \delta b + \delta^2)w(t) - \delta qe^{\delta r}w(t - r) = 0 \tag{21}$$

for $t \geq 0$. (Problem 4.) This is a special case of Eq. (16), and the conditions of Example 4 will be fulfilled if

$$b - 2\delta > 0, \quad k - \delta b + \delta^2 > 0, \quad \delta qe^{\delta r} < k - \delta b + \delta^2,$$

$$\text{and} \quad |q|e^{\delta r} + r\delta|q|e^{\delta r} \leq b - 2\delta. \tag{22}$$

Thanks to the strict inequalities $b > 0$, $k > 0$, and $|q| < b$, all the conditions in (22) can be met by making δ a sufficiently small positive number. Then it follows that $w(t)$ is bounded, say $|w(t)| \leq c$, so that

$$|z(t)| \leq ce^{-\delta t} \to 0 \quad \text{as} \quad t \to \infty.$$

It should be noted that the conditions given in Examples 1, 2, 4, and 5 for solutions to be bounded or to approach zero as $t \to \infty$ are all sufficient conditions. They are not necessary conditions. Other criteria can be found in [1], [4], [6], [7], and [9].

Example 6. Let us try to analyze the behavior of solutions of

$$x''(t) + x'(t) + x'(t - 1) + \pi^2 x(t) = \cos \pi t \quad \text{for} \quad t \geq 0 \tag{23}$$

with

$$x(t) = \theta(t) \quad \text{for} \quad -1 \leq t \leq 0, \tag{24}$$

where θ is a given continuously differentiable function.

It is easy to verify directly that if x and $\tilde{x}$ are any two solutions of Eq. (23), then $z = x - \tilde{x}$ is a solution of the homogeneous equation

$$z''(t) + z'(t) + z'(t - 1) + \pi^2 z(t) = 0 \quad \text{for} \quad t \geq 0. \tag{25}$$

Now the coefficients in Eq. (25) satisfy conditions (17) but not (20). So we can only conclude that z is bounded. In other words, if we find any particular solution $\tilde{x}$ of Eq. (23) and if x is the unique solution of Eqs. (23) and (24), then

$$x(t) = \tilde{x}(t) + z(t) \quad \text{for} \quad t \geq 0, \tag{27}$$

where $z(t)$ is bounded.

In an attempt to find a particular solution of (23) it is natural to try $\tilde{x}(t) = A \cos \pi t + B \sin \pi t$. But, as the reader should verify, this function satisfies the homogeneous equation (25), and hence cannot satisfy (23). A little reflection on similar cases in the study of ordinary differential equations might lead one to try

$$\tilde{x}(t) = At \cos \pi t + Bt \sin \pi t. \tag{28}$$

Carry out the substitution and you should find that this works with $A = 0$ and $B = 1/3\pi$. So the unique solution of Eqs. (23) and (24) has the form

$$x(t) = \frac{1}{3\pi} t \sin \pi t + z(t)$$

where $z(t)$ is bounded.

This would be an unsatisfactory situation if, as in Eq. 30-(15), (25) represented the equation for the angle of roll of a ship. For then, if the ship were riding in periodic waves which provided a forcing function $\cos \pi t$, regardless of the initial conditions, the ship would begin oscillating strongly and would soon be upside down. (Of course, the mathematical model would not apply for large oscillations.)

We conclude with two examples of what not to do. Occasionally one encounters a published book or research paper in which the author proposes to analyze a delay differential equation by replacing $x(t - r)$ with the first few terms of a Taylor's (or Maclaurin) series, say

$$x(t) - rx'(t) + \frac{1}{2} r^2 x''(t) - \cdots + (-1)^m \frac{1}{m!} r^m x^{(m)}(t).$$

Beware of such papers!

Example 7. Every solution of

$$x'(t) = -2x(t) + x(t - r)$$

is bounded and tends to zero as $t \to \infty$. However, replacement of $x(t - r)$ with $x(t) - rx'(t) + \frac{1}{2}r^2 x''(t)$ produces the ordinary differential equation

$$x'(t) = -2x(t) + x(t) - rx'(t) + \frac{1}{2}r^2 x''(t).$$

The latter has exponentially increasing solutions, $ce^{\lambda t}$ with $\lambda > 0$, regardless of how small the value of $r > 0$. (Problem 9.)

Example 8. Every solution of

$$x'(t) = -3x(t) - 2x(t-1)$$

is bounded and tends to zero as $t \to \infty$. But the ordinary differential equation obtained by replacing $x(t-1)$ by $x(t) - x'(t)$ has exponentially increasing solutions $x(t) = ce^{5t}$. (Problem 10).

Despite Examples 7 and 8, it should be remarked that for some equations with "sufficiently small" delay one does get useful approximations by replacing $x(t-r)$ by $x(t)$ or by $x(t) - rx'(t)$. See, for example, the *American Mathematical Monthly*, **80**, 990–995 (1973).

PROBLEMS

1. In the proof of Theorem A, what theorems should you invoke to prove existence and uniqueness on $[0, r_2]$?
2. Prove Theorem C.
*3. In Example 28-3, for the nonlinear equation of a simple pendulum, we defined $v(t) = x_2^2(t)/2 + (g/l)[1 - \cos x_1(t)]$ which was essentially the energy in the pendulum. We showed that $v'(t) \leq 0$. Now show that condition (4) is also satisfied if $|x_1(t)| \leq \pi/2$. [*Hint:* Assume $|\xi| \leq \pi/2$ and show that $|\sin \xi| \geq (2/\pi)|\xi|$. Then use the fact that $1 - \cos \xi = \sin^2 \xi/(1 + \cos \xi) \geq \frac{1}{2}\sin^2 \xi$.]
4. Derive Eq. (21).
5. Show that the trivial solution of Eq. (16) is asymptotically stable if $p > -k$ and $|q| + \tau|p| < b$.
6. Use your imagination (and analogy to ordinary differential equations) to find a particular solution, $\tilde{x}$ on $[-r, \infty)$, of each of the following scalar equations.

   (a) $x'(t) + 2x(t) - x\left(t - \dfrac{\pi}{4}\right) = \sin t$.

   (b) $x'(t) + x\left(t - \dfrac{\pi}{2}\right) = \sin t$.

   (c) $x'(t) = -2x(t) + x(t-1) + t$.
   (d) $x'(t) = -x(t) + x(t-1) + t$.
   (e) $x'(t) = x(t) - x(t-1) + t$.

7. Describe (if possible) the behavior as $t \to \infty$ of all solutions of

$$x''(t) + 2x'(t) + x'\left(t - \dfrac{\pi}{3}\right) + x(t) = \cos t.$$

*8. Consider the equation

$$y''(t) - p(t)y(t-r) = 0, \quad \text{for} \quad t \geq 0. \tag{29}$$

where $p(t) \geq 0$ is continuous and $r \geq 0$ is constant. This equation may or may not have nontrivial bounded solutions. For example, $y''(t) - y(t - \pi) = 0$ has nontrivial bounded solutions $c_1 \cos t + c_2 \sin t$, while $y''(t) - y(t) = 0$ does not. Argue, nevertheless, that in any case if Eq. (29) represented a real-world problem, the bounded solutions would never be encountered. [*Hint:* If y is any bounded solution of Eq. (29), show that the addition of an *arbitrarily small* error $e(t) > 0$ with $e'(0_+) > 0$ to the initial function $\theta(t)$ on $[-r, 0]$ would lead to an unbounded solution.]

9. Verify the assertions of Example 7.
10. Verify the assertions of Example 8.

Chapter Seven

POWER SERIES SOLUTION OF LINEAR ORDINARY EQUATIONS

We return now to the study of linear nth order ordinary differential equations.

The methods of Chapter Three do not by any means give the solution of every such equation. The only reason we were able to solve so many equations in Chapter Three is that the equations were especially chosen so as to be amenable to our simple methods. If one considers a linear equation of the form

$$x^{(n)} + a_{n-1}(t)x^{(n-1)} + \cdots + a_1(t)x' + a_0(t)x = h(t),$$

with arbitrarily chosen continuous coefficients, one must be prepared to accept a less tidy form for the solution than we have encountered thus far. In general, one either seeks a solution in the form of an infinite series or else accepts numerical approximations to the solution. The present chapter introduces the series method.

In principle, we can always solve the linear equation (by variation of parameters) if we know n linearly independent solutions of the associated homogeneous equation. Thus we shall concentrate our attention on the latter.

34. CONVERGENCE AND ALGEBRA OF POWER SERIES

To motivate the theoretical discussion which will follow we shall first present a "formal" calculation using power series to solve a simple differential equation—one for which we already have a much easier method. In other words, we will perform various calculations without worrying about their mathematical justification.

Example 1 (Formal computations). Solve

$$x'' - 4x = 0 \quad \text{with} \quad x(0) = 1, \quad x'(0) = 2.$$

We shall seek a solution in the form of an infinite series

$$x(t) = \sum_{k=0}^{\infty} c_k t^k = c_0 + c_1 t + c_2 t^2 + \cdots,$$

where $c_0, c_1, c_2, \ldots$ are constants.

Assuming such a series can be differentiated term by term we find

$$x'(t) = \sum_{k=1}^{\infty} c_k k t^{k-1} \quad \text{and} \quad x''(t) = \sum_{k=2}^{\infty} c_k k(k-1) t^{k-2}.$$

Substitution into the differential equation yields

$$\sum_{k=2}^{\infty} c_k k(k-1) t^{k-2} - 4 \sum_{k=0}^{\infty} c_k t^k = 0,$$

or, letting $j = k - 2$ in the first of these series,

$$\sum_{j=0}^{\infty} c_{j+2}(j+2)(j+1) t^j - 4 \sum_{k=0}^{\infty} c_k t^k = 0.$$

Since j is only a " dummy index " we can now relabel it as we please. Let us replace j by k. Then, *assuming* we can algebraically combine the two series, we have

$$\sum_{k=0}^{\infty} [c_{k+2}(k+2)(k+1) - 4c_k] t^k = 0.$$

Clearly a sufficient condition for this to hold is that each coefficient be zero, that is,

$$c_{k+2}(k+2)(k+1) - 4c_k = 0 \quad \text{for} \quad k = 0, 1, 2, \ldots.$$

(It is true, but not obvious, that this is also a necessary condition.) What we have obtained is a " recursion relation,"

$$c_{k+2} = \frac{4}{(k+2)(k+1)} c_k \quad \text{for} \quad k = 0, 1, 2, \ldots,$$

which enables us to determine all the c_k's if c_0 and c_1 are specified. Indeed we find

$$c_2 = \frac{4}{2 \cdot 1} c_0, \quad c_4 = \frac{4}{4 \cdot 3} c_2 = \frac{4^2}{4!} c_0, \quad \ldots, \quad c_{2j} = \frac{4^j}{(2j)!} c_0, \quad \ldots$$

and

$$c_3 = \frac{4}{3 \cdot 2} c_1, \quad c_5 = \frac{4}{5 \cdot 4} c_3 = \frac{4^2}{5!} c_1, \quad \ldots, \quad c_{2j+1} = \frac{4^j}{(2j+1)!} c_1, \quad \ldots.$$

But we have not yet applied the initial conditions. These give

$$c_0 = x(0) = 1 \quad \text{and} \quad c_1 = x'(0) = 2.$$

Thus our series becomes

$$x(t) = 1 + 2t + \frac{4}{2!} t^2 + \frac{4}{3!} \cdot 2t^3 + \frac{4^2}{4!} t^4 + \frac{4^2}{5!} \cdot 2t^5 + \cdots = \sum_{k=0}^{\infty} \frac{(2t)^k}{k!}.$$

Perhaps the reader will recognize this as the Maclaurin series for e^{2t}. And indeed $x(t) = e^{2t}$ is the unique solution we would have obtained using the easier method of Chapter Three.

We presented this example, not because we will be using series methods to solve such simple equations, but in order to illustrate the computations involved and to raise a number of questions about power series: When does a power series converge? Is the sum of a convergent power series a continuous function? Is it differentiable? Can we differentiate termwise? When does the differentiated series converge? Are we justified in algebraically combining the various series after substituting into the differential equation? If a power series sums to zero does it follow that every term is zero? Can we always expect to recognize a power series solution of a differential equation as a familiar function (such as e^{2t})?

In order to answer these questions we shall use some elementary properties of series of constants. The reader is assumed to be familiar with the definition of convergence and with the basic properties of series of constants. The assumed prerequisites are summarized in Theorem A2-H.

We shall discuss *power series* of the form

$$\sum_{k=0}^{\infty} c_k(t - t_0)^k = c_0 + c_1(t - t_0) + c_2(t - t_0)^2 + \cdots, \tag{1}$$

where $t_0, c_0, c_1, c_2, \ldots$ are constants (real or complex). (If the basic convergence properties and the algebra of power series are also familiar, then proceed to Section 35.)

Remarks. The symbol $(t - t_0)^0$ will always be interpreted as 1. Thus series (1) must always converge to c_0 when $t = t_0$.

We shall often consider the special case of (1) with $t_0 = 0$.

Even though we are interested in real solutions of differential equations with real coefficients and real initial conditions, it will often be helpful to admit complex numbers in the series (1). Moreover, this does not make the statements or proofs of the theorems any more difficult. The definitions and theorems listed in Appendix 2 for series of constants remain valid for complex series.

LEMMA A

If the power series (1) converges when $t = t_1$ for some constant $t_1 \neq t_0$, then (1) converges absolutely whenever $|t - t_0| < |t_1 - t_0|$. If (1) diverges when $t = t_2$, then (1) also diverges whenever $|t - t_0| > |t_2 - t_0|$.

Proof. If (1) converges when $t = t_1 \neq t_0$, then we conclude that $\lim_{k \to \infty} c_k(t_1 - t_0)^k = 0$. Thus there exists K such that

$$|c_k(t_1 - t_0)^k| \leq 1 \quad \text{for all} \quad k \geq K.$$

Hence if $|t - t_0| < |t_1 - t_0|$ and $k \geq K$,

$$|c_k(t - t_0)^k| = |c_k(t_1 - t_0)^k| \left| \frac{t - t_0}{t_1 - t_0} \right|^k \leq \left(\frac{|t - t_0|}{|t_1 - t_0|} \right)^k.$$

It follows, by comparison with the geometric series $\sum_{k=0}^{\infty} r^k$ where $r = |t - t_0|/|t_1 - t_0| < 1$, that the series $\sum_{k=0}^{\infty} c_k(t - t_0)^k$ converges.

Now assume that (1) diverges when $t = t_2$. Then (1) cannot converge if $|t - t_0| > |t_2 - t_0|$ without contradicting the first assertion of the lemma. ∎

Example 2. We know that

$$\frac{1}{1 - t} = \sum_{k=0}^{\infty} t^k \quad \text{whenever} \quad |t| < 1.$$

Moreover, since $(1 - t)^{-1}$ is undefined when $t = 1$, we can naturally suspect that the series $\sum_{k=0}^{\infty} t^k$ will diverge at $t = 1$. This is easily verified. But then it follows from Lemma A that $\sum_{k=0}^{\infty} t^k$ must also diverge, for example, when $t = -2$ or $t = 1 + i$, even though $(1 - t)^{-1}$ is well defined for those values of t.

THEOREM B (RADIUS OF CONVERGENCE)

For every power series (1) there exists a unique $R \in [0, \infty]$ (called the *radius of convergence* of the series) such that (1) converges absolutely whenever $|t - t_0| < R$ and diverges whenever $|t - t_0| > R$. (This theorem does not say what happens when $|t - t_0| = R$.)

*Proof. We define R to be the least upper bound of the values of $|t_1 - t_0|$ such that (1) converges for $t = t_1$. In symbols,

$$R \equiv \sup \left\{ |t_1 - t_0| : \sum_{k=0}^{\infty} c_k(t_1 - t_0)^k \text{ converges} \right\}.$$

Then if $|t - t_0| > R$ it follows, from the definition of R, that (1) diverges. And if $|t - t_0| < R$, then, by definition of supremum, there must exist *some* t_1 such that $|t - t_0| < |t_1 - t_0| < R$ and (1) converges at t_1. Thus, by Lemma A, (1) converges at t.

To prove the uniqueness of R let us suppose (for contradiction) that there are two numbers R_1 and R_2 such that $0 \leq R_1 < R_2 \leq \infty$ and both R_1 and R_2 have the claimed properties. Then consider $t \equiv t_0 + (R_1 + R_2)/2$. We find that $R_1 < |t - t_0| < R_2$ which means that (1) must converge and diverge at t—a contradiction. ∎

We remark again that everything which has been said and proved is valid for complex numbers as well as real.

When dealing with real values of t the possibilities permitted by Theorem B are as follows:

(i) If $R = 0$, series (1) converges only for $t = t_0$.
(ii) If $0 < R < \infty$, (1) converges for $t_0 - R < t < t_0 + R$ and diverges when $t < t_0 - R$ or $t > t_0 + R$. The interval $(t_0 - R, t_0 + R)$ is called the *interval of convergence*. The series may converge or it may diverge at the endpoints, $t_0 - R$ and $t_0 + R$.
(iii) If $R = \infty$, (1) converges for all real t.

When t is considered complex the three cases are as follows:

(i) If $R = 0$, series (1) converges only for $t = t_0$.
(ii) If $0 < R < \infty$, (1) converges inside the circle of radius R with center at t_0 and diverges outside that circle. The circle $\{t \in \mathbf{C} : |t - t_0| = R\}$ is called the *circle of convergence*. The series may either converge or diverge at points on that circle, that is, when $|t - t_0| = R$.
(iii) If $R = \infty$, (1) converges for all t in the complex plane.

Example 3. Suppose we could find a series of the form (1) with $t_0 = 1$ which (when it converges) represents the function $1/(t^2 - 2t + 5)$. Then we conclude that this series must diverge for all $t > 3$ and all $t < -1$ or, more generally, for all complex t such that $|t - 1| > 2$. Why?

The conclusion that $R \le 2$ follows from the fact that the function $1/(t^2 - 2t + 5)$ itself is undefined at $t = 1 \pm 2i$ and the distances $|(1 + 2i) - 1|$ and $|(1 - 2i) - 1|$ each equal 2 (Figure 1). (We might say that the misbehavior of the series at $1 \pm 2i$ is the "cause" of its divergence when, say, $t < -1$.)

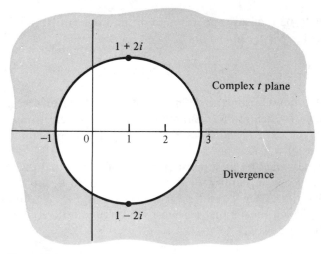

FIGURE 1

As a matter of fact we can easily and explicitly find the required series in this example using our knowledge of geometric series:

$$\frac{1}{t^2 - 2t + 5} = \frac{1}{4 + (t-1)^2} = \frac{1/4}{1 + [(t-1)/2]^2} = \sum_{j=0}^{\infty} \frac{(-1)^j}{4} \left(\frac{t-1}{2}\right)^{2j}.$$

One can now verify directly that this series converges when $|t - 1| < 2$ and diverges when $|t - 1| > 2$ so that the radius of convergence *is* $R = 2$. [Problem 1(b).]

Example 4. Returning to the function $(1 - t)^{-1}$ of Example 2, suppose this is represented as a power series in powers of $(t + 2)$ instead of powers of t; that is, suppose

$$\frac{1}{1 - t} = \sum_{k=0}^{\infty} c_k(t + 2)^k$$

whenever this series converges. Then we must have $R \leq 3$ because the series cannot converge at $t = 1$.

Direct computation gives

$$\frac{1}{1-t} = \frac{1}{3 - (t+2)} = \frac{1/3}{1 - (t+2)/3} = \sum_{k=0}^{\infty} \frac{1}{3} \left(\frac{t+2}{3}\right)^k$$

from which we find $R = 3$. [Problem 1(c).]

The most useful simple tool for determining R for a given power series is the ratio test.

Example 5. Determine the radius of convergence for

$$\sum_{k=0}^{\infty} \frac{2^k}{k(k+1)} (t - \pi)^k.$$

This is $\sum_{k=0}^{\infty} a_k$, where

$$a_k = \frac{2^k(t - \pi)^k}{k(k+1)} \quad \text{and} \quad a_{k+1} = \frac{2^{k+1}(t - \pi)^{k+1}}{(k+1)(k+2)}.$$

To apply the ratio test we compute

$$\lim_{k \to \infty} \frac{|a_{k+1}|}{|a_k|} = \lim_{k \to \infty} \frac{2|t - \pi|k}{k+2} = \lim_{k \to \infty} \frac{2|t - \pi|}{1 + 2/k} = 2|t - \pi|.$$

The series converges when $2|t - \pi| < 1$ and diverges when $2|t - \pi| > 1$. Thus $R = \frac{1}{2}$.

PROBLEMS

1. Find the best condition of the form $|t - t_0| < R$ which assures convergence of the following power series.

(a) $\displaystyle\sum_{k=0}^{\infty} \frac{(2t)^k}{k!}$ (from Example 1).

(b) $\displaystyle\sum_{j=0}^{\infty} \frac{(-1)^j}{4} \left(\frac{t-1}{2}\right)^{2j}$ (from Example 3).

(c) $\displaystyle\sum_{k=0}^{\infty} \frac{1}{3} \left(\frac{t+2}{3}\right)^k$ (from Example 4).

(d) $\displaystyle\sum_{k=0}^{\infty} \frac{k!}{(2k)!} t^k$.

(e) $\displaystyle\sum_{j=0}^{\infty} \frac{1 \cdot 3 \cdot 5 \cdots (2j-3)}{2^j \cdot j!} t^{2j+1}$.

2. Find power series which converge to the following expressions. Use the form (1) with t_0 as indicated, and find a region of convergence

(a) $\dfrac{1}{1+t}$, $t_0 = 0$.

(b) $\dfrac{1}{1+t}$, $t_0 = 1$.

(c) $\dfrac{1}{1+t^2}$, $t_0 = 0$.

(d) $\dfrac{t+1}{t^2 - 3t + 2}$, $t_0 = 0$.

$$\left[Hint:\ \text{First expand in partial fractions } \frac{A}{2-t} + \frac{B}{1-t} \cdot \right]$$

35. CALCULUS OPERATIONS ON POWER SERIES

If a power series in powers of $t - t_0$ has radius of convergence $R > 0$, then it represents some function of t for $|t - t_0| < R$. Let

$$f(t) \equiv \sum_{k=0}^{\infty} c_k(t - t_0)^k \tag{1}$$

wherever the series converges.

We shall see that this function f is continuous and that it can be integrated and differentiated termwise. That is, one can integrate or differentiate $f(t)$ by performing the corresponding integration or differentiation on the individual terms of the series as long as $t_0 - R < t < t_0 + R$.

The ability to differentiate a power series term by term is essential for the series solution of differential equations. But the proof that this is justified is seldom presented in a first course on calculus. So this and some related proofs are given below. A full understanding of this material may take too much time for the reader who has not studied "advanced calculus." Thus you may prefer to pass over Lemma A and simply accept Theorems B and C without proof.

*LEMMA A (UNIFORM CONVERGENCE)

Let the series (1) converge when $|t - t_0| < R$ for some $R \in (0, \infty]$, and consider the partial sums

$$f_m(t) \equiv \sum_{k=0}^{m} c_k(t - t_0)^k.$$

Then, for any $R_1 \in (0, R)$, the sequence $\{f_m\}$ converges uniformly to f on $[t_0 - R_1, t_0 + R_1]$. (See Appendix 2 for the definition of uniform convergence.)

Proof. Choose any $R_2 \in (R_1, R)$ (Figure 1). Then, since (1) converges at $t = t_0 + R_2$, it follows that $\lim_{k \to \infty} |c_k| R_2^k = 0$. Thus there exists a number K such that $|c_k| R_2^k \leq 1$ for all $k \geq K$. So, if we define

$$B = \max \{|c_0|, |c_1| R_2, \ldots, |c_{K-1}| R^{K-1}, 1\},$$

then $|c_k| R_2^k \leq B$ for all $k = 0, 1, 2, \ldots$.

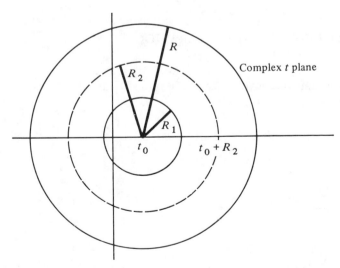

FIGURE 1

Now let $|t - t_0| \leq R_1$ and $p > m \geq 0$. Then

$$|f_p(t) - f_m(t)| = \left| \sum_{k=m+1}^{p} c_k(t - t_0)^k \right| \leq \sum_{k=m+1}^{p} |c_k| R_1^k$$

$$\leq \sum_{k=m+1}^{p} B\left(\frac{R_1}{R_2}\right)^k < \frac{B(R_1/R_2)^{m+1}}{1 - R_1/R_2}.$$

Since $R_1 < R_2$, it follows that for each $\varepsilon > 0$ there exists M so large that $B(R_1/R_2)^{M+1}(1 - R_1/R_2)^{-1} < \varepsilon$. So, letting $p \to \infty$, we find

$$|f(t) - f_m(t)| < \varepsilon \quad \text{whenever} \quad |t - t_0| \leq R_1 \quad \text{and} \quad m \geq M. \quad \blacksquare$$

Remark. Lemma A and the following results are phrased for real t since this will be sufficient for our needs. With the addition of appropriate definitions, the results can be shown to remain true when t is complex.

THEOREM B (CONTINUITY OF THE SUM OF A POWER SERIES)

If series (1) converges whenever $|t - t_0| < R$ for some $R \in (0, \infty]$, then f is continuous on $(t_0 - R, t_0 + R)$.

*Proof. It will suffice to establish continuity at an arbitrary but fixed point $t_1 \in (t_0 - R, t_0 + R)$. For any such t_1 we can choose $R_1 \in (0, R)$ such that $t_1 \in (t_0 - R_1, t_0 + R_1)$. Now $\{f_m\}$ is a sequence of continuous functions (polynomials in fact) which converges uniformly to f on $[t_0 - R_1, t_0 + R_1]$. So the continuity of f at t_1 follows from Theorem A2-I(i). $\quad \blacksquare$

THEOREM C (TERM-BY-TERM INTEGRATION AND DIFFERENTIATION)

If series (1) has radius of convergence $R > 0$ and if t_1 and t are any two real numbers in $(t_0 - R, t_0 + R)$, then

$$\int_{t_1}^{t} f(s)\, ds = \sum_{k=0}^{\infty} \int_{t_1}^{t} c_k(s - t_0)^k\, ds. \tag{2}$$

In particular,

$$\int_{t_0}^{t} f(s)\, ds = \sum_{k=0}^{\infty} \frac{c_k}{k + 1}(t - t_0)^{k+1} \quad \text{for} \quad t_0 - R < t < t_0 + R. \tag{3}$$

Also

$$f'(t) = \sum_{k=1}^{\infty} k c_k(t - t_0)^{k-1} \quad \text{for} \quad t_0 - R < t < t_0 + R. \tag{4}$$

Remarks. The above assertions carry the implication that the new series,

$$\sum_{k=0}^{\infty} \frac{c_k}{k+1}(t-t_0)^{k+1} \quad \text{and} \quad \sum_{k=1}^{\infty} kc_k(t-t_0)^{k-1},$$

also converge whenever $|t - t_0| < R$. Note that we can then respectively differentiate and integrate these new series term by term to recover (1). Thus, from Theorem C, we also conclude that the new series must diverge whenever $|t - t_0| > R$. In other words, the result of term-by-term integration or differentiation of (1) is a new power series having *exactly the same radius of convergence* as (1). Note that we have not been making any general assertions about convergence when $|t - t_0| = R$.

*__Proof of Theorem C.__ Let t_1, $t \in (t_0 - R, t_0 + R)$ and choose $R_1 \in (0, R)$ such that

$$t_1, t \in (t_0 - R_1, t_0 + R_1).$$

Then $\{f_m\}$ converges uniformly to f on $[t_0 - R_1, t_0 + R_1]$ and, by Theorem A2-I(ii),

$$\int_{t_1}^{t} f(s)\, ds = \lim_{m \to \infty} \int_{t_1}^{t} f_m(s)\, ds = \lim_{m \to \infty} \sum_{k=0}^{m} \int_{t_1}^{t} c_k(s - t_0)^k\, ds,$$

which is equivalent to (2).

Specialization to the case $t_1 = t_0$ easily gives (3).

Now consider the possibility of term-by-term differentiation. As in the Proof of Lemma A, the convergence of (1) at $t_0 + R_1$ provides the existence of a number, B, such that $|c_k| \le BR_1^{-k}$ for all $k = 0, 1, 2, \ldots$. In addition to (1), let us now consider the series $\sum_{k=1}^{\infty} kc_k(s - t_0)^{k-1}$. We shall prove, in three easy steps, that this too converges whenever $|s - t_0| < R_1$.

(i) For all $k = 1, 2, \ldots$, $|kc_k(s - t_0)^{k-1}| \le k \dfrac{B}{R_1} \left(\dfrac{|s - t_0|}{R_1} \right)^{k-1}$.

(ii) $\displaystyle\sum_{k=1}^{\infty} \frac{B}{R_1} k \left(\frac{|s - t_0|}{R_1} \right)^{k-1}$ converges by the ratio test, since

$$\lim_{k \to \infty} \left(\frac{k+1}{k} \frac{|s - t_0|}{R_1} \right) = \frac{|s - t_0|}{R_1} < 1.$$

(iii) By the comparison test, $\sum_{k=1}^{\infty} kc_k(s - t_0)^{k-1}$ converges.

We now define $g(s) = \sum_{k=1}^{\infty} kc_k(s - t_0)^{k-1}$ for $|s - t_0| < R_1$. Then by the first part of this theorem we find

$$\int_{t_0}^{t} g(s)\, ds = \sum_{k=1}^{\infty} c_k(t - t_0)^k = f(t) - c_0.$$

Differentiation then gives $f'(t) = g(t)$—the desired result. ∎

Example 1. Starting with the elementary result

$$\frac{1}{1-s} = \sum_{k=0}^{\infty} s^k \quad \text{for} \quad |s| < 1,$$

we can integrate term by term from 0 to t, and immediately obtain the more interesting series

$$\ln(1-t) = -\sum_{k=0}^{\infty} \frac{t^{k+1}}{k+1} \quad \text{for} \quad -1 < t < 1.$$

Example 2. Term-by-term differentiation of $(1-t)^{-1} = \sum_{k=0}^{\infty} t^k$ gives

$$\frac{1}{(1-t)^2} = \sum_{k=1}^{\infty} k t^{k-1} \quad \text{for} \quad -1 < t < 1.$$

COROLLARY D

If series (1) has radius of convergence $R > 0$, then f has derivatives of all orders on $(t_0 - R, t_0 + R)$ and these can be computed by repeated application of Theorem C. Thus, for $t_0 - R < t < t_0 + R$,

$$f''(t) = \sum_{k=2}^{\infty} k(k-1)c_k(t-t_0)^{k-2},$$

$$f^{(3)}(t) = \sum_{k=3}^{\infty} k(k-1)(k-2)c_k(t-t_0)^{k-3},$$

and in general, for $p = 1, 2, \ldots,$

$$f^{(p)}(t) = \sum_{k=p}^{\infty} k(k-1)\cdots(k-p+1)c_k(t-t_0)^{k-p}. \tag{5}$$

The proof consists of a straightforward mathematical induction using the result of Theorem C.

THEOREM E

If, for some $R > 0$,

$$f(t) = \sum_{k=0}^{\infty} c_k(t-t_0)^k \quad \text{for} \quad t_0 - R < t < t_0 + R,$$

then $c_k = f^{(k)}(t_0)/k!$ for $k = 0, 1, 2, \ldots$. (We understand $f^{(0)} = f$ and $0! = 1$.)

Proof. Substitution of $t = t_0$ in the series (1) yields $f(t_0) = c_0$. Substitution of $t = t_0$ in the series (5) yields

$$f^{(p)}(t_0) = p! c_p \quad \text{for} \quad p = 1, 2, \ldots. \quad \blacksquare$$

Remarks. The reader will probably recognize the result,

$$f(t) = \sum_{k=0}^{\infty} \frac{f^{(k)}(t_0)}{k!} (t - t_0)^k, \tag{6}$$

as a Taylor's series. The related result, Taylor's theorem with remainder is stated in Appendix 2, Theorem B. It is also pointed out in a remark following that theorem that not every function f can be represented by a Taylor's series even though f may be infinitely differentiable.

COROLLARY F

If, for some $R > 0$,

$$\sum_{k=0}^{\infty} c_k(t - t_0)^k = \sum_{k=0}^{\infty} d_k(t - t_0)^k \quad \text{for} \quad t_0 - R < t < t_0 + R,$$

then $c_k = d_k$ for $k = 0, 1, 2, \ldots$. In particular if series (1) converges to zero with some positive radius of convergence, then every coefficient $c_k = 0$.

Proof. Apply Theorem E. ■

DEFINITION

Let a function f be given. If, for some $R > 0$,

$$f(t) = \sum_{k=0}^{\infty} c_k(t - t_0)^k \quad \text{for} \quad t_0 - R < t < t_0 + R,$$

then we will say f is *analytic* at t_0.

Example 3. If $f(t) \equiv \sin t$ for all real t, then f is analytic at 0. (Actually f is analytic at every point t_0.) This is proved as follows.

For all real t and all $m = 1, 2, \ldots$, Taylor's theorem (Theorem A2-B) gives

$$f(t) = \sum_{k=0}^{m-1} \frac{f^{(k)}(0)}{k!} t^k + \frac{f^{(m)}(\theta)}{m!} t^m$$

for some θ between 0 and t. Since $|f^{(m)}(\theta)| = $ either $|\sin \theta|$ or $|\cos \theta|$ we conclude that

$$\left| f(t) - \sum_{k=0}^{m-1} \frac{f^{(k)}(0)}{k!} t^k \right| \leq \frac{|t|^m}{m!}.$$

For each fixed t, $\lim_{m \to \infty} |t|^m/m! = 0$. Why? So it follows that

$$\sin t = f(t) = \sum_{k=0}^{\infty} \frac{f^{(k)}(0)}{k!} t^k = \sum_{j=0}^{\infty} \frac{(-1)^j}{(2j + 1)!} t^{2j+1}$$

for all t.

PROBLEMS

1. Integrate the geometric series

$$\frac{1}{1+t^2} = \sum_{j=0}^{\infty} (-1)^j t^{2j} \quad \text{for} \quad -1 < t < 1,$$

obtained in Problem 34-2(c), to find a series for Arctan t.

2. Use the series you obtained in Problem 1 to evaluate $\pi/6$ to within an error of 0.005.

3. Defining $f(t) \equiv \sum_{k=0}^{\infty} t^k/k!$, prove that $f'(t) = f(t)$ for all real t and that $f(0) = 1$. Using these results and Theorem 8-A, conclude that $f(t) = e^t$. Thus we obtain $e^t = \sum_{k=0}^{\infty} t^k/k!$—the well-known "Maclaurin series" for e^t. Find the radius of convergence of this series.

4. (a) Let us take the following as *definitions* of $\cos t$ and $\sin t$:

$$\cos t = \sum_{j=0}^{\infty} \frac{(-1)^j}{(2j)!} t^{2j} \quad \text{and} \quad \sin t = \sum_{j=0}^{\infty} \frac{(-1)^j}{(2j+1)!} t^{2j+1}$$

wherever these series converge. Prove that the series converge for *all t*.

(b) Prove, using these definitions, that $\cos 0 = 1$, $\sin 0 = 0$, $\dfrac{d}{dt} \sin t = \cos t$, and

$\dfrac{d}{dt} (\cos t) = -\sin t$ for all real t.

(c) Prove the identity

$$\sin(t_1 + t_2) = \sin t_1 \cos t_2 + \cos t_1 \sin t_2$$

for all real t_1 and t_2. [*Hint:* For fixed t_2 consider the function defined by

$$f(t) = \sin(t + t_2) - \sin t \cos t_2 - \cos t \sin t_2.$$

Prove that $f''(t) = f(t)$ for all t and that $f(0) = 0$ and $f'(0) = 0$. Then invoke uniqueness.]

(d) Prove the identity

$$\cos(t_1 + t_2) = \cos t_1 \cos t_2 - \sin t_1 \sin t_2.$$

For a continuation of this development of the sine and cosine functions, see Landau [10], Chapter 16.

5. Using the series for e^t, $\cos t$, and $\sin t$ from Problems 3 and 4, verify that $e^{it} = \cos t + i \sin t$ for all t.

36. SOLUTION OF A DIFFERENTIAL EQUATION NEAR AN ORDINARY POINT

We now have enough machinery to solve certain linear ordinary differential equations by the power-series method and to *justify* the procedure.

Example 1. Find nontrivial solutions, valid near $t = 0$, for the equation

$$y'' + ty' + 2y = 0. \tag{1}$$

Note that this is an equation to which (apparently) none of our previous methods apply.

Let us conjecture tentatively that there is a solution of the form

$$y(t) = \sum_{k=0}^{\infty} c_k t^k \quad \text{valid for} \quad -R < t < R,$$

where $R > 0$. (We will confirm this conjecture later.) We then conclude that

$$y'(t) = \sum_{k=1}^{\infty} k c_k t^{k-1} \quad \text{for} \quad -R < t < R,$$

and

$$y''(t) = \sum_{k=2}^{\infty} k(k-1) c_k t^{k-2} \quad \text{for} \quad -R < t < R.$$

The known theorems on addition and multiplication by scalars enable us to now substitute the above series for y, y', and y'' into Eq. (1) and write

$$\sum_{k=2}^{\infty} k(k-1) c_k t^{k-2} + \sum_{k=0}^{\infty} [k c_k t^k + 2 c_k t^k] = 0 \quad \text{for} \quad -R < t < R.$$

Now let us introduce a new dummy index $j = k - 2$ in the first of these series and write

$$\sum_{j=0}^{\infty} (j+2)(j+1) c_{j+2} t^j + \sum_{k=0}^{\infty} (k+2) c_k t^k = 0.$$

Then it is convenient to relabel j as k and combine the two series to get

$$\sum_{k=0}^{\infty} [(k+2)(k+1) c_{k+2} + (k+2) c_k] t^k = 0.$$

In order that this be true we should set

$$(k+2)(k+1) c_{k+2} + (k+2) c_k = 0 \quad \text{for} \quad k = 0, 1, 2, \ldots.$$

Thus we obtain the recursion relation

$$c_{k+2} = -\frac{k+2}{(k+2)(k+1)} c_k = -\frac{1}{k+1} c_k \quad \text{for} \quad k = 0, 1, 2, \ldots.$$

Taking $k = 0, 2, 4, \ldots$, this gives

$$c_2 = -\frac{c_0}{1}, \quad c_4 = -\frac{c_2}{3} = \frac{c_0}{1 \cdot 3}, \quad c_6 = -\frac{c_4}{5} = -\frac{c_0}{1 \cdot 3 \cdot 5}, \quad \ldots,$$

$$c_{2j} = (-1)^j \frac{c_0}{1 \cdot 3 \cdots (2j-1)}, \quad \ldots.$$

Then, taking $k = 1, 3, 5, \ldots$, the recursion relation gives

$$c_3 = -\frac{c_1}{2}, \quad c_5 = -\frac{c_3}{4} = \frac{c_1}{2 \cdot 4}, \quad c_7 = -\frac{c_5}{6} = -\frac{c_1}{2 \cdot 4 \cdot 6}, \quad \ldots,$$

$$c_{2j+1} = (-1)^j \frac{c_1}{2 \cdot 4 \cdots (2j)}, \quad \ldots.$$

Thus, *if* the two series converge, we have

$$y(t) = c_0 \left[1 + \sum_{j=1}^{\infty} \frac{(-1)^j}{1 \cdot 3 \cdots (2j-1)} t^{2j} \right] + c_1 \sum_{j=0}^{\infty} \frac{(-1)^j}{2^j j!} t^{2j+1}. \qquad (2)$$

In Problem 1 the reader is asked to verify that each of these two series does converge for *all* t.

Now we are ready to justify all the above computations.

Since the two power series in Eq. (2) converge for all *t*, their sums represent functions which are infinitely often differentiable for all real *t* and whose derivatives are obtained by term-by-term differentiation. Thus, for *that function y defined by* Eq. (2), all the steps which we have already performed would be justified. And, in fact, our computations show that this *y* is a solution for any values of the constants c_0 and c_1.

We can assert further that *y* defined by Eq. (2) is the *general* solution of Eq. (1). To show this, let

$$y_1(t) = 1 + \sum_{j=1}^{\infty} \frac{(-1)^j}{1 \cdot 3 \cdots (2j+1)} t^{2j} \quad \text{and} \quad y_2(t) = \sum_{j=0}^{\infty} \frac{(-1)^j}{2^j j!} t^{2j+1}.$$

Then $y = c_0 y_1 + c_1 y_2$ will be the general solution if y_1 and y_2 are linearly independent solutions. The fact that y_1 and y_2 are each solutions of Eq. (1) follows by setting $c_0 = 1$, $c_1 = 0$, and then $c_0 = 0$, $c_1 = 1$ in (2). The linear independence is most easily demonstrated by evaluating

$$W(y_1, y_2)(0) = \begin{vmatrix} y_1(0) & y_2(0) \\ y_1'(0) & y_2'(0) \end{vmatrix} = \begin{vmatrix} 1 & 0 \\ 0 & 1 \end{vmatrix} = 1 \neq 0.$$

As a guide to what one should expect from the power-series method of solution we state (without proof) the following theorem.

THEOREM A

Consider the equation

$$y^{(n)} + a_{n-1}(t)y^{(n-1)} + \cdots + a_1(t)y' + a_0(t)y = 0 \qquad (3)$$

with initial conditions

$$y(t_0) = b_0, \quad y'(t_0) = b_1, \quad \ldots, \quad y^{(n-1)}(t_0) = b_{n-1}, \qquad (4)$$

where $t_0, b_0, b_1, \ldots, b_{n-1}$ are given real numbers. If the coefficients $a_{n-1}, \ldots, a_1$, and a_0 can each be represented by a convergent power series in powers of $t - t_0$ for

$$|t - t_0| < R \quad \text{for some} \quad R > 0,$$

then Eqs. (3) and (4) have a (unique) solution of the form

$$y(t) = \sum_{k=0}^{\infty} c_k(t - t_0)^k \qquad (5)$$

valid for $|t - t_0| < R$. Moreover,

$$c_k = \frac{b_k}{k!} \quad \text{for} \quad k = 0, 1, \ldots, n - 1,$$

and $c_n, c_{n+1}, \ldots$ can be computed from a recursion relation obtained by substituting (5) into Eq. (3).

(If the power series for $a_{n-1}, \ldots, a_0$ have different radii of convergence, then R is the smallest of these radii.)

The proof of this theorem can be found in Coddington [2], Chapter 3, Section 9 for the important special case $n = 2$.

When Eq. (3) satisfies the hypotheses of Theorem A we shall refer to t_0 as an *ordinary point* for Eq. (3).

In most of the problems to be considered, as in Example 1, we can give an independent proof of the convergence of our final power-series solutions, and hence of their validity.

Note, incidentally, that Theorem A says that the series (5) will converge *at least* for all t such that $|t - t_0| < R$. In some cases the series might converge and represent a solution on an even larger set.

Under the hypotheses of Theorem A we can also assert the following. There will be n particular solutions of Eq. (3), called $y_1, y_2, \ldots, y_n$, respectively satisfying the initial conditions:

$$y_1(t_0) = 1, \quad y_1'(t_0) = 0, \quad \ldots, \quad y_1^{(n-1)}(t_0) = 0,$$

$$y_2(t_0) = 0, \quad y_2'(t_0) = 1, \quad \ldots, \quad y_2^{(n-1)}(t_0) = 0,$$

$$\vdots$$

$$y_n(t_0) = 0, \quad y_n'(t_0) = 0, \quad \ldots, \quad y_n^{(n-1)}(t_0) = 1.$$

These n solutions are linearly independent, as the reader should verify by evaluating the Wronskian

$$W(y_1, \ldots, y_n)(t_0) = 1.$$

Thus Theorem A also asserts the existence of n linearly independent solutions of the homogeneous Eq. (3) on $(t_0 - R, t_0 + R)$.

Example 2 (Legendre's equation). The equation

$$(1 - t^2)y'' - 2ty' + p(p + 1)y = 0 \tag{6}$$

is called Legendre's equation of order p. This equation arises (with various values of the constant p) in many physical problems.

As long as $t \neq \pm 1$ we can write

$$y'' - \frac{2t}{1 - t^2} y' + \frac{p(p + 1)}{1 - t^2} y = 0, \tag{7}$$

which is in the form of Eq. (3). Now if $t_0 = 0$, then Theorem A asserts that

Legendre's equation, with any value of p, has power-series solutions in powers of t which are valid for $-1 < t < 1$. Why? (Problem 2.) The actual series substitution is more conveniently done in Eq. (6) rather than (7).

Some examples of Legendre's equation occur in the problems at the end of this section. We remark, without proof, that whenever p is a nonnegative integer Eq. (7) will have some solution which is a polynomial. Particular polynomials which satisfy these equations are called Legendre polynomials. An example is encountered in Problem 6(c).

The next example does not work out so nicely as Example 1. Theorem A then plays an essential role.

Example 3. Consider power-series solutions of

$$(1 + 2t^2)y'' + y' + (1 + t^2)y = 0 \quad \text{with} \quad t_0 = 0. \tag{8}$$

We again seek a solution of the form

$$y = \sum_{k=0}^{\infty} c_k t^k \quad \text{valid for} \quad -R < t < R, \tag{9}$$

where $R > 0$. Proceeding as in Example 1, one finds

$$\sum_{k=2}^{\infty} k(k-1)c_k t^{k-2} + \sum_{k=0}^{\infty} 2k(k-1)c_k t^k + \sum_{k=1}^{\infty} kc_k t^{k-1}$$

$$+ \sum_{k=0}^{\infty} c_k t^k + \sum_{k=0}^{\infty} c_k t^{k+2} = 0.$$

In the first of these series let us introduce $j = k - 2$, in the third let $j = k - 1$, and in the fifth let $j = k + 2$. Then, replacing each j with k, we find

$$\sum_{k=0}^{\infty} [(k+2)(k+1)c_{k+2} + 2k(k-1)c_k + (k+1)c_{k+1} + c_k]t^k + \sum_{k=2}^{\infty} c_{k-2} t^k = 0.$$

This leads to the recursion relations

$$2c_2 + c_1 + c_0 = 0,$$

$$6c_3 + 2c_2 + c_1 = 0,$$

and

$$(k+2)(k+1)c_{k+2} + (k+1)c_{k+1} + (2k^2 - 2k + 1)c_k + c_{k-2} = 0, \tag{10}$$

for $k = 2, 3, \ldots$. Equations (10) will be satisfied if, having chosen c_0 and c_1 arbitrarily, one takes

$$c_2 = -\tfrac{1}{2}c_0 - \tfrac{1}{2}c_1,$$

$$c_3 = -\tfrac{1}{3}c_2 - \tfrac{1}{6}c_1 = \tfrac{1}{6}c_0,$$

$$c_4 = -\tfrac{1}{4}c_3 - \tfrac{5}{12}c_2 - \tfrac{1}{12}c_0 = \tfrac{1}{12}c_0 + \tfrac{5}{24}c_1,$$

and so on. In other words, it is possible, in principle, to express all the co-efficients c_k in terms of c_0 and c_1. However, it is not at all easy to find a general formula for c_k for $k = 2, 3, \ldots$. Thus we cannot explicitly write down the series (9) and thereby determine its radius of convergence. This difficulty arises when the recursion relations involve more than the two different c_k's, as is the case in Eqs. (10).

Nevertheless we can invoke Theorem A to find that the series (9) with appropriate values of c_k for $k = 2, 3, \ldots$ will be a solution of Eq. (8) for $-1/\sqrt{2} < t < 1/\sqrt{2}$. (Problem 4.) The beginning of the series solution, using the coefficients we have found, is

$$y(t) = c_0(1 - \tfrac{1}{2}t^2 + \tfrac{1}{6}t^3 + \tfrac{1}{12}t^4 + \cdots) + c_1(t - \tfrac{1}{2}t^2 + \tfrac{5}{24}t^4 + \cdots).$$

The problems given below [except 8(a)] have been designed so that the difficulty encountered in Example 3 will not arise. Fortunately, many of the equations which occur in mathematical physics are of the special types selected here.

Another problem which we have ignored so far is the case when some coefficients in Eq. (3) are analytic at t_0 but are neither polynomials nor ratios of polynomials. Such equations, for example,

$$y'' + (\sin t)y = 0,$$

are encompassed by Theorem A. They can be solved by the power series method using multiplication of two power series, say to compute $(\sin t)y(t)$. [See Theorem A2-H(vi).] Or one can use the Taylor-series method outlined in Problem 10. In fact, using these tools, it is also possible to consider power-series solutions of certain nonlinear equations.

It was asserted, in the introduction to this chapter, that if we can solve a linear homogeneous equation, such as Eq. (3), then the related nonhomo-geneous equations can be solved by variation of parameters. In practice, however, it is usually easier to seek a series solution directly for a nonhomo-geneous equation. This is illustrated for a simple equation in Problem 5(c).

PROBLEMS

1. Prove that each of the two power series in Eq. (2) converges for all t.
2. Verify that the coefficients in Eq. (7) satisfy the hypotheses of Theorem A with $t_0 = 0$ and $R = 1$.
3. If we take $t_0 = 3$ what does Theorem A say about Eq. (7)?
4. Apply Theorem A to Eq. (8) to verify that the series (9) will converge for $|t| < 1/\sqrt{2}$.
5. Find power-series solutions of each of the following using $t_0 = 0$. Determine an interval on which your solution is valid.

   (a) $y'' + y = 0$, $y(0) = 1$, $y'(0) = -2$. Compare with the solution obtained by the methods of Chapter Three.

(b) $y'' + ty' - y = 0$, $y(0) = -1$, $y'(0) = 3$.

(c) $x'' + tx' - x = 1 + t$.

(d) $(1 - t^2)y'' - 2ty' + y = 0$, $y(0) = 17$, $y'(0) = \pi$—Legendre's equation with $p = (-1 \pm \sqrt{5})/2$.

(e) $y' + ty = 0$, $y(0) = 1$. Compare with the solution obtained by the method of Section 2.

6. Find two linearly independent solutions of each of the following equations as power series in t (using $t_0 = 0$). Determine where your solutions are valid.

(a) $(1 - t^2)y'' - ty' + p^2 y = 0$, where p is a (real) constant—Chebyshev's equation. For simplicity, set $p = 2$.

(b) $y'' - ty = 0$—Airy's equation.

(c) $(1 - t^2)y'' - 2ty' + 2y = 0$—Legendre's equation of order 1.

(d) $(1 - t^2)y'' - 2ty' + \frac{3}{4}y = 0$—Legendre's equation of order $\frac{1}{2}$.

(e) $y'' - 2ty' + 2py = 0$ where p is a (real) constant—Hermite's equation.

7. Apply a method of Section 19 to find the general solution of $y'' + ty' - y = 0$, the equation of Problem 5(b).

*8. (a) Find two linearly independent solutions in powers of $t - 1$ for $t^2 y'' - ty' + y = 0$.

(b) Using another method of solution, indirectly find the sums of the series you obtained in (a).

9. Transform Hermite's equation [Problem 6(e)] using Eq. 19-(22) to remove the first derivative term. This leads to an equation arising in quantum mechanics, $z'' + (2p + 1 - t^2)z = 0$. Begin the solution of this latter equation using a power series with $t_0 = 0$, and discover why it is more difficult than solving Hermite's equation.

10. (The Taylor-Series Method). With $t_0 = 0$, find the terms through $c_6 t^6$ in the power series solution (5) of the following problems. For what values of t does Theorem A assert the convergence and validity of the power series solution?

(a) $y'' + ty' - y = 0$ with $y(0) = -1$, $y'(0) = 3$—the same as Problem 5(b). *Outline:* From $y'' = y - ty'$ find $y''(0) = -1$. Then, from $y''' = -ty''$, find $y'''(0)$, and so on. Put the values obtained into Eq. 35-(6).

(b) $y'' + (\sin t)y = 0$ with $y(0) = 1$, $y'(0) = 2$.

*37. SOLUTION OF A DIFFERENTIAL EQUATION NEAR A REGULAR SINGULAR POINT

The equation

$$a_n(t)y^{(n)} + a_{n-1}(t)y^{(n-1)} + \cdots + a_1(t)y' + a_0(t)y = 0 \tag{1}$$

with coefficients analytic at t_0 is said to have a *singular point* or *singularity* at t_0 if $a_n(t_0) = 0$.

In general, one should not expect to be able to find a unique solution of Eq. (1) satisfying the usual type of initial conditions at t_0 if t_0 is a singular point. (Note that we have no existence or uniqueness theorems to cover that situation.) Nevertheless, equations with singular points do arise in real-world physical problems, and then it is the study of solutions *near* the singular points which often gives valuable information about the physical problem. For this reason one tries to express solutions near a singular point t_0 in the most convenient form possible—often in powers of $(t - t_0)$.

Example 1. The first-order equation

$$ty' + ay = 0, \tag{2}$$

where a is a real constant, has a singularity at $t_0 = 0$.

Note what happens when one tries to solve by the method of Section 36 with $t_0 = 0$. Substitution of

$$y(t) = \sum_{k=0}^{\infty} c_k t^k \tag{3}$$

into Eq. (2) leads to the trivial "recursion relation"

$$(k + a)c_k = 0 \quad \text{for} \quad k = 0, 1, 2, \ldots,$$

which implies $c_k = 0$ whenever $k \neq -a$.

Case (i). If a is a negative integer or zero one finds solutions of the form

$$y(t) = c_{-a} t^{-a},$$

for arbitrary constant c_{-a}.

Case (ii). If $a \neq 0$, or -1, or -2, ..., then $c_k = 0$ for all k, and one gets by this method only the trivial solution $y(t) = 0$. This does *not* imply that Eq. (2) has no nontrivial solutions. The trouble is that any solution of the form (3) must be analytic at 0, and in this case Eq. (2) has no solutions which are analytic at $t_0 = 0$ except the trivial solution. Indeed, it is easy to solve Eq. (2) rewritten in the form

$$y' + \frac{a}{t} y = 0 \quad \text{for} \quad t \neq 0.$$

One simply multiplies through by the integrating factor $e^{\int (a/t)\, dt} = e^{a \ln |t|} = |t|^a$ to find $(d/dt)(|t|^a y) = 0$. This implies

$$y(t) = ct^{-a} \quad \text{for} \quad t > 0.$$

Now if $a > 0$, then $y(t) \to \pm\infty$ as $t \to 0+$ (unless $c = 0$). And, on the other hand, if $a < 0$ (and a is not an integer), then some derivative of y approaches $\pm\infty$ as $t \to 0+$ (unless $c = 0$). [Problem 1(a).] So in these cases y certainly cannot represent a function which is analytic at $t = 0$.

Note that (unless $a = 0$) there will not be any solution of Eq. (2) satisfying the initial condition $y(0) = y_0$ if $y_0 \neq 0$. [Problem 1(b).] However, if $a < 0$, there are infinitely many solutions satisfying $y(0) = 0$. [Problem 1(c).]

Example 2. The equation

$$t^2 y' + ay = 0, \quad \text{with} \quad a \neq 0, \tag{4}$$

also has a singularity at $t_0 = 0$. The solutions, obtained with the aid of an integrating factor, are

$$y(t) = c e^{a/t} \quad \text{for} \quad t > 0 \quad \text{or} \quad t < 0, \tag{5}$$

where c is an arbitrary constant. Verify this.

In case $a < 0$ we can define a solution on **R** by

$$y(t) = \begin{cases} 0 & \text{for} \quad t \leq 0 \\ c e^{a/t} & \text{for} \quad t > 0. \end{cases}$$

This function *is* infinitely often differentiable; but it is *not* analytic at 0. (A similar function is mentioned in Appendix 2.) In this example one can show that, regardless of the value of $a \neq 0$, the only solution of Eq. (4) which is analytic at $t = 0$ is the trivial solution. Hence, the substitution of (3) into Eq. (4) will yield only $y(t) = 0$. (Compare Problem 2.)

The differential equations (2) and (4) both have singularities at $t_0 = 0$. However we shall make a distinction between them. In fact we can say that the singularity of Eq. (2) is "milder," and in a sense more manageable, than that of Eq. (4).

DEFINITION

We shall say that Eq. (1) has a *regular singular point* at t_0 if the equation is equivalent to

$$(t - t_0)^n y^{(n)} + (t - t_0)^{n-1} q_{n-1}(t) y^{(n-1)} + \cdots + (t - t_0) q_1(t) y' + q_0(t) y = 0, \tag{6}$$

where the functions $q_{n-1}, \ldots, q_0$ are analytic at t_0.

Note that Euler equations, such as Eq. (2), are of this type with constant $q_{n-1}, \ldots, q_0$. Moreover, if t_0 is an ordinary point for Eq. (1) with $a_n(t) = 1$, as described in Theorem 36-A, then t_0 is also a regular singular point for that equation. To see this, one merely multiplies Eq. (1) by $(t - t_0)^n$ throughout. Thus the concept of regular singular point includes the concept of ordinary point as a special case.

The reader should verify that, while Eq. (2) has a regular singular point at $t_0 = 0$, Eq. (4) does not. A singular point which is not regular, such as $t_0 = 0$ for Eq. (4), is called an *irregular singular point*.

Procedures for the series solution of Eq. (6) near its regular singular point, t_0, are described and illustrated below. The solution should not be expected to be valid, in general, *at* $t = t_0$, but only for values of t *near* t_0 ($t \neq t_0$).

Example 3. The equation

$$2(t-1)^2 y'' - 2(t-1)y' + ty = 0 \qquad (7)$$

has a regular singular point at $t_0 = 1$. Verify this (Problem 3). We shall now make an *attempt* to find solutions of Eq. (7), as in Section 36, in the form

$$y(t) = \sum_{k=0}^{\infty} c_k(t-1)^k. \qquad (8)$$

When this attempt fails we will have good reason to try something else.

It is convenient to rewrite Eq. (7) as

$$2(t-1)^2 y'' - 2(t-1)y' + [1 + (t-1)]y = 0, \qquad (7')$$

that is, to write all coefficients in powers of $(t-1)$. Then, assuming that the series in (8) converges and defines a solution of Eq. (7) when $|t-1| < R$ for some $R > 0$, we substitute (8) into (7'). This gives

$$\sum_{k=0}^{\infty} 2k(k-1)c_k(t-1)^k - \sum_{k=0}^{\infty} 2kc_k(t-1)^k + \sum_{k=0}^{\infty} c_k(t-1)^k$$

$$+ \sum_{k=0}^{\infty} c_k(t-1)^{k+1} - 0.$$

It then follows, in the usual way, that

$$c_0 = 0$$

and, for $k = 1, 2, \ldots,$

$$(2k^2 - 4k + 1)c_k + c_{k-1} = 0.$$

But these equations imply that every $c_k = 0$ $(k = 0, 1, 2, \ldots)$. Hence the only solution we have obtained is the trivial solution $y(t) \equiv 0$.

The basic trick that will enable us to find nontrivial solutions of Eq. (6) for $t > t_0$ is to set

$$y(t) = (t-t_0)^\lambda \sum_{k=0}^{\infty} c_k(t-t_0)^k, \qquad (9)$$

or

$$y(t) = \sum_{k=0}^{\infty} c_k(t-t_0)^{\lambda+k}, \qquad (10)$$

where λ is a further constant to be determined. The constant λ will, in general, not be an integer. We shall concern ourselves primarily with the case $t > t_0$. Then, for any λ,

$$(t-t_0)^\lambda \equiv e^{\lambda \ln(t-t_0)}.$$

If λ is complex, say $\lambda = \mu + i\omega$, then for $t > t_0$ we would write

$$(t - t_0)^\lambda = e^{(\mu + i\omega)\ln(t - t_0)}$$

$$= e^{\mu \ln(t - t_0)} \cos[\omega \ln(t - t_0)] + ie^{\mu \ln(t - t_0)} \sin[\omega \ln(t - t_0)].$$

In case $t < t_0$ one should modify (10) by replacing $(t - t_0)^{\lambda + k}$ by $|t - t_0|^{\lambda + k}$ $= (t_0 - t)^{\lambda + k}$.

The series in Eq. (10)—called a Frobenius series—is not a power series as previously defined unless λ happens to be a nonnegative integer. Thus it is not obvious that we have the right to differentiate term by term.

LEMMA A

For any series of the form (10) there will exist a radius of convergence $R \geq 0$ such that the series converges whenever $0 < t - t_0 < R$ and diverges whenever $t - t_0 > R$. Moreover, if $R > 0$, the series in (10) can be differentiated termwise giving

$$y'(t) = \sum_{k=0}^{\infty} (\lambda + k)c_k(t - t_0)^{\lambda + k - 1} \quad \text{for} \quad 0 < t - t_0 < R.$$

Similarly for y'', y''', ...

Proof. The series involved in Eq. (9) *is* a power series. Thus it converges when $|t - t_0| < R$ and diverges when $|t - t_0| > R$ for some $R \geq 0$. Hence, if $t - t_0 > 0$, so that $(t - t_0)^\lambda$ is well defined, and if $t - t_0 < R$, then the series in Eq. (10) also converges and the two expressions of Eqs. (9) and (10) are equal. Moreover, assuming $R > 0$ and using the properties of the power series in (9), we can compute for $0 < t - t_0 < R$:

$$y'(t) = \lambda(t - t_0)^{\lambda - 1} \sum_{k=0}^{\infty} c_k(t - t_0)^k + (t - t_0)^\lambda \sum_{k=1}^{\infty} kc_k(t - t_0)^{k-1}$$

$$= \sum_{k=0}^{\infty} (\lambda + k)c_k(t - t_0)^{\lambda + k - 1}. \quad \blacksquare$$

In case $t < t_0$, so that (10) becomes

$$y(t) = \sum_{k=0}^{\infty} c_k|t - t_0|^{\lambda + k} \quad \text{for} \quad -R < t - t_0 < 0, \tag{10'}$$

one finds

$$y'(t) = -\sum_{k=0}^{\infty} (\lambda + k)c_k|t - t_0|^{\lambda + k - 1}.$$

Example 4. Let us reconsider Eq. (7) for $t > 1$, seeking a solution of the form given in Eq. (10) (with $t_0 = 1$). Substituting into (7'), with the aid of

Lemma A, we obtain

$$\sum_{k=0}^{\infty} 2(\lambda + k)(\lambda + k - 1)c_k(t - 1)^{\lambda+k} - \sum_{k=0}^{\infty} 2(\lambda + k)c_k(t - 1)^{\lambda+k}$$

$$+ \sum_{k=0}^{\infty} c_k(t - 1)^{\lambda+k} + \sum_{k=0}^{\infty} c_k(t - 1)^{\lambda+k+1} = 0. \tag{11}$$

Note that, since we are restricting ourselves to values of $t > 1$, it is now possible to divide out $(t - 1)^{\lambda}$ from Eq. (11) and consider the resulting equation involving ordinary power series. From this, we find the recursion relations

$$[2\lambda(\lambda - 1) - 2\lambda + 1]c_0 = 0, \tag{12}$$

and, for $k = 1, 2, \ldots$,

$$[2(\lambda + k)(\lambda + k - 2) + 1]c_k + c_{k-1} = 0. \tag{13}$$

Since the object is to get a nontrivial solution of Eq. (7) we want to avoid requiring $c_0 = 0$. Thus, to satisfy Eq. (12), we shall set

$$2\lambda(\lambda - 1) - 2\lambda + 1 = 0 \tag{14}$$

or

$$2\lambda^2 - 4\lambda + 1 = 0.$$

Equation (14) is called the *indicial equation* for (7). It has two solutions

$$\lambda_1 = 1 + \frac{\sqrt{2}}{2} \quad \text{and} \quad \lambda_2 = 1 - \frac{\sqrt{2}}{2}.$$

Putting $\lambda = \lambda_1$ in the recursion relation (13) for $k = 1, 2, \ldots$. we find

$$\left[2\left(1 + \frac{\sqrt{2}}{2} + k\right)\left(-1 + \frac{\sqrt{2}}{2} + k\right) + 1\right]c_k + c_{k-1} = 0,$$

which gives

$$c_k = -\frac{c_{k-1}}{2k(k + \sqrt{2})} \quad \text{for} \quad k = 1, 2, \ldots.$$

Thus, with arbitrary c_0,

$$c_1 = -\frac{c_0}{2 \cdot (1 + \sqrt{2})},$$

$$c_2 = \frac{c_0}{2^2 \cdot 2!(1 + \sqrt{2})(2 + \sqrt{2})},$$

$$\vdots$$

$$c_k = \frac{(-1)^k c_0}{2^k \cdot k!(1 + \sqrt{2})(2 + \sqrt{2}) \cdots (k + \sqrt{2})}, \ldots.$$

We shall set $c_0 = 1$ for convenience. The resulting series, the series in $y_1(t) =$

$$(t-1)^{1+\sqrt{2}/2}\left[1 + \sum_{k=1}^{\infty} \frac{(-1)^k}{2^k \cdot k!(1+\sqrt{2})(2+\sqrt{2})\cdots(k+\sqrt{2})}(t-1)^k\right],$$

is easily verified, by the ratio test, to converge for all t. Thus all the preceding computations are justified for the case $\lambda = \lambda_1$, and it follows that y_1 is a solution of Eq. (7) valid for all $t > 1$.

Similarly we obtain a solution of the form

$$\sum_{k=0}^{\infty} d_k(t-1)^{\lambda_2+k},$$

namely, setting $d_0 = 1$,

$y_2(t) =$

$$(t-1)^{1-\sqrt{2}/2}\left[1 + \sum_{k=1}^{\infty} \frac{(-1)^k}{2^k \cdot k!(1-\sqrt{2})(2-\sqrt{2})\cdots(k-\sqrt{2})}(t-1)^k\right]$$

also valid for all $t > 1$.

If we could show y_1 and y_2 to be linearly independent, then the problem would be completely solved. Unfortunately we cannot simply compute the Wronskian of y_1 and y_2 at $t = 1$ since y_2' is not even defined at that point. Instead, let us suppose that for some constants, C_1 and C_2,

$$C_1 y_1(t) + C_2 y_2(t) = 0$$

for $t > 1$. Then divide through by $(t-1)^{1-\sqrt{2}/2}$ and let $t \to 1$ to conclude that

$$C_1 \cdot 0 + C_2 \cdot 1 = 0.$$

This shows that $C_2 = 0$. It then follows that $C_1 = 0$ also.

When this method is applied to a given differential equation of second order, it may turn out that the "indicial equation" has a double root, $\lambda_2 = \lambda_1$. Then apparently only one linearly independent solution will be obtained. More generally, if $\lambda_1 - \lambda_2$ is any integer it *may* happen that our procedure fails to produce two linearly independent solutions. When such a shortage or deficiency of linearly independent solutions occurs we could always resort to the method of "reduction of order" (variation of parameters) described in Section 19. However it is usually easier to follow the procedures to be described in the next section.

PROBLEMS

1. (a) If $y(t) = t^{-a}$ for $t > 0$, where $a < 0$ and a is not an integer, show that, for some integer $k > 0$, $|y^{(k)}(t)| \to \infty$ as $t \to 0+$.
   (b) Let $y(t)$ be a solution of Eq. (2) for $t > 0$. Show that if $a < 0$, then

$\lim_{t \to 0^+} y(t) = 0$ [so that we would naturally define $y(0) = 0$]; and if $a > 0$, then in order that $\lim_{t \to 0^+} y(t)$ exist one must have $y(t) \equiv 0$. Thus one could never solve Eq. (2) for $t > 0$ (or $t \geq 0$) with an initial condition of the form $y(0) = y_0 \neq 0$.

(c) If $a < 0$, show that Eq. (2) has infinitely many solutions for $t > 0$ satisfying $\lim_{t \to 0^+} y(t) = 0$.

2. What happens if one tries to find a solution of Eq. (4) with $a \neq 0$ by substituting $y(t) = \sum_{k=0}^{\infty} c_k t^k$ or $y(t) = \sum_{k=0}^{\infty} c_k t^{\lambda + k}$?

3. Prove that Eq. (7) has a regular singular point at $t_0 = 1$.

4. Show that each of the following equations (important in applications) has a regular singular point at $t_0 = 0$. In particular, determine the functions $q_1(t)$ and $q_0(t)$ which make each equation a special case of (6). (p, q, and r are constants.)

   (a) $t^2 y'' + t y' + (t^2 - p^2) y = 0$—Bessel's equation.
   (b) $t(1 - t) y'' + [r - (1 + p + q) t] y' - pq y = 0$—the hypergeometric equation.

5. Find two linearly independent solutions expressed in powers of $t > 0$ for the following special cases of the equations of Problem 4. Determine the domain of validity of the solutions.

   (a) $t^2 y'' + t y' + (t^2 - \frac{1}{9}) y = 0$—Bessel's equation with $p = \frac{1}{3}$.
   (b) $t(1 - t) y'' + (2 - 3t) y' - y = 0$—the hypergeometric equation with $p = 1$, $q = 1$, $r = 2$.

6. Use the simpler solution you found in Problem 5(b) to obtain a second linearly independent solution by "reduction of order" (variation of parameters). Compare your answer with the other solution of Problem 5(b).

*38. THE MORE COMPLICATED CASES FOR A REGULAR SINGULAR POINT

At the end of Section 37 we noted that near a regular singular point t_0 the equation

$$(t - t_0)^2 y'' + (t - t_0) q_1(t) y' + q_0(t) y = 0 \tag{1}$$

might not have two linearly independent solutions of the form

$$y(t) = \sum_{k=0}^{\infty} c_k (t - t_0)^{\lambda + k} \quad \text{for} \quad t - t_0 > 0. \tag{2}$$

The following theorem describes the possible nature of the solutions near a regular singular point in *all* cases.

THEOREM A

Consider Eq. (1) with a regular singular point at t_0, where q_1 and q_0 can be represented by convergent power series in powers of $t - t_0$ for

$$|t - t_0| < R \quad \text{for some} \quad R > 0.$$

Let the *indicial equation*

$$\lambda(\lambda - 1) + q_1(t_0)\lambda + q_0(t_0) = 0, \tag{3}$$

obtained by substituting (2) into Eq. (1) and requiring $c_0 \neq 0$, have roots λ_1 and λ_2.

If $\lambda_1 - \lambda_2$ is not an integer, then Eq. (1) has two linearly independent solutions valid for $0 < t - t_0 < R$ of the form

$$y_1(t) = \sum_{k=0}^{\infty} c_k(t - t_0)^{\lambda_1 + k} \tag{4}$$

and

$$y_2(t) = \sum_{k=0}^{\infty} d_k(t - t_0)^{\lambda_2 + k}. \tag{5}$$

Without loss of generality we can set $c_0 = d_0 = 1$.

If $\lambda_1 - \lambda_2$ is an integer, then, assuming $\lambda_1 - \lambda_2 \geq 0$, Eq. (1) has two linearly independent solutions for $0 < t - t_0 < R$ either of the form (4) and (5) or of the form (4) and

$$y_2(t) = \sum_{k=0}^{\infty} d_k(t - t_0)^{\lambda_2 + k} + y_1(t) \ln (t - t_0). \tag{5'}$$

(Here we cannot assume that $d_0 = 1$.)

Remarks. In case solutions are desired for $t < t_0$, it is useful to note that the same theorem holds for $-R < t - t_0 < 0$ when $t - t_0$ is replaced with $|t - t_0| = t_0 - t$.

Theorem A can also be generalized to higher-order equations. However, the most important equations arising in applications seem to be of second order.

We give no proof of the above theorem, but refer the interested reader once again to Coddington [2], Chapter 4. Our examples and problems will all be such that it is possible to give a complete proof of the validity of the solution without depending on Theorem A. In other words, Theorem A will be used only as a guide to the type of solutions we should seek.

To illustrate the assertions of Theorem A when $\lambda_1 - \lambda_2$ is an integer we shall consider two special cases of Bessel's equation,

$$t^2 y'' + ty' + (t^2 - p^2)y = 0, \tag{6}$$

where p is a constant. Bessel's equation, previously encountered in Problems 19-12, 37-4, and 37-5, arises with various values of p in many problems of applied mathematics.

Bessel's equation has a regular singular point at $t = 0$, regardless of the value of p. (Problem 37-4.) Thus we seek solutions of the form

$$y(t) = \sum_{k=0}^{\infty} c_k t^{\lambda + k} \quad \text{for} \quad t > 0. \tag{7}$$

Example 1 (Bessel's Equation with $p^2 = \frac{1}{4}$). Find two linearly independent solutions of

$$t^2 y'' + t y' + (t^2 - \tfrac{1}{4}) y = 0 \quad \text{for} \quad t > 0. \tag{8}$$

If $\mathbf{y}(t) = \sum_{k=0}^{\infty} c_k t^{\lambda+k}$, then

$$y'(t) = \sum_{k=0}^{\infty} c_k(\lambda + k) t^{\lambda+k-1}$$

and

$$y''(t) = \sum_{k=0}^{\infty} c_k(\lambda + k)(\lambda + k - 1) t^{\lambda+k-2}.$$

Substitution of these expressions into the differential equation gives

$$\sum_{k=0}^{\infty} c_k[(\lambda + k)(\lambda + k - 1) + (\lambda + k) - \tfrac{1}{4}] t^{\lambda+k} + \sum_{k=0}^{\infty} c_k t^{\lambda+k+2} = 0. \tag{9}$$

We shall equate the coefficient of each power of t to zero. For the lowest power, t^λ, this gives

$$[\lambda(\lambda - 1) + \lambda - \tfrac{1}{4}] c_0 = 0.$$

In order not to require $c_0 = 0$, we shall require that λ satisfy the indicial equation

$$\lambda(\lambda - 1) + \lambda - \tfrac{1}{4} = 0,$$

which has solutions $\lambda_1 = \frac{1}{2}$ and $\lambda_2 = -\frac{1}{2}$.

According to Theorem A, we should seek two linearly independent solutions of the form

$$y_1(t) = t^{1/2} \sum_{k=0}^{\infty} c_k t^k \quad \text{and} \quad y_2(t) = t^{-1/2} \sum_{k=0}^{\infty} d_k t^k + C y_1(t) \ln t, \tag{10}$$

where $C = 0$ or 1 [for (5) or (5′), respectively]. Now if it should happen that $C = 0$, then y_1 and y_2 will have very similar forms. Considering this possibility, we *might* be lucky enough to save some work if, instead of substituting y_1 into (8) we begin by seeking a solution of the form

$$y(t) = t^{-1/2} \sum_{k=0}^{\infty} c_k t^k. \tag{11}$$

In other words, when the difference between λ_1 and λ_2 (real) is an integer, it is best to use the smaller of these two indices first. In effect both $t^{\lambda_1} \sum_{k=0}^{\infty} c_k t^k$ and $t^{\lambda_2} \sum_{k=0}^{\infty} d_k t^k$ are being considered simultaneously, and *if it turns out* that $C = 0$ in Eq. (10) we will have the complete solution at once.

Using $\lambda = -\frac{1}{2}$, (9) now becomes

$$t^{-1/2} \left\{ \sum_{k=0}^{\infty} c_k k(k - 1) t^k + \sum_{k=0}^{\infty} c_k t^{k+2} \right\} = 0.$$

Introducing $j = k + 2$ in the last sum, this becomes

$$t^{-1/2} \left\{ \sum_{k=2}^{\infty} c_k k(k-1)t^k + \sum_{j=2}^{\infty} c_{j-2} t^j \right\} = 0,$$

which simplifies to

$$\sum_{k=2}^{\infty} [k(k-1)c_k + c_{k-2}]t^k = 0.$$

Hence c_0 and c_1 are arbitrary and

$$c_k = \frac{-c_{k-2}}{k(k-1)} \quad \text{for} \quad k = 2, 3, \ldots.$$

This gives

$$c_2 = \frac{-1}{2 \cdot 1} c_0, \quad c_4 = \frac{-1}{4 \cdot 3} c_2 = \frac{(-1)^2}{4!} c_0, \quad \ldots, \quad c_{2j} = \frac{(-1)^j}{(2j)!} c_0, \quad \ldots$$

and

$$c_3 = \frac{-1}{3 \cdot 2} c_1, \quad c_5 = \frac{-1}{5 \cdot 4} c_3 = \frac{(-1)^2}{5!} c_1, \quad \ldots, \quad c_{2j+1} = \frac{(-1)^j}{(2j+1)!} c_1, \quad \ldots$$

Thus

$$y(t) = c_0 t^{-1/2} \sum_{j=0}^{\infty} \frac{(-1)^j}{(2j)!} t^{2j} + c_1 t^{-1/2} \sum_{j=0}^{\infty} \frac{(-1)^j}{(2j+1)!} t^{2j+1}. \tag{12}$$

Left as an exercise for the reader is the proof that both of the series in (12) converge for all t. (Problem 2.) So, thanks to Lemma 37-A, the above calculations are validated and y is a solution for all $t > 0$.

For this problem we define *Bessel functions of order* $\frac{1}{2}$ (of the first and second kinds) by

$$J_{1/2}(t) = \sqrt{\frac{2}{\pi}} t^{1/2} \sum_{j=0}^{\infty} \frac{(-1)^j}{(2j+1)!} t^{2j}$$

and

$$J_{-1/2}(t) = \sqrt{\frac{2}{\pi}} t^{-1/2} \sum_{j=0}^{\infty} \frac{(-1)^j}{(2j)!} t^{2j}. \tag{13}$$

The multiplier $\sqrt{2/\pi}$ is a conventional (but to us arbitrary) choice.

The reader should verify that $J_{1/2}$ and $J_{-1/2}$ are two linearly independent solutions of (8). [Problem 3(a).] Thus, in this case, one *does have* $C = 0$ in Eq. (10).

It so happens that we can recognize the infinite series in (13) as familiar functions

$$J_{1/2}(t) = \left(\frac{2}{\pi t}\right)^{1/2} \sin t \quad \text{and} \quad J_{-1/2}(t) = \left(\frac{2}{\pi t}\right)^{1/2} \cos t. \tag{13'}$$

(Problem 4.) Another derivation of these solutions was outlined in Problem 19-12.

Example 2 (Bessel's equation with $p=0$). Find two linearly independent solutions of

$$t^2 y'' + t y' + t^2 y = 0 \quad \text{for} \quad t > 0. \tag{14}$$

(This equation could be rewritten as $t y'' + y' + t y = 0$.) Again we substitute $y(t) = \sum_{k=0}^{\infty} c_k t^{\lambda+k}$ into the equation. We find

$$\sum_{k=0}^{\infty} c_k(\lambda+k)(\lambda+k-1)t^{\lambda+k} + \sum_{k=0}^{\infty} c_k(\lambda+k)t^{\lambda+k} + \sum_{k=0}^{\infty} c_k t^{\lambda+k+2} = 0. \tag{15}$$

The indicial equation, obtained by requiring $c_0 \neq 0$, is

$$\lambda(\lambda-1) + \lambda = 0,$$

which gives $\lambda = 0, 0$. We can now rewrite (15) (introducing $j = k + 2$ in the last sum) as

$$\sum_{k=0}^{\infty} k^2 c_k t^k + \sum_{j=2}^{\infty} c_{j-2} t^j = 0.$$

This simplifies to

$$c_1 t + \sum_{k-2}^{\infty} (k^2 c_k + c_{k-2})t^k = 0. \tag{16}$$

Thus we require

$$c_1 = 0 \quad \text{and} \quad c_k = -\frac{1}{k^2} c_{k-2} \quad \text{for} \quad k = 2, 3, \ldots.$$

Hence $c_3 = c_5 = c_7 = \cdots = 0$ and

$$c_2 = -\frac{1}{2^2} c_0, \quad c_4 = -\frac{1}{4^2} c_2 = \frac{1}{4^2 2^2} c_0, \ldots,$$

and, by mathematical induction,

$$c_{2j} = \frac{(-1)^j}{(2j)^2(2j-2)^2 \cdots 4^2 2^2} c_0 = \frac{(-1)^j}{2^{2j}(j!)^2} c_0,$$

where c_0 is arbitrary. We shall take $c_0 = 1$.
    The resulting series is

$$y_1(t) = \sum_{j=0}^{\infty} \frac{(-1)^j}{2^{2j}(j!)^2} t^{2j}. \tag{17}$$

To prove convergence let us apply the ratio test. We find

$$\left| \frac{c_{2j+2} t^{2j+2}}{c_{2j} t^{2j}} \right| = \left| \frac{2^{2j}(j!)^2 t^2}{2^{2j+2}[(j+1)!]^2} \right| = \frac{|t|^2}{2^2(j+1)^2} \to 0 \quad \text{as} \quad j \to \infty$$

for every t. Thus our series (17) converges everywhere and hence all the above calculations are valid. This means that y_1 is a solution of Eq. (14) for all $t > 0$. (In fact, in this case, it is a solution for all t, including $t = 0$.)

The function defined by the series in (17) is given a name:

$$J_0(t) \equiv \sum_{j=0}^{\infty} \frac{(-1)^j}{(j!)^2} \left(\frac{t}{2} \right)^{2j}$$

is called the *Bessel function of order zero of the first kind*. Its values can be found in various published tables, and many properties of this function are known. To those who work regularly with such functions, J_0 is not a mysterious infinite series. Rather it is just another familiar function like the exponential, sine, or cosine.

To find a second linearly independent solution of (14) we should, according to Theorem A, try

$$y(t) = \sum_{k=0}^{\infty} d_k t^k + C y_1(t) \ln t. \tag{18}$$

Now we can not have $C = 0$. For, with $C = 0$, substitution of (18) into (14) would just be a rerun of the calculations performed above, and we would obtain again some multiple of J_0. Thus we set $C = 1$ in (18). Furthermore, we can subtract from y the function $d_0 y_1$, or equivalently, without loss of generality, we can set $d_0 = 0$ in this example.

Let $z(t) = \sum_{k=1}^{\infty} d_k t^k$ and, instead of (18), we will use

$$y_2(t) = z(t) + y_1(t) \ln t. \tag{19}$$

Then

$$y_2'(t) = z'(t) + y_1'(t) \ln t + \frac{y_1(t)}{t},$$

$$y_2''(t) = z''(t) + y_1''(t) \ln t + \frac{2y_1'(t)}{t} - \frac{y_1(t)}{t^2}.$$

Substituting into (14), and remembering that y_1 is itself a solution, we have

$$t^2 z''(t) + t z'(t) + t^2 z(t) + 2t y_1'(t) = 0. \tag{20}$$

This is the equation which is going to yield the numbers $d_1, d_2, \ldots$. We can avoid repeating some calculations by noting the similarity between the substitution of $z(t) = \sum_{k=1}^{\infty} d_k t^k$ and the earlier substitution of $\sum_{k=0}^{\infty} c_k t^k$. Thus, by comparison with (16), Eq. (20) becomes

$$d_1 t + 2^2 d_2 t^2 + \sum_{k=3}^{\infty} (k^2 d_k + d_{k-2}) t^k + 2t y_1'(t) = 0. \tag{21}$$

But from (17) it follows that

$$2ty_1'(t) = \sum_{j=1}^{\infty} \frac{(-1)^j j}{2^{2(j-1)}(j!)^2} t^{2j}.$$

Equating to zero the coefficients of the various powers of t in (21) we find

$$d_1 = 0, \qquad 2^2 d_2 = 1,$$

and, for $k = 3, 5, 7, \ldots$,

$$d_k = -\frac{d_{k-2}}{k^2} \quad \text{so that} \quad d_3 = d_5 = d_7 = \cdots = 0,$$

and, for $k = 2j$,

$$d_{2j} = -\frac{d_{2j-2}}{(2j)^2} - \frac{(-1)^j j}{2^{2(j-1)}(j!)^2(2j)^2}.$$

Thus

$$d_2 = \frac{1}{2^2}, \quad d_4 = -\left(\frac{d_2}{4^2} + \frac{2}{2^2(2!)^2 4^2}\right) = -\frac{1}{2^2 4^2}(1 + \tfrac{1}{2})$$

and, by mathematical induction, one can verify that

$$d_{2j} = \frac{(-1)^{j-1}}{(j!)^2 2^{2j}}\left(1 + \frac{1}{2} + \cdots + \frac{1}{j}\right) \quad \text{for} \quad j = 1, 2, \ldots.$$

The resulting function y_2 is customarily denoted by K_0 and is called the Bessel function of zero order of the second kind:

$$K_0(t) = y_2(t) = \sum_{j=1}^{\infty} \frac{(-1)^{j-1}}{(j!)^2}\left(1 + \frac{1}{2} + \cdots + \frac{1}{j}\right)\left(\frac{t}{2}\right)^{2j} + J_0(t)\ln t$$

$$\text{for} \quad t > 0. \quad (22)$$

The reader should verify, by the ratio test, that the new series involved in K_0 also converges for all t. [Problem 5(a).] Because of the occurrence of $\ln t$, the function K_0 is not defined at $t = 0$ and, of course, cannot satisfy the differential equation there. The function K_0, like J_0, has also been tabulated and studied.

PROBLEMS

1. The Euler equation $t^2 y'' + b_1 t y' + b_0 y = 0$ is a special case of Eq. (1) when $t_0 = 0$. Compare the corresponding indicial equation (3) with the equation obtained when one seeks solutions of the form t^λ. Compare the solutions of the Euler equation in the cases $(b_1 - 1)^2 > 4b_0$ and $(b_1 - 1)^2 = 4b_0$ with the assertions of Theorem A. [This problem suggests the reason for the logarithm term in Eq. (5').]

2. Prove that the two series in Eq. (12) converge for all t.

3. (a) Using the series definitions of Eqs. (13), prove that $J_{1/2}$ and $J_{-1/2}$ are linearly independent solutions of Eq. (8).

 (b) If y is a bounded solution of Eq. (8) on $(0, \beta)$, what else can you say about it? How does $y'(t)$ behave as $t \rightarrow 0+$?

4. Prove that $J_{1/2}$ and $J_{-1/2}$ as defined in (13) have the simple forms given in Eqs. (13').

5. (a) Prove that the first infinite series on the right-hand side of Eq. (22) converges for all t.

 (b) Show that J_0 and K_0 are linearly independent solutions of Eq. (14).

 (c) If y is a bounded solution of Eq. (14) on $(0, \beta)$, what else can you say about it? How does $y'(t)$ behave as $t \rightarrow 0+$?

6. Show that each of the following equations has a regular singular point at t_0 (p, q, and r are constants). Then apply Theorem A to find $R > 0$ such that the equation has two linearly independent solutions for $0 < t - t_0 < R$. Find the roots, λ_1 and λ_2, of the indicial equation in each case. (These differential equations all have applications in physics.)

 (a) $t^2 y'' + t y' + (t^2 - p^2)y = 0$, $t_0 = 0$—Bessel's equation.

 (b) $t y'' + (1 - t)y' + py = 0$, $t_0 = 0$—Laguerre's equation.

 (c) $(1 - t^2)y'' - 2t y' + p(r + 1)y = 0$, $t_0 = 1$—Legendre's equation. [*Hint:* Begin by expressing the coefficients $(1 - t^2)$ and $-2t$ as polynomials in $(t - 1)$.]

 (d) $t(1 - t)y'' + [r - (1 + p + q)t]y' - pqy = 0$, $t_0 = 0$—the hypergeometric equation.

7. Find two linearly independent solutions expressed in powers of $t - t_0 > 0$ for each of the following special cases of equations from Problem 6. Determine the domain of validity for each and compare with $0 < t - t_0 < R$ as predicted by Theorem A.

 (a) $t^2 y'' + t y' + (t^2 - \frac{9}{4})y = 0$, $t_0 = 0$—Bessel's equation with $p = \frac{3}{2}$.

 (b) $t y'' + (1 - t)y' = 0$, $t_0 = 0$—Laguerre's equation with $p = 0$.

 *(c) $t y'' + (1 - t)y' + y = 0$, $t_0 = 0$—Laguerre's equation with $p = 1$.

 *(d) $t^2 y'' + t y' + (t^2 - 1)y = 0$, $t_0 = 0$—Bessel's equation with $p = 1$.

 *(e) $(1 - t^2)y'' - 2t y' + 2y = 0$, $t_0 = 1$—Legendre's equation with $p = 1$.

8. Use the simplest solution you found for Problem 7(e) to obtain the second linearly independent solution by reduction of order.

Chapter Eight

THE LAPLACE TRANSFORM METHOD FOR LINEAR ORDINARY EQUATIONS

This chapter gives a brief introduction to another important method for solving linear equations, especially those with constant coefficients but complicated forcing functions.

In this method one "transforms" the given differential equation into an algebraic equation. Solving the resulting algebraic equation is usually an elementary computation. Then one tries to convert the solution of the algebraic equation into a solution of the original differential equation.

39. THE LAPLACE TRANSFORM

If g is a real- or complex-valued function on $[0, \infty)$, the Laplace transform of g, denoted by $\mathscr{L}[g]$ or G, is defined by

$$\mathscr{L}[g](s) = G(s) = \int_0^\infty e^{-st} g(t) \, dt, \tag{1}$$

provided this improper integral converges. The Laplace transform of g is thus a new function (of s) for those values of s for which the integral converges.

Example 1. Let $g(t) = e^{\lambda t}$ for $t \geq 0$, where λ is any given constant (real or complex). Then if $s > \operatorname{Re} \lambda$, the integral in Eq. (1) converges since

$$\int_0^\infty e^{-st} g(t) \, dt = \lim_{b \to \infty} \int_0^b e^{(\lambda - s)t} \, dt$$

$$= \lim_{b \to \infty} \left[\frac{e^{(\lambda - s)b}}{\lambda - s} - \frac{1}{\lambda - s} \right] = \frac{1}{s - \lambda}.$$

Justify these steps. Thus

$$\mathscr{L}[g](s) = \frac{1}{s - \lambda} \quad \text{for all} \quad s > \operatorname{Re} \lambda.$$

It will be convenient at times to think of the parameter s as being complex-valued. Then $\mathscr{L}[g]$ becomes a complex-valued function of a "complex variable." Note that the computation above is unchanged if s is complex with $\operatorname{Re} s > \operatorname{Re} \lambda$. In particular, if $\lambda - s = \mu + i\omega$ where μ and ω are real with $\mu = \operatorname{Re} \lambda - \operatorname{Re} s < 0$, then

$$\left| \frac{e^{(\lambda - s)b}}{\lambda - s} \right| = \frac{e^{\mu b}}{|\lambda - s|} \to 0 \quad \text{as} \quad b \to \infty.$$

If you prefer, you can regard s as real throughout the chapter; but some computations then become more complicated.

Example 2. Let $g(t) = t$ for $t \geq 0$ and let $\operatorname{Re} s > 0$. Then integration by parts gives

$$\mathscr{L}[g](s) = \int_0^\infty e^{-st} t \, dt$$

$$= \lim_{b \to \infty} \left[-\frac{t}{s} e^{-st} + \int \frac{1}{s} e^{-st} \, dt \right]_0^b$$

$$= \lim_{b \to \infty} \left[-\frac{b}{s} e^{-sb} - \frac{1}{s^2} e^{-sb} + \frac{1}{s^2} \right] = \frac{1}{s^2}.$$

These steps should all be familiar in the case of real positive s. You should convince yourself that they are also valid when s is a complex number with $\operatorname{Re} s > 0$. In particular, show that integration by parts is valid for complex-valued functions of a real variable, t. (Problem 1.)

Examples 1 and 2 show that the Laplace transforms of certain simple functions can be easily computed directly from the definition, Eq. (1). But sometimes the function g will be too complicated to permit easy evaluation of the integral in (1), and sometimes g will not be specifically known. In these cases we will welcome a criterion which assures that $\mathscr{L}[g]$ does at least exist.

Unless otherwise specified, g will be a real- or complex-valued function on $[0, \infty)$. A useful set of sufficient conditions for the existence of $\mathscr{L}[g]$ (to be given in Theorem A) uses the following two concepts.

DEFINITION

We shall say that g is *piecewise continuous* if, for every finite $b > 0$, g is continuous on $[0, b]$ except possibly at a finite number of points, say t_1, $t_2, \ldots, t_k$, and at each of these points the left- and right-hand limits,

$$\lim_{t \to t_i-} g(t) \quad \text{and} \quad \lim_{t \to t_i+} g(t),$$

exist. See Figure 1.

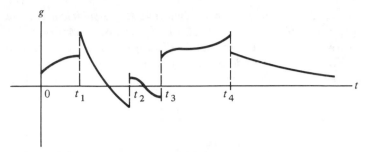

FIGURE 1 **A piecewise continuous function.**

DEFINITION

We shall say that g is of *exponential order* if there exist real numbers M and ρ such that

$$|g(t)| \le Me^{\rho t} \quad \text{for all} \quad t \ge 0. \tag{2}$$

THEOREM A

If g is piecewise continuous and of exponential order, $|g(t)| \le Me^{\rho t}$ for $t \ge 0$, then $\mathcal{L}[g](s)$ exists and

$$|\mathcal{L}[g](s)| \le \frac{M}{\text{Re } s - \rho} \quad \text{whenever} \quad \text{Re } s > \rho.$$

Proof. The piecewise continuity of g is enough to assure the existence of

$$\int_0^b e^{-st} g(t)\, dt$$

for each $b > 0$ and each complex s.

For Re $s > \rho$ and each $b > 0$

$$\int_0^b |e^{-st} g(t)|\, dt \le \int_0^b e^{-(\text{Re } s - \rho)t} M\, dt = \frac{M}{\text{Re } s - \rho}[1 - e^{-(\text{Re } s - \rho)b}] < \frac{M}{\text{Re } s - \rho}.$$

But this proves the convergence of the improper integral

$$\int_0^\infty |e^{-st} g(t)|\, dt.$$

Hence, by the comparison test for improper integrals [the analog of Theorem A2-H(ii) for series], it follows that

$$\mathcal{L}[g](s) = \int_0^\infty e^{-st} g(t)\, dt$$

also converges and

$$|\mathcal{L}[g](s)| \le \int_0^\infty |e^{-st} g(t)|\, dt \le \frac{M}{\text{Re } s - \rho}. \quad \blacksquare$$

Note that the conditions on g given in Theorem A are *sufficient* for the existence of $\mathscr{L}[g]$. It can be shown that they are not necessary conditions. It does, however, turn out to be the case that if $\mathscr{L}[g](s)$ exists for some s, then it will exist whenever Re $s > \rho$ for some real number ρ.

The next theorem asserts that the "Laplace transform operator" $\mathscr{L}$ is linear.

THEOREM B

Let g and h be two functions mapping $[0, \infty) \to \mathbf{C}$ such that $\mathscr{L}[g](s)$ exists whenever Re $s > \rho_1$ and $\mathscr{L}[h](s)$ exists whenever Re $s > \rho_2$. Let c_1 and c_2 be any two constants (real or complex). Then

$$\mathscr{L}[c_1 g + c_2 h](s) = c_1 \mathscr{L}[g](s) + c_2 \mathscr{L}[h](s) \quad \text{for} \quad \text{Re } s > \rho, \qquad (3)$$

where $\rho = \max\{\rho_1, \rho_2\}$.

Proof. Clearly $\mathscr{L}[g](s)$ and $\mathscr{L}[h](s)$ both exist when Re $s > \rho$. For Re $s > \rho$ and for each $b > 0$,

$$\int_0^b e^{-st}[c_1 g(t) + c_2 h(t)] \, dt = c_1 \int_0^b e^{-st} g(t) \, dt + c_2 \int_0^b e^{-st} h(t) \, dt.$$

But now, letting $b \to \infty$, we get Eq. (3). ∎

Another basic property of the Laplace transform is its effect on derivatives. This, we will see, is the key feature which makes Laplace transforms useful for solving differential equations.

THEOREM C

Let $x: [0, \infty) \to \mathbf{C}$ be continuous and of exponential order, $|x(t)| \leq Me^{\rho t}$ for $t \geq 0$, and let x' be piecewise continuous. Then $\mathscr{L}[x]$ and $\mathscr{L}[x']$ exist and

$$\mathscr{L}[x'](s) = s\mathscr{L}[x](s) - x(0) \quad \text{for} \quad \text{Re } s > \rho. \qquad (4)$$

Proof. Let Re $s > \rho$ and let $b > 0$. Let the discontinuities of x' in $[0, b]$, if any, occur at $t_1, t_2, \ldots, t_k$ where $0 \equiv t_0 < t_1 < t_2 < \cdots < t_k < t_{k+1} \equiv b$. Then, using integration by parts,

$$\int_0^b e^{-st} x'(t) \, dt = \sum_{i=1}^{k+1} \int_{t_{i-1}}^{t_i} e^{-st} x'(t) \, dt$$

$$= \sum_{i=1}^{k+1} \left[e^{-st_i} x(t_i) - e^{-st_{i-1}} x(t_{i-1}) + s \int_{t_{i-1}}^{t_i} e^{-st} x(t) \, dt \right]$$

$$= e^{-sb} x(b) - x(0) + s \int_0^b e^{-st} x(t) \, dt.$$

But now

$$|e^{-sb}x(b)| \le Me^{-(\mathrm{Re}\,s-\rho)b} \to 0 \quad \text{as} \quad b \to \infty.$$

Thus Eq. (4) follows. ■

Before any further discussion of the theory of Laplace transforms, let us illustrate their use in solving differential equations by treating a simple example.

Example 3. Solve

$$x' + 2x = e^{-t} \quad \text{with} \quad x(0) = 3. \tag{5}$$

If we assume, for the moment, that the unknown solution x is of exponential order, then its transform $\mathscr{L}[x] = X$ must exist. Moreover, by Theorems B and C and Example 1, we can take the Laplace transform of both sides of the differential equation and find

$$\mathscr{L}[x'](s) + 2X(s) = \frac{1}{s+1}$$

or

$$sX(s) - x(0) + 2X(s) = \frac{1}{s+1} \quad \text{for} \quad \mathrm{Re}\,s > \rho,$$

for some ρ. Substituting the initial value $x(0) = 3$ and solving for $X(s)$ we obtain

$$X(s) = \frac{3s+4}{(s+2)(s+1)}. \tag{6}$$

Expansion of the right-hand side of this equation in partial fractions then gives

$$X(s) = \frac{1}{s+1} + \frac{2}{s+2}.$$

But now we observe, by Example 1 and Theorem B, that this X is just the Laplace transform of

$$x(t) = e^{-t} + 2e^{-2t}. \tag{7}$$

One easily verifies, by substitution, that this is a solution of Eqs. (5). So, by uniqueness, it must be *the* solution.

Remarks. At this stage in our discussion of the Laplace transform, a substitution check of the solution (7) is *essential*. Of course, we know from Chapters Two and Three (or even from Section 2) that Eqs. (5) have a unique solution valid on **R**. However, if we merely followed the above calculations leading to Eq. (7)—and did *not* check by substitution into (5)—we could not conclude that (7) gives the solution of (5). Two logical defects would keep us from this conclusion. What are they? (Problem 7.) We shall resolve these difficulties in Section 40.

When we attempt to apply the method illustrated in Example 3 to more complicated problems we also encounter practical difficulties. First of all we must be able to compute the transform $\mathscr{L}[h]$ of a given forcing function h. Then, when we have found $X = \mathscr{L}[x]$, the Laplace transform of the solution, we must be able to find x.

As an aid in finding these it is convenient to have available a table of Laplace transforms of some functions which arise frequently. A brief list of such results will be given in Table 1. We derive some of these results and then ask the reader to check the rest.

Example 4. If $g(t) = t^n$, where $n = 0, 1, 2, \ldots$, and Re $s > 0$, then

$$\mathscr{L}[g](s) = \int_0^\infty e^{-st}t^n \, dt = \frac{n!}{s^{n+1}}, \tag{8}$$

or, using a loose but descriptive notation, we will sometimes write

$$\mathscr{L}[t^n](s) = \frac{n!}{s^{n+1}}. \tag{8'}$$

The proof of (8) is by mathematical induction. For $n = 0$ it is easily verified that

$$\mathscr{L}[1](s) = \int_0^\infty e^{-st} \, dt = \frac{1}{s} = \frac{0!}{s^{0+1}}.$$

For $n = 1$, Eq. (8) was obtained in Example 2. Now assume the validity of (8) for some n and compute for $b > 0$, using integration by parts,

$$\int_0^b e^{-st}t^{n+1} \, dt = -\frac{1}{s} e^{-sb}b^{n+1} + \frac{n+1}{s} \int_0^b e^{-st}t^n \, dt.$$

Since Re $s > 0$, we can take the limit as $b \to \infty$ and find

$$\mathscr{L}[t^{n+1}](s) = \frac{n+1}{s} \mathscr{L}[t^n](s)$$

or, using the induction hypothesis,

$$\mathscr{L}[t^{n+1}](s) = \frac{(n+1)!}{s^{n+2}}.$$

This completes the proof of (8).

The following theorem provides a useful tool for finding new Laplace transforms from those which are already known.

THEOREM D

Let g have Laplace transform $G(s)$ for Re $s > \rho$, and let $\lambda \in C$. Then

$$\mathscr{L}[e^{\lambda t}g(t)](s) = \mathscr{L}[g(t)](s - \lambda) = G(s - \lambda) \quad \text{for} \quad \text{Re } s > \rho + \text{Re } \lambda. \qquad (9)$$

Proof. For Re $s > \rho + \text{Re } \lambda$, $\text{Re}(s - \lambda) > \rho$ and hence,

$$\mathscr{L}[e^{\lambda t}g(t)](s) = \int_0^\infty e^{-(s-\lambda)t}g(t)\, dt = \mathscr{L}[g(t)](s - \lambda). \quad \blacksquare$$

Example 5. Combining Example 4 and Theorem D we find at once that, for $n = 0, 1, 2, \ldots$,

$$\mathscr{L}[e^{\lambda t}t^n](s) = \frac{n!}{(s - \lambda)^{n+1}} \quad \text{for} \quad \text{Re } s > \text{Re } \lambda.$$

Example 6. Compute $\mathscr{L}[e^{\mu t} \cos \omega t]$ where μ and ω are real numbers. From Problem 13-12 we can write

$$e^{\mu t} \cos \omega t = \tfrac{1}{2}e^{(\mu + i\omega)t} + \tfrac{1}{2}e^{(\mu - i\omega)t}.$$

But then Example 1 easily gives for Re $s > \mu$

$$\mathscr{L}[e^{\mu t} \cos \omega t](s) = \frac{1}{2}\left[\frac{1}{s - \mu - i\omega} + \frac{1}{s - \mu + i\omega}\right] = \frac{s - \mu}{(s - \mu)^2 + \omega^2}.$$

Note that we could not have used the result of Example 1 in this computation if we had not allowed λ to be complex in Example 1. And we could not have allowed λ to be complex in Theorem D if we had not also been allowing s to be complex.

The energetic reader should undertake to compute $\mathscr{L}[e^{\mu t} \cos \omega t](s)$ for real s without any use of complex functions. (Problem 2.) This tends to increase one's appreciation of complex function methods.

Example 7. Compute $\mathscr{L}[t \cos \omega t]$ where ω is real. Again we shall make use of complex functions, writing

$$t \cos \omega t = \tfrac{1}{2}te^{i\omega t} + \tfrac{1}{2}te^{-i\omega t}.$$

Then, for Re $s > 0$, Example 5 gives

$$\mathscr{L}[t \cos \omega t](s) = \frac{1}{2}\left[\frac{1}{(s - i\omega)^2} + \frac{1}{(s + i\omega)^2}\right] = \frac{s^2 - \omega^2}{(s^2 + \omega^2)^2}.$$

Another tool which is useful in computing Laplace transforms is Theorem E below.

TABLE 1 **A Short List of Laplace Transforms***
(λ is a complex number; μ and ω are real)

$g(t)$	$G(s) = \mathscr{L}[g](s)$
1. $e^{\lambda t}$	$\dfrac{1}{s - \lambda}$ $(\mathrm{Re}\, s > \mathrm{Re}\, \lambda)$
2. t^n $(n = 0, 1, 2, \ldots)$	$\dfrac{n!}{s^{n+1}}$ $(\mathrm{Re}\, s > 0)$
3. $t^n e^{\lambda t}$ $(n = 0, 1, 2, \ldots)$	$\dfrac{n!}{(s - \lambda)^{n+1}}$ $(\mathrm{Re}\, s > \mathrm{Re}\, \lambda)$
4. $\cos \omega t$	$\dfrac{s}{s^2 + \omega^2}$ $(\mathrm{Re}\, s > 0)$
5. $\sin \omega t$	$\dfrac{\omega}{s^2 + \omega^2}$ $(\mathrm{Re}\, s > 0)$
6. $e^{\mu t} \cos \omega t$	$\dfrac{s - \mu}{(s - \mu)^2 + \omega^2}$ $(\mathrm{Re}\, s > \mu)$
7. $e^{\mu t} \sin \omega t$	$\dfrac{\omega}{(s - \mu)^2 + \omega^2}$ $(\mathrm{Re}\, s > \mu)$
8. $t \cos \omega t$	$\dfrac{s^2 - \omega^2}{(s^2 + \omega^2)^2}$ $(\mathrm{Re}\, s > 0)$
9. $t \sin \omega t$	$\dfrac{2\omega s}{(s^2 + \omega^2)^2}$ $(\mathrm{Re}\, s > 0)$
10. $\sin \omega t - \omega t \cos \omega t$	$\dfrac{2\omega^3}{(s^2 + \omega^2)^2}$ $(\mathrm{Re}\, s > 0)$
11. $\sin \omega t + \omega t \cos \omega t$	$\dfrac{2\omega s^2}{(s^2 + \omega^2)^2}$ $(\mathrm{Re}\, s > 0)$

* Extensive tables of Laplace transforms are available:
  Erdelyi, A., Magnus, W., Oberhettinger, F., and Tricomi, F., *Tables of Integral Transforms* (Bateman Manuscript Project), Vol. 1. New York: McGraw-Hill Book Company, Inc., 1954.
  Roberts, G. E., and Kaufman, H., *Table of Laplace Transforms*. Philadelphia: W. B. Saunders Company, 1966.

THEOREM E

Let $g(t)$ be piecewise continuous and of exponential order, with $|g(t)| \leq M e^{\rho t}$ for $t \geq 0$, so that $G(s) = \mathscr{L}[g](s)$ exists for all $s > \rho$. Then

$$G'(s) = -\mathscr{L}[tg(t)](s) \quad \text{for all (real)} \quad s > \rho.$$

The proof would be trivial if we could be sure that

$$\frac{d}{ds} \int_0^\infty e^{-st} g(t)\, dt = \int_0^\infty \frac{\partial}{\partial s} [e^{-st} g(t)]\, dt.$$

But, passing a derivative across an integral sign (and other operations which amount to interchanging the order of limits) should never be considered obvious. For example, the reader can verify that

$$\int_0^\infty se^{-s^2t}\,dt = \begin{cases} \dfrac{1}{s} & \text{if } s \neq 0 \\[2mm] 0 & \text{if } s = 0. \end{cases}$$

So here is an example of an integral which exists for each value of the parameter s, but is not even continuous (let alone differentiable) with respect to s at $s = 0$.

The proof of Theorem E is outlined in Problem 8.

Note how Theorem E could have been used in Example 7. We know from Example 6 that $\mathscr{L}[\cos \omega t](s) = s/(s^2 + \omega^2)$ for $s > 0$. So

$$\mathscr{L}[t \cos \omega t](s) = -\frac{d}{ds}\frac{s}{s^2 + \omega^2} = \frac{s^2 - \omega^2}{(s^2 + \omega^2)^2} \quad \text{for} \quad s > 0.$$

The examples given thus far have verified lines 1, 2, 3, 4, 6, and 8 of Table 1. The verification of the rest of this table is left as Problem 4.

We are going to use Table 1 not only to determine $\mathscr{L}[g]$ when g is given, but also to find a function x such that $\mathscr{L}[x] = X$ when X is given.

Example 8. Find a function x such that

$$\mathscr{L}[x](s) = X(s) = \frac{s^3 - 2s^2 + 4s - 6}{s^4 + 8s^2 + 16}.$$

Since this function does not resemble anything in the second column of the table, let us first break X down into partial fractions. The appropriate partial fraction form is

$$X(s) = \frac{As + B}{s^2 + 4} + \frac{Cs + D}{(s^2 + 4)^2}.$$

Evaluating the constants A, B, C, and D, one finds

$$X(s) = \frac{s - 2}{s^2 + 4} + \frac{2}{(s^2 + 4)^2}.$$

Now referring to lines 4, 5, and 10 of the table, we get

$$x(t) = \cos 2t - \sin 2t + \tfrac{1}{8}(\sin 2t - 2t \cos 2t)$$

$$= \cos 2t - \tfrac{7}{8}\sin 2t - \tfrac{1}{4}t \cos 2t.$$

We conclude this section by observing that the bound for $|\mathscr{L}[g](s)|$ given in Theorem A actually represents a growth estimate for Laplace transforms. This tells us that certain types of functions X cannot be Laplace transforms.

Example 9. The functions defined by

$$X(s) = \frac{s^2 - 4}{s^2 + 4} \quad \text{and by} \quad X(s) = \frac{1}{\sqrt{\text{Re } s}}$$

cannot be Laplace transforms of piecewise continuous functions of exponential order. Why?

PROBLEMS

1. Let u and v be two continuous complex-valued functions with piecewise continuous derivatives on $[a, b]$. Then show that

$$\int_a^b u(t)dv(t) = u(b)v(b) - u(a)v(a) - \int_a^b v(t)du(t),$$

that is, integration by parts is valid for complex-valued functions. [*Hint:* Consider the real and imaginary parts of the above equation.]

2. Compute $\mathcal{L}[e^{\mu t} \cos \omega t](s)$, where μ, ω, and s are real, by direct integration of

$$\int_0^\infty e^{-st}e^{\mu t} \cos \omega t \, dt.$$

Do not use complex functions. Compare your work with that in Example 6.

3. Use Theorem E to give another proof of Eq. (8).

4. Verify Table 1, lines 5, 7, 9, 10, and 11.

5. Find the Laplace transforms, if they exist, of each of the following functions. You may use Table 1 when it helps. (μ and ω are real constants, and a is a positive constant.)

(a) $g(t) = \cos 3t + 2te^{2t}$.

(b) $g(t) = \sinh \mu t$.

(c) $g(t) = \begin{cases} 1 & \text{for } 0 \le t < a \\ 0 & \text{for } t \ge a. \end{cases}$

(d) $g(t) = \begin{cases} t & \text{for } 0 \le t < a \\ 0 & \text{for } t \ge a. \end{cases}$

(e) $g(t) = e^{t^2}$.

(f) $g(t) = te^{\mu t} \cos \omega t$.

(g) $g(t) = te^{\mu t} \sin \omega t$.

(h) $g(t) = t^2 \cos \omega t$.

6. Find continuous functions x of exponential order, if they exist, such that $\mathcal{L}[x] = X$ for each of the following given functions X.

(a) $X(s) = \dfrac{s^2 + 3s + 4}{(s + 2)^3(s + 3)} \quad \text{for } s > -2.$

(b) $X(s) = \dfrac{3s}{s^2 + 2s + 5} \quad \text{for } s > -1.$

[*Hint:* Rewrite the denominator as $(s + 1)^2 + 2^2$. Then use Table 1 lines 6 and 7.]

(c) $X(s) = \dfrac{1}{2}s - \dfrac{2}{s} \quad \text{for } s > 0.$

(d) $X(s) = \dfrac{s^2 + 1}{s^3 + s^2 - 2} \quad \text{for } s > 10.$

7. Find two reasons why the calculations leading from Eqs. (5) to Eq. (7) do *not* prove (without further checking) that (7) gives a solution of (5).

*8. Prove Theorem E by completing the details in the following outline: If $G(s) = \int_0^\infty e^{-st} g(t)\, dt$ for $s > \rho$, we must show that

$$\lim_{\Delta s \to 0} \frac{\int_0^\infty e^{-(s+\Delta s)t} g(t)\, dt - \int_0^\infty e^{-st} g(t)\, dt}{\Delta s} = -\int_0^\infty e^{-st} t g(t)\, dt.$$

This is equivalent to showing that

$$H(\Delta s) \equiv \int_0^\infty \left[\frac{e^{-t\Delta s} - 1}{\Delta s} + t\right] e^{-st} g(t)\, dt \to 0 \quad \text{as} \quad \Delta s \to 0.$$

Since $s > \rho$, there exists a number $\eta > 0$ such that $s - \rho \geq 2\eta$. Show that, for $0 < |\Delta s| < \eta$,

$$|H(\Delta s)| \leq \int_0^\infty \left|\frac{1}{\Delta s}\right| \cdot \left|1 - t\Delta s + \frac{e^{-\theta} t^2 \Delta s^2}{2} - 1 + t\Delta s\right| M e^{-2\eta t}\, dt,$$

where $\theta = \theta(t\Delta s)$ lies between 0 and $t\Delta s$. From this

$$|H(\Delta s)| \leq \int_0^\infty \frac{1}{2} t^2 |\Delta s| M e^{-\eta t}\, dt = \frac{M}{\eta^3} |\Delta s| \to 0 \quad \text{as} \quad \Delta s \to 0.$$

*9. Find $\mathscr{L}[t^{-1} \sin t] = G$. [*Hint;* First prove that $G(s)$ exists for $s > 0$ (if $t^{-1} \sin t$ is defined to be, say, 0 at $t = 0$). Then find $G'(s)$ by using Theorem E. Integrate, and determine the constant of integration, to get $G(s)$.]

40. SOLUTION OF SIMPLE LINEAR EQUATIONS

Example 39-3 illustrated the Laplace transform method for solving a linear differential equation with constant coefficients. But the argument given there had two defects. We shall soon remove these.

In order to also treat equations of higher order than the first, we will need the following result on the Laplace transforms of higher derivatives of x.

THEOREM A

Let $x: [0, \infty) \to \mathbf{C}$ be such that $x^{(n-1)}$ is continuous and of exponential order and $x^{(n)}$ is piecewise continuous. Then, for some $\rho > 0$, $\mathscr{L}[x](s)$, $\mathscr{L}[x'](s), \ldots, \mathscr{L}[x^{(n)}](s)$ exist for $s > \rho$ and, for each $k = 1, 2, \ldots, n$,

$$\mathscr{L}[x^{(k)}](s) = s^k \mathscr{L}[x](s) - s^{k-1} x(0) - s^{k-2} x'(0) - \cdots - x^{(k-1)}(0). \quad (1)$$

Proof. The continuity of $x^{(n-1)}$ assures the continuity also of $x, x', x'', \ldots, x^{(n-2)}$.

Let M and ρ be positive numbers such that

$$|x^{(n-1)}(t)| \leq M e^{\rho t} \quad \text{for} \quad t \geq 0.$$

Then

$$x^{(n-2)}(t) = x^{(n-2)}(0) + \int_0^t x^{(n-1)}(\sigma) \, d\sigma \quad \text{for} \quad t \geq 0.$$

Hence

$$|x^{(n-2)}(t)| \leq |x^{(n-2)}(0)| + \int_0^t M e^{\rho\sigma} \, d\sigma$$

$$= |x^{(n-2)}(0)| + \frac{M}{\rho} (e^{\rho t} - 1)$$

$$\leq \left(|x^{(n-2)}(0)| + \frac{M}{\rho} \right) e^{\rho t} \quad \text{for} \quad t \geq 0,$$

that is, $x^{(n-2)}$ is also of exponential order with the same value of ρ. It follows, by mathematical induction, that $x, x', \ldots, x^{(n-2)}$ are all of exponential order with the same value of ρ. Thus each of these has a Laplace transform for $s > \rho$.

In case $k = 1$ the desired result,

$$\mathscr{L}[x'](s) = s\mathscr{L}[x](s) - x(0) \quad \text{for} \quad s > \rho, \tag{2}$$

is found in Theorem 39-C.

Now consider $k = 2$. It then follows by applying (2) to x' (instead of x) that

$$\mathscr{L}[x''](s) = s\mathscr{L}[x'](s) - x'(0) \quad \text{for} \quad s > \rho. \tag{3}$$

Combining this with (2) itself gives

$$\mathscr{L}[x''](s) = s^2 \mathscr{L}[x](s) - sx(0) - x'(0) \quad \text{for} \quad s > \rho, \tag{4}$$

which is Eq. (1) for $k = 2$.

The general proof of Eq. (1) requires a straightforward mathematical induction which is left to the reader. (Problem 1.) ∎

Example 1. Solve

$$x'' + 4x' + 4x = \sin 2t \quad \text{for} \quad t \geq 0 \tag{5}$$

with

$$x(0) = 1, \qquad x'(0) = -2. \tag{6}$$

We know that this problem has a solution. (In fact we *could* explicitly solve by the methods of Chapter Three.) It follows that x, x', and x'' are continuous. Let us *assume* that x' is of exponential order, so that we can apply Theorem A. The Laplace transform $X = \mathscr{L}[x]$ will exist on (ρ, ∞) for some

$\rho > 0$, and transforming the differential equation (5) we find (with the aid of Table 39-1)

$$s^2 X(s) - sx(0) - x'(0) + 4sX(s) - 4x(0) + 4X(s) = \frac{2}{s^2 + 4}.$$

Now apply the initial conditions (6) to get

$$(s^2 + 4s + 4)X(s) = s + 2 + \frac{2}{s^2 + 4}, \tag{7}$$

or

$$X(s) = \frac{s^3 + 2s^2 + 4s + 10}{(s + 2)^2(s^2 + 4)}. \tag{8}$$

To find x such that $\mathscr{L}[x] = X$ we first rewrite $X(s)$ in terms of partial fractions

$$X(s) = \frac{A}{s + 2} + \frac{B}{(s + 2)^2} + \frac{Cs + D}{s^2 + 4}.$$

Evaluating the unknown coefficients A, B, C, and D one gets

$$X(s) = \frac{9}{8} \frac{1}{s + 2} + \frac{1}{4} \frac{1}{(s + 2)^2} - \frac{1}{8} \frac{s}{s^2 + 4}.$$

Now we can refer again to Table 39-1 to find

$$x(t) = \tfrac{9}{8}e^{-2t} + \tfrac{1}{4}te^{-2t} - \tfrac{1}{8}\cos 2t. \tag{9}$$

Note that the coefficient of $X(s)$ in Eq. (7) is just the characteristic polynomial for Eq. (5). It is not difficult to see that this will *always* be the case when the Laplace transform method is applied to linear equations with constant coefficients.

The two important questions that arise after this argument, just as for Example 39-3 (Problem 39-7), are: 1. How do we know that the solution x has the desired properties of continuity and exponential order so that Theorem A is applicable? and 2. While (9) certainly defines a function whose Laplace transform is given by Eq. (8), how do we know that other functions do not also have this same Laplace transform? [If there were several functions with Laplace transforms given by Eq. (8), at most one of them could be the solution of (5) and (6). But how could we tell which one?]

Of course we could resolve any doubts raised by these questions simply by checking our apparent solution (9) through substitution into (4) and (5). If it checks, then the appropriate uniqueness theorem from Chapter Two assures us that the problem is finished. (Problem 2.) But in more complicated cases it may be quite inconvenient to carry out the substitution check.

We proceed as follows: Consider a given linear constant-coefficient differential equation

$$x^{(n)}(t) + a_{n-1}x^{(n-1)}(t) + \cdots + a_0 x(t) = h(t) \quad \text{for} \quad t \geq 0 \tag{10}$$

with h continuous on $[0, \infty)$, and with initial conditions

$$x(0) = b_0, \quad x'(0) = b_1, \quad \ldots, \quad x^{(n-1)}(0) = b_{n-1}. \tag{11}$$

For convenience define $h(t) = h(0)$ on $(\alpha, 0]$ for some $\alpha < 0$. Then variation of parameters guarantees that Eqs. (10) and (11) have a solution on (α, ∞), and in particular on the half line $[0, \infty)$. We also know that this solution is unique. It follows that $x, x', \ldots,$ and $x^{(n)}$ are continuous on $[0, \infty)$. Why?

But to assure the existence of the required Laplace transforms we also need to be sure that $x, x', \ldots, x^{(n-1)}$ are of exponential order.

THEOREM B

If h is continuous and of exponential order and if x is the solution of Eqs. (10) and (11), then $x, x', \ldots, x^{(n-1)}$ are also of exponential order.

*Proof (Depending on Section 12). For $t \geq 0$ we have

$$x(t) = b_0 + \int_0^t x'(\sigma) \, d\sigma,$$

$$x'(t) = b_1 + \int_0^t x''(\sigma) \, d\sigma,$$

$$\vdots$$

$$x^{(n-2)}(t) = b_{n-2} + \int_0^t x^{(n-1)}(\sigma) \, d\sigma,$$

$$x^{(n-1)}(t) = b_{n-1} + \int_0^t [-a_0 x(\sigma) - a_1 x'(\sigma) - \cdots - a_{n-1}x^{(n-1)}(\sigma) + h(\sigma)] \, d\sigma.$$

Letting

$$v(t) = |x(t)| + |x'(t)| + \cdots + |x^{(n-1)}(t)|, \tag{12}$$

the above equations yield, for $t \geq 0$,

$$v(t) \leq |b_0| + |b_1| + \cdots + |b_{n-1}| + \int_0^t v(\sigma) \, d\sigma \tag{13}$$

$$+ \max\{|a_0|, |a_1|, \ldots, |a_{n-1}|\} \int_0^t v(\sigma) \, d\sigma + \int_0^t |h(\sigma)| \, d\sigma.$$

But now, since h is of exponential order,

$$|h(t)| \leq Me^{\rho t} \quad \text{for} \quad t \geq 0.$$

Without loss of generality we can assume $\rho > 0$. Thus inequality (13) gives, with appropriate constants $B > 0$ and $K > 0$,

$$v(t) \leq B + K \int_0^t v(\sigma)\, ds + \int_0^t M e^{\rho\sigma}\, d\sigma$$

$$\leq B + \frac{M}{\rho} e^{\rho t} + K \int_0^t v(\sigma)\, d\sigma.$$

To this we apply Corollary 12-E to find

$$v(t) \leq \left(B + \frac{M}{\rho} e^{\rho t}\right) e^{Kt} \leq \left(B + \frac{M}{\rho}\right) e^{(\rho + K)t} \quad \text{for} \quad t \geq 0.$$

This shows that x, x', $\ldots$, and $x^{(n-1)}$ are each of exponential order. ∎

Thus $X = \mathscr{L}[x]$ exists, Theorem A is applicable, and (10) and (11) give

$$(s^n + a_{n-1}s^{n-1} + \cdots + a_0)X(s) = \text{a known function of } s,$$

as in Eq. (7).

The question remains: Is there a *unique* continuous function x of exponential order on $[0, \infty)$ having Laplace transform equal to the specified $X(s)$? If so, then x must be the solution of our differential equation and initial conditions.

THEOREM C (UNIQUENESS OF THE INVERSE LAPLACE TRANSFORM)

Given a continuous function $X: (\rho, \infty) \to \mathbf{C}$, there is at most one continuous function $x: [0, \infty) \to \mathbf{C}$ of exponential order such that $\mathscr{L}[x] = X$.

We call such an x, if it exists, the *inverse Laplace transform* of X and write

$$x = \mathscr{L}^{-1}[X].$$

The proof of Theorem C (Lerch's Theorem) is outlined in Problem 4.

Note that one only needs to know the values of $X(s)$ for all *real* values of $s > \rho$ in order to conclude that there is at most one continuous function x of exponential order with $\mathscr{L}[x] = X$. Of course, if we also knew $X(s)$ for complex s, that would not hurt anything.

Note also that Theorem C does not assert the existence of $x = \mathscr{L}^{-1}[X]$. As a matter of fact there *is* a systematic procedure for finding $\mathscr{L}^{-1}[X]$ given the values of $X(s)$ for Re $s > \rho$. But this uses the theory of complex functions of a complex variable which we do not assume of the reader.

Subject to proving Theorem C, we have now fully justified the Laplace transform method for solving Eqs. (10) and (11), or in particular Eqs. (5) and (6).

Problem 3 gives some further simple equations which can now be rigorously solved by the use of the Laplace transform. But each of these examples and

Eqs. (5) and (6) can also be solved just about as easily by the method of Section 17.

Section 41 will contain examples in which the Laplace transform method begins to show its real advantage.

The following theorem, which is clearly related to Theorem A, is sometimes of use in computing Laplace transforms and inverse transforms.

THEOREM D

Let $x: [0, \infty) \to \mathbf{C}$ be piecewise continuous and of exponential order, $|x(t)| \leq Me^{\rho t}$ for $t \geq 0$ where $\rho > 0$. Then

$$\mathscr{L}\left[\int_0^t x(\sigma)\, d\sigma\right](s) = \frac{1}{s} \mathscr{L}[x](s) \quad \text{for} \quad \text{Re } s > \rho.$$

Proof. Introduce

$$y(t) \equiv \int_0^t x(\sigma)\, d\sigma \quad \text{for} \quad t \geq 0.$$

Then, just as in the proof of Theorem A, we find that $|y(t)| \leq M_1 e^{\rho t}$ for $t \geq 0$. Moreover, y is continuous and has a piecewise continuous derivative, x. Now apply Theorem A (or Theorem 39-C) to y to find

$$\mathscr{L}[x](s) = \mathscr{L}[y'](s) = s\mathscr{L}[y](s) - y(0)$$

$$= s\mathscr{L}\left[\int_0^t x(\sigma)\, d\sigma\right](s) \quad \text{for} \quad \text{Re } s > \rho. \quad \blacksquare$$

Example 2. Find

$$\mathscr{L}^{-1}\left[\frac{1}{s(s - \lambda)}\right].$$

This can, of course, be done easily by partial fractions. But another approach is to first find, from Table 39-1,

$$\mathscr{L}^{-1}\left[\frac{1}{s - \lambda}\right](t) = e^{\lambda t}$$

and then apply Theorem D to compute

$$\mathscr{L}^{-1}\left[\frac{1}{s(s - \lambda)}\right](t) = \int_0^t e^{\lambda \sigma}\, d\sigma = \frac{1}{\lambda}(e^{\lambda t} - 1).$$

The two examples given thus far to demonstrate the solution of differential equations using the Laplace transform have had initial data conveniently given at $t_0 = 0$. We now present just one example to illustrate the procedure when the initial data is specified at some other value of t_0.

Example 3. Solve

$$x'' + 2x' + 5x = e^{-t} \quad \text{for} \quad t \geq 3$$

with

$$x(3) = 2, \qquad x'(3) = 1.$$

We convert this to a problem with data at zero by introducing $\sigma = t - 3$ and $z(\sigma) = x(\sigma + 3)$. Then

$$\frac{d^2z}{d\sigma^2} + 2\frac{dz}{d\sigma} + 5z = e^{-3}e^{-\sigma} \quad \text{for} \quad \sigma \geq 0$$

with

$$z(0) = 2, \qquad \frac{dz}{d\sigma}(0) = 1.$$

Taking Laplace transforms in this new problem, we find

$$s^2Z(s) - 2s - 1 + 2[sZ(s) - 2] + 5Z(s) = \frac{e^{-3}}{s+1}$$

or

$$Z(s) = \frac{2s + 5}{s^2 + 2s + 5} + \frac{e^{-3}}{(s^2 + 2s + 5)(s + 1)}.$$

And, upon expanding the last fraction into partial fractions, one gets

$$Z(s) = \frac{(2 - e^{-3}/4)s + (5 - e^{-3}/4)}{(s + 1)^2 + 2^2} + \frac{e^{-3}/4}{s + 1},$$

from which

$$z(\sigma) = \left(2 - \frac{e^{-3}}{4}\right)e^{-\sigma}\cos 2\sigma + \frac{3}{2}e^{-\sigma}\sin 2\sigma + \frac{e^{-3}}{4}e^{-\sigma},$$

for $\sigma \geq 0$. Finally, putting $\sigma = t - 3$ and $z(\sigma) = x(t)$, we have

$$x(t) = (2e^3 - \tfrac{1}{4})e^{-t}\cos(2t - 6) + \tfrac{3}{4}e^3 e^{-t}\sin(2t - 6) + \tfrac{1}{4}e^{-t} \text{ for } t \geq 3.$$

PROBLEMS

1. Complete the proof of Theorem A by mathematical induction.
2. (a) Verify by substitution that Eq. (9) defines a solution of Eqs. (5) and (6).
   (b) Which uniqueness theorem from Chapter Two would you now invoke to show that the problem (5) and (6) is completely solved?
3. Solve by use of the Laplace transform:

   (a) $x'' + 4x = 0, \; x(0) = 1, \; x'(0) = -2.$
   (b) $x'' + 4x = \sin 2t, \; x(0) = 1, \; x'(0) = -2.$

(c) $x''' - 4x'' + 4x' = e^{3t}$, $x(0) = 2$, $x'(0) = 1$, $x''(0) = -5$.
(d) $x'' + x' - 2x = 2e^{-2t} + 3te^{-2t}$, $x(0) = 1$, $x'(0) = 2$.
(e) $x'' + 2x' + 5x = 0$, $x(0) = 2$, $x'(0) = -1$.

*4. Prove Theorem C by filling in the details in the following argument (adapted from Ritger and Rose [14]):

Suppose that x and $\tilde{x}$ are two continuous functions of exponential order mapping $[0, \infty) \to \mathbf{C}$ such that $\mathcal{L}[x](s) = \mathcal{L}[\tilde{x}](s)$ for $s > \rho$ for some $\rho > 0$. We can assume $|x(t)| \le M_1 e^{\rho t}$ and $|\tilde{x}(t)| \le M_2 e^{\rho t}$. (Why?) Then, defining $y(t) = x(t) - \tilde{x}(t)$ we have

$$\int_0^\infty e^{-st} y(t)\, dt = 0 \quad \text{for all} \quad s > \rho.$$

Since e^{-st} is real, this same equation must hold when y is replaced by either of the real-valued functions $\operatorname{Re} y$ or $\operatorname{Im} y$. Thus, it will suffice to prove that $y(t) \equiv 0$ for $t \ge 0$ *under the assumption that y is real valued.*

Taking $s = \rho + n + 2$ for $n = 0, 1, 2, \ldots$ the integral condition on y becomes

$$\int_0^\infty (e^{-t})^{n+1}(e^{-t})^{\rho+1} y(t)\, dt = 0.$$

Make a change of variables in this integral by introducing $u = e^{-t}$. This yields

$$\int_0^1 u^n [u^{\rho+1} y(-\ln u)]\, du = 0 \quad \text{for} \quad n = 0, 1, 2, \ldots.$$

Now define $z(u) = u^{\rho+1} y(-\ln u)$ on $(0, 1]$ and $z(0) = 0$. Show that z is continuous on $[0, 1]$, paying special attention to its behavior near 0. It will suffice to prove that $z(u) \equiv 0$.

Since the last integral condition holds for $n = 0, 1, 2, \ldots$, it follows that

$$\int_0^1 P(u)z(u)\, du = 0 \tag{14}$$

for every polynomial P.

Suppose (for contradiction) that $z(u) \ne 0$ for some $u \in [0, 1]$. Then, since z is continuous, there must exist numbers a, b, a_1, b_1, and c such that $0 \le a_1 < a < b < b_1 \le 1$, $c > 0$,

$$|z(u)| \ge c \quad \text{for} \quad a \le u \le b,$$

and

$$|z(u)| > 0 \quad \text{for} \quad a_1 \le u \le b_1.$$

Without loss of generality we can assume

$$z(u) > 0 \quad \text{for} \quad a_1 \le u \le b_1.$$

Again invoking the continuity of z on $[0, 1]$, there must exist M such that

$$|z(u)| \le M \quad \text{for} \quad 0 \le u \le 1.$$

Now consider the quadratic polynomial defined by

$$Q(u) = 1 + (u - a)(b - u) \quad \text{for} \quad 0 \le u \le 1.$$

Show that

$$Q(u) \geq 1 \quad \text{for} \quad u \in [a, b]$$

and

$$0 < Q(u) < 1 \quad \text{for} \quad u \notin [a, b].$$

Choose an integer m so large that

$$[Q(u)]^m < \frac{c(b - a)}{2M} \quad \text{for} \quad u \notin [a_1, b_1].$$

Why is this possible? Then

$$\int_0^1 [Q(u)]^m z(u) \, du \geq c(b - a) - \frac{c(b - a)}{2M} M > 0,$$

which contradicts (14).

[Note that the above argument actually proves more than was asserted in Theorem C. Instead of the values of $X(s)$ for all real $s > \rho$, we have used only the values at $s = \rho + 2, \rho + 3, \ldots$]

41. EQUATIONS WITH MORE COMPLICATED FORCING FUNCTION, h

Up to this point, when we have considered the equation

$$x^{(n)} + a_{n-1} x^{(n-1)} + \cdots + a_0 x = h(t), \tag{1}$$

we have always assumed h to be continuous. Now we are going to extend the concept of "solution" to the case of piecewise continuous h.

As usual let us impose initial conditions

$$x(0) = b_0, \quad x'(0) = b_1, \quad \ldots, \quad x^{(n-1)}(0) = b_{n-1}. \tag{2}$$

DEFINITION

Let h be piecewise continuous on $[0, \infty)$. Then a *solution* of Eqs. (1) and (2) is a function $x: [0, \beta) \to \mathbf{C}$ for some $\beta \in (0, \infty]$ such that

(i) $x, x', \ldots,$ and $x^{(n-1)}$ are continuous and satisfy Eq. (2), and
(ii) $x^{(n)}$ is piecewise continuous and satisfies Eq. (1) on $[0, \beta)$ except at points of discontinuity of h.

DEFINITION

If Eqs. (1) and (2) have a solution x on $[0, \beta)$, we shall say the solution is *unique* if, for every other solution $\tilde{x}$, $\tilde{x}(t) = x(t)$ as far as both solutions are defined.

THEOREM A

If h is piecewise continuous on $[0, \infty)$, then Eqs. (1) and (2) have a unique solution x on the entire interval $[0, \infty)$. Moreover, if h is of exponential order, then $x, x', \ldots, x^{(n-1)}$ are also of exponential order.

Proof. Choose any (finite) number $b > 0$. Then h is continuous on $[0, b]$ except possibly at a finite number of points, $t_1 < t_2 < \cdots < t_p$, and at these points the left- and right-hand limits of h exist. (At 0 and b, h has right- and left-hand limits, respectively.)

Thus on $[0, t_1]$ we can treat Eqs. (1) and (2) just as if h were continuous on $\mathbf{R}$. Indeed, one could imagine replacing h with the continuous function $\tilde{h}$ defined by

$$\tilde{h}(t) = \begin{cases} \lim_{\sigma \to 0+} h(\sigma) & \text{for} \quad t \leq 0 \\ h(t) & \text{for} \quad 0 < t < t_1 \\ \lim_{\sigma \to t_1-} h(\sigma) & \text{for} \quad t \geq t_1. \end{cases}$$

The unique solution obtained on $[0, t_1]$ using the function $\tilde{h}$ will also be valid when the original h is used. Why? Since $x, x', \ldots, x^{(n-1)}$ are continuous on $[0, t_1]$ it follows from Eq. (1) that $x^{(n)}$ is continuous on $(0, t_1)$ and has a (finite) right-hand limit at 0 and left-hand limit at t_1.

Now consider Eq. (1) on the interval $[t_1, t_2]$ and require as initial conditions

$$x(t_1) = \lim_{t \to t_1-} x(t), \quad x'(t_1) = \lim_{t \to t_1-} x'(t), \quad \ldots, \quad x^{(n-1)}(t_1) = \lim_{t \to t_1-} x^{(n-1)}(t).$$

Subject to these new initial conditions Eq. (1) has a unique solution on $[t_1, t_2]$. And this provides a continuation of the solution of Eqs. (1) and (2) to $[0, t_2]$.

Using mathematical induction one then establishes the existence of a unique solution of Eqs. (1) and (2) on $[0, b]$. But since $b > 0$ was arbitrary one can argue that the unique solution actually continues to $[0, \infty)$.

If we further assume h to be of exponential order, then the proof that x, $x', \ldots$, and $x^{(n-1)}$ are of exponential order is word-for-word what it was for Theorem 40-B. ∎

This theorem shows that the Laplace transform method is applicable to Eqs. (1) and (2) for the case when h is piecewise continuous and of exponential order—just as in the case of continuous h. Before solving equations with these more complicated functions h, we compute some further Laplace transforms which will be useful.

DEFINITION

The *unit step function* (or Heaviside function) is defined by

$$u(t) = \begin{cases} 0 & \text{for} \quad t < 0 \\ 1 & \text{for} \quad t \geq 0. \end{cases}$$

Note that, given any number $a \geq 0$,

$$u(t - a) = \begin{cases} 0 & \text{for} \quad t < a \\ 1 & \text{for} \quad t \geq a \end{cases}$$

is a piecewise continuous function of exponential order. It is easy to verify (Problem 1) that

$$\mathscr{L}[u(t - a)](s) = \frac{1}{s} e^{-as} \quad \text{for} \quad \text{Re } s > 0. \tag{3}$$

THEOREM B

Let g be a piecewise continuous function of exponential order, $|g(t)| \leq Me^{pt}$ for $t \geq 0$, and let $a \geq 0$. If a new function is defined by

$$f(t) = u(t - a)g(t - a) \quad \text{for} \quad t \geq 0$$

(see Figure 1), then

$$\mathscr{L}[u(t - a)g(t - a)](s) = e^{-as}\mathscr{L}[g](s) \quad \text{for} \quad \text{Re } s > p.$$

Proof. For $t \geq 0$,

$$|f(t)| \leq Me^{p(t-a)} = Me^{-pa}e^{pt}.$$

Thus f is also of exponential order, with the same p, and clearly f is also piecewise continuous. This establishes the existence of $\mathscr{L}[f](s)$ for $\text{Re } s > p$,

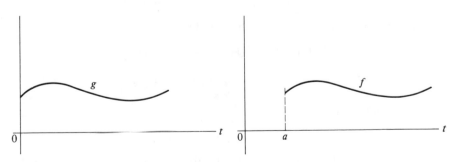

FIGURE 1

and we can now compute

$$\mathscr{L}[f](s) = \int_0^\infty e^{-st}u(t-a)g(t-a)\,dt$$

$$= \int_a^\infty e^{-st}g(t-a)\,dt$$

$$= \int_0^\infty e^{-s(\sigma+a)}g(\sigma)\,d\sigma$$

$$= e^{-as}\mathscr{L}[g](s). \quad\blacksquare$$

Example 1. Compute the Laplace transform of the function defined by

$$g(t) = \begin{cases} 0 & \text{for} \quad 0 \le t < a, \\ 1 & \text{for} \quad a \le t < b, \\ 0 & \text{for} \quad t \ge b, \end{cases}$$

a single "square pulse" (Figure 2).

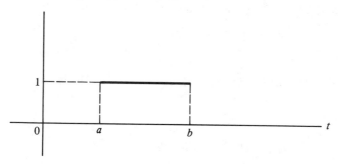

FIGURE 2

The transform is easily computed by noting that

$$g(t) = u(t-a) - u(t-b).$$

Thus, using Theorem B and linearity, we find

$$\mathscr{L}[g](s) = e^{-as}\mathscr{L}[1](s) - e^{-bs}\mathscr{L}[1](s)$$

$$= \frac{1}{s}(e^{-as} - e^{-bs}) \quad \text{for} \quad \text{Re } s > 0.$$

Example 2. Solve for x:

$$Lx' + Rx = v(t) = \begin{cases} 0 & \text{for} \quad 0 \le t < a, \\ E & \text{for} \quad a \le t < b, \\ 0 & \text{for} \quad t \ge b. \end{cases}$$

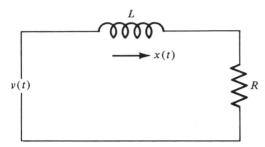

FIGURE 3

This is the equation for the current $x(t)$ in amperes in a series circuit contain-ing inductance L henries and resistance R ohms. (Figure 3.) The applied voltage is $v(t) = Eg(t)$ volts, where g is the single pulse of Example 1.

Let us assume the circuit is inactive at $t = 0$, that is,

$$x(0) = 0.$$

Using the result of Example 1, the equation for $X = \mathscr{L}[x]$ is

$$Ls X(s) + R X(s) = \mathscr{L}[v](s) = \frac{E}{s}(e^{-as} - e^{-bs})$$

or

$$X(s) = \frac{E}{Ls(s + R/L)}(e^{-as} - e^{-bs}) \quad \text{for} \quad s > 0.$$

Using either partial fractions or the method of Example 40-2, one easily finds

$$\mathscr{L}^{-1}\left[\frac{E}{Ls(s + R/L)}\right](t) = \frac{E}{R}(1 - e^{-Rt/L}).$$

But then Theorem B gives

$$x(t) = \frac{E}{R}[1 - e^{-R(t-a)/L}]u(t - a) - \frac{E}{R}[1 - e^{-R(t-b)/L}]u(t - b)$$

$$= \begin{cases} 0, & \text{for} \quad 0 \le t < a \\[2mm] \dfrac{E}{R}[1 - e^{-R(t-a)/L}], & \text{for} \quad a \le t < b \\[2mm] \dfrac{E}{R}[e^{-R(t-b)/L} - e^{-R(t-a)/L}], & \text{for} \quad t \ge b, \end{cases}$$

as illustrated in Figure 4.

Theorem C which follows is sometimes useful in computing the Laplace

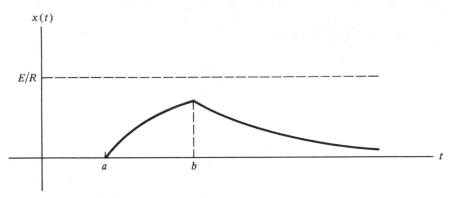

FIGURE 4

transform of a periodic function. Theorem D will be useful in computing Laplace transforms of certain functions, both periodic and nonperiodic.

THEOREM C

Let $g: [0, \infty) \to \mathbf{C}$ be piecewise continuous and periodic, that is, $g(t + T) = g(t)$ for all $t \geq 0$ for some $T > 0$, where T is called a "period" of g. Then $\mathscr{L}[g]$ exists and

$$\mathscr{L}[g](s) = \frac{\int_0^T e^{-st} g(t)\, dt}{1 - e^{-sT}} \quad \text{for} \quad \text{Re } s > 0. \tag{4}$$

Proof. Since g is piecewise continuous, it must be bounded on $[0, T]$. Why? Thus $|g(t)| \leq M$ for all $t \geq 0$, that is, g is of exponential order with $\rho = 0$. So it follows that $\mathscr{L}[g](s)$ exists whenever Re $s > 0$.

Now note that

$$g(t) - u(t - T)g(t - T) = \begin{cases} g(t) & \text{for} \quad 0 \leq t < T \\ 0 & \text{for} \quad t \geq T. \end{cases}$$

Thus, taking the Laplace transform of both sides of this last equation, we get

$$\mathscr{L}[g](s) - e^{-Ts}\mathscr{L}[g](s) = \int_0^T e^{-st} g(t)\, dt \quad \text{for} \quad \text{Re } s > 0,$$

which is equivalent to (4). ∎

Example 3. Compute the Laplace transform of the "square wave" function defined by

$$g(t) = \begin{cases} 1 & \text{for} \quad nT \leq t < (n + \tfrac{1}{2})T \\ -1 & \text{for} \quad (n + \tfrac{1}{2})T \leq t < (n + 1)T \end{cases} \quad n = 0, 1, 2, \ldots \tag{5}$$

(Figure 5.)

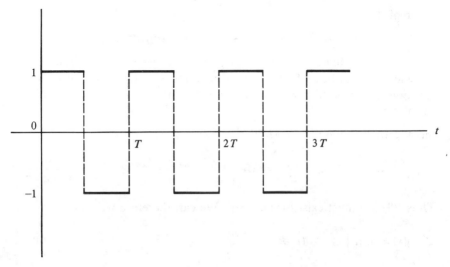

FIGURE 5

With the aid of Theorem C we easily find

$$\mathscr{L}[g](s) = \frac{1}{1 - e^{-sT}}\left[\int_0^{T/2} e^{-st}\,dt - \int_{T/2}^{T} e^{-st}\,dt\right]$$

$$= \frac{1}{1 - e^{-sT}}\left[\frac{1}{s}(1 - e^{-sT/2}) - \frac{1}{s}(e^{-sT/2} - e^{-sT})\right]$$

$$= \frac{(1 - e^{-sT/2})^2}{s(1 - e^{-sT})} = \frac{1 - e^{-sT/2}}{s(1 + e^{-sT/2})} \quad \text{for} \quad \text{Re } s > 0.$$

THEOREM D

Let $f: [0, \infty) \to \mathbf{R}$ be piecewise continuous and of exponential order, $|f(t)| \le Me^{\rho t}$ where $\rho > 0$, and let $\mathscr{L}[f](s) = F(s)$ for $s > \rho$. Then, given $a > 0$,

$$\mathscr{L}[f(t) - u(t - a)f(t - a) + u(t - 2a)f(t - 2a) - \cdots](s)$$

$$= F(s)\{1 - e^{-as} + e^{-2as} - \cdots\} = \frac{F(s)}{1 + e^{-as}} \quad (6)$$

and

$$\mathscr{L}[f(t) + u(t - a)f(t - a) + u(t - 2a)f(t - 2a) + \cdots](s)$$

$$= F(s)\{1 + e^{-as} + e^{-2as} + \cdots\} = \frac{F(s)}{1 - e^{-as}} \quad (7)$$

for all $s > \rho$.

Proof. Let

$$g(t) = f(t) - u(t-a)f(t-a) + u(t-2a)f(t-2a) - \cdots$$

for $t \geq 0$. Note that there is no problem of convergence for this infinite series since at any t all but a finite number of terms are zero. Moreover, the new function g will be piecewise continuous and of exponential order. Indeed, for all $t \geq 0$,

$$|g(t)| \leq Me^{\rho t} + Me^{\rho(t-a)} + Me^{\rho(t-2a)} + \cdots$$

$$= Me^{\rho t}(1 + e^{-\rho a} + e^{-2\rho a} + \cdots) = \frac{M}{1 - e^{-\rho a}} e^{\rho t}.$$

Thus $\mathscr{L}[g](s)$ must exist for all $s > \rho$. We can therefore write

$$\mathscr{L}[g](s) = \lim_{n \to \infty} \int_0^{na} e^{-st} g(t)\, dt$$

$$= \lim_{n \to \infty} \sum_{k=0}^{n-1} \int_0^{na} e^{-st}(-1)^k u(t-ka)f(t-ka)\, dt$$

$$= \lim_{n \to \infty} \sum_{k=0}^{n-1} (-1)^k \bigg\{ \mathscr{L}[u(t-ka)f(t-ka)](s)$$

$$- \int_{na}^{\infty} e^{-st} u(t-ka)f(t-ka)\ dt \bigg\}$$

$$= \lim_{n \to \infty} \sum_{k=0}^{n-1} (-1)^k \bigg\{ e^{-kas}F(s) - \int_{na}^{\infty} e^{-st}f(t-ka)\, dt \bigg\}.$$

To show that this gives Eq. (6) it suffices to note that, for $s > \rho$,

$$\sum_{k=0}^{n-1} \bigg| \int_{na}^{\infty} e^{-st}f(t-ka)\, dt \bigg| \leq \sum_{k=0}^{n-1} \int_{na}^{\infty} Me^{-st}e^{\rho t - \rho ka}\, dt$$

$$= \sum_{k=0}^{n-1} Me^{-k\rho a} \frac{e^{(\rho-s)na}}{s-\rho} < \frac{M}{1-e^{-\rho a}} \frac{e^{(\rho-s)na}}{s-\rho} \to 0 \quad \text{as} \quad n \to \infty.$$

The similar proof of Eq. (7) is left to the reader as Problem 2. ∎

Example 4. Compute the Laplace transform of the function defined by

$$g(t) = \begin{cases} E \sin \omega t & \text{for } \dfrac{2n\pi}{\omega} \leq t < \dfrac{(2n+1)\pi}{\omega}, \\[2ex] -E \sin \omega t & \text{for } \dfrac{(2n+1)\pi}{\omega} \leq t < \dfrac{(2n+2)\pi}{\omega} \end{cases}$$

for $n = 0, 1, 2, \ldots$, the output of an idealized "full wave rectifier," Figure 6.

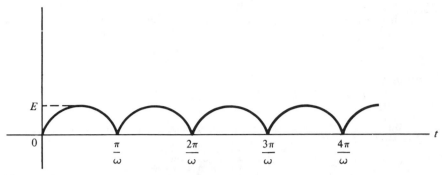

FIGURE 6

If we define $f(t) = E \sin \omega t$, then

$$g(t) = f(t) + 2u\left(t - \frac{\pi}{\omega}\right) f\left(t - \frac{\pi}{\omega}\right) + 2u\left(t - \frac{2\pi}{\omega}\right) f\left(t - \frac{2\pi}{\omega}\right) + \cdots$$

$$= 2\left[f(t) + u\left(t - \frac{\pi}{\omega}\right) f\left(t - \frac{\pi}{\omega}\right) + \cdots\right] - f(t).$$

Thus, using Eq. (7), for $s > 0$

$$\mathscr{L}[g](s) = \mathscr{L}[E \sin \omega t](s)\left[\frac{2}{1 - e^{-\pi s/\omega}} - 1\right]$$

$$= \frac{E\omega}{s^2 + \omega^2} \cdot \frac{1 + e^{-\pi s/\omega}}{1 - e^{-\pi s/\omega}}.$$

PROBLEMS

1. Verify Eq. (3) by direct integration.
2. Prove Eq. (7) of Theorem D.
3. Use the theorems of this section to verify the following Laplace transforms.

(a) If $g(t) = \begin{cases} t & \text{for} \quad 0 \le t < \pi \\ 0 & \text{for} \quad t \ge \pi, \end{cases}$

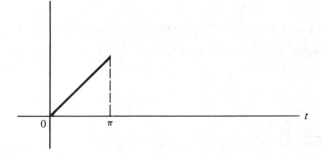

then $G(s) = (1 - e^{-\pi s})\dfrac{1}{s^2} - \pi e^{-\pi s}\dfrac{1}{s}$. Compare Problem 39-5(d).

(b) $g(t) = \begin{cases} E \sin \omega t & \text{for} \quad \dfrac{2n\pi}{\omega} \le t < \dfrac{(2n+1)\pi}{\omega} \\[3mm] 0 & \text{for} \quad \dfrac{(2n+1)\pi}{\omega} \le t < \dfrac{(2n+2)\pi}{\omega}, \end{cases}$

for $n = 0, 1, 2, \ldots$ (the output of an idealized "half-wave rectifier"), then

$$G(s) = \frac{E\omega}{s^2 + \omega^2} \cdot \frac{1}{1 - e^{-\pi s/\omega}}.$$

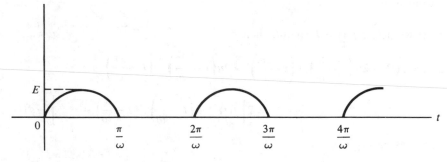

(c) If $g(t) = t/T - n$ for $nT \le t < (n+1)T$ for $n = 0, 1, \ldots$ (a "saw tooth" wave), then

$$G(s) = \frac{1}{Ts^2} - \frac{e^{-sT}}{s(1 - e^{-sT})}.$$

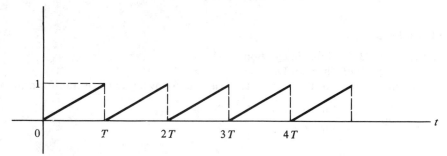

(d) If g is the square wave of Example 3, then recompute $G(s)$ using Theorem D.

4. Find the inverses (if they exist) of the following Laplace transforms and sketch the resulting functions x.

(a) $X(s) = \dfrac{1}{as^2}(1 - 2e^{-as} + e^{-2as})$ for $s > 0$, where $a > 0$.

(b) $X(s) = \dfrac{1}{as^2} \cdot \dfrac{1 - e^{-as}}{1 + e^{-as}}$, for $s > 0$, where $a > 0$.

(c) $X(s) = \dfrac{1}{s(s-\lambda)} \cdot \dfrac{1}{1+e^{-as}}$, for $s > 0$, where $\lambda < 0$ and $a > 0$.

(d) $X(s) = \dfrac{1}{1+e^{-as}}$ for $s > 0$, where $a > 0$.

5. Find the current x in the circuit of Figure 3 if

(a) $v(t) = \begin{cases} t & \text{for } 0 \le t < \pi \\ 0 & \text{for } t \ge \pi \end{cases}$ [the g of Problem 3(a)],

and $x(0) = 1/R$.

(b) $v(t) \quad \begin{cases} E & \text{for } nT \le t < (n+\tfrac{1}{2})T \\ 0 & \text{for } (n+\tfrac{1}{2})T \le t < (n+1)T \end{cases}$

for $n = 0, 1, 2, \ldots$, and $x(0) = 0$.

42. LINEAR SYSTEMS

The Laplace transform is sometimes a useful tool for solving linear differential systems—especially in the case of constant coefficients.

For the system

$$\mathbf{x}' = A\mathbf{x} + \mathbf{h}(t) \tag{1}$$

with initial condition

$$\mathbf{x}(0) = \mathbf{x}_0, \tag{2}$$

the Laplace transform method is a completely natural extension of the method described for scalar equations. We will consider Eq. (1) on $J = (\alpha, \infty)$ where $\alpha < 0$.

Let a vector-valued function $\mathbf{g}$ be defined on $[0,\infty)$. We shall say $\mathbf{g}$ is *piecewise continuous* and of *exponential order* if each component of $\mathbf{g}$ is piecewise continuous and of exponential order. For such a vector-valued function the Laplace transform is the vector-valued function defined by

$$\mathcal{L}[\mathbf{g}](s) = \mathbf{G}(s) = \int_0^\infty e^{-st}\mathbf{g}(t)\,dt = \begin{pmatrix} \displaystyle\int_0^\infty e^{-st}g_1(t)\,dt \\ \vdots \\ \displaystyle\int_0^t e^{-st}g_n(t)\,dt \end{pmatrix}$$

where Re s is sufficiently large to assure convergence of these improper integrals.

In order to take the Laplace transform of both sides of Eq. (1) we must know, in addition to the behavior of $\mathbf{h}$, that $\mathbf{x}$ exists and is of exponential order. If we consider only the case of continuous $\mathbf{h}$, then the existence of a unique solution $\mathbf{x}$ of (1) and (2) follows from Theorems 24-B and 26-B. If we now assume also that $\mathbf{h}$ is of exponential order, then it follows from Theorem 27-F that $\mathbf{x}$ is also of exponential order. Why?

Invoking Theorem 39-C, we now take the Laplace transform of both sides of (1). Using the initial condition (2), we find

$$sX(s) - \mathbf{x}_0 = AX(s) + H(s), \tag{3}$$

where X and H are vector-valued Laplace transforms.

It is then a matter of solving for $X(s)$ and finally finding the unique continuous function $\mathbf{x}$ such that $\mathscr{L}[\mathbf{x}] = X$.

One simple example should suffice to illustrate the method. The example is a familiar one, but the Laplace transform method appears to give a more automatic solution than the method of Example 24-2.

Example 1. Solve

$$\mathbf{y}' = \begin{pmatrix} -1 & -1 \\ 1 & -3 \end{pmatrix} \mathbf{y} \quad \text{with} \quad \mathbf{y}(0) = \mathbf{y}_0. \tag{4}$$

Taking the Laplace transform gives

$$sY(s) - \mathbf{y}_0 = \begin{pmatrix} -1 & -1 \\ 1 & -3 \end{pmatrix} Y(s)$$

or

$$\begin{pmatrix} 1 + s & 1 \\ -1 & 3 + s \end{pmatrix} \begin{pmatrix} Y_1(s) \\ Y_2(s) \end{pmatrix} = \begin{pmatrix} y_{01} \\ y_{02} \end{pmatrix}.$$

This pair of algebraic equations can be solved for $Y_1(s)$ and $Y_2(s)$ as long as the determinant of the coefficients is not zero, that is, as long as

$$s^2 + 4s + 4 = (s + 2)^2 \neq 0.$$

Taking $s > -2$, therefore, the solution is

$$Y(s) = \begin{pmatrix} Y_1(s) \\ Y_2(s) \end{pmatrix} = \frac{1}{(s + 2)^2} \begin{pmatrix} (3 + s)y_{01} - y_{02} \\ y_{01} + (1 + s)y_{02} \end{pmatrix}.$$

Now refer to Table 39-1 to find the inverse transform of each component of Y:

$$y_1(t) = y_{01}e^{-2t} + (y_{01} - y_{02})te^{-2t},$$

$$y_2(t) = y_{02}e^{-2t} + (y_{01} - y_{02})te^{-2t}.$$

(Problem 1.)

The forcing functions involved in the problems below are all elementary functions. To handle more complicated forcing functions one would proceed as in Section 41.

PROBLEMS

1. Verify the computations of Example 1, especially the computation of y_1 and y_2 from Y_1 and Y_2.

2. Solve the following systems using the Laplace transform method.

(a) $\mathbf{x}' = \begin{pmatrix} -1 & -1 \\ 1 & -3 \end{pmatrix} \mathbf{x} + \begin{pmatrix} 0 \\ e^{-2t} \end{pmatrix}$, $\mathbf{x}(0) = \mathbf{x}_0$.

(b) $\mathbf{x}' = \begin{pmatrix} 1 & 2 \\ -1 & -1 \end{pmatrix} \mathbf{x} + \begin{pmatrix} 1 \\ -2 \end{pmatrix}$, $\mathbf{x}(0) = \mathbf{x}_0$.

(c) $\mathbf{x}' = \begin{pmatrix} 1 & 2 \\ -1 & -1 \end{pmatrix} \mathbf{x} + \begin{pmatrix} 2 \cos t \\ 0 \end{pmatrix}$, $\mathbf{x}(0) = \mathbf{x}_0$.

APPENDIXES

1. NOTATION FOR SETS, FUNCTIONS, AND DERIVATIVES

Some notation used throughout the text is assembled here for ready reference.

Sets

A set can be described either by listing its elements (or members or points), regardless of order, or by stating some properties which characterize the elements. For example, the set of all integers from one through four can be written

$$X = \{1, 2, 3, 4\},$$

$$X = \{x : x \text{ an integer, } 1 \le x \le 4\},$$

$$X = \{x \text{ an integer} : 0 < x^3 < 117\},$$

or any number of other ways.

The symbol **R** stands for the set of all real numbers and **C** stands for the set of all complex numbers. For intervals of real numbers we use the notation

$$[\alpha, \beta] = \{x \text{ real} : \alpha \le x \le \beta\}, \quad \text{a closed interval,}$$

$$(\alpha, \beta) = \{x \text{ real} : \alpha < x < \beta\}, \quad \text{an open interval,}$$

$$\left. \begin{array}{l} [\alpha, \beta) = \{x \text{ real} : \alpha \le x < \beta\}, \\ (\alpha, \beta] = \{x \text{ real} : \alpha < x \le \beta\}, \end{array} \right\} \quad \text{half-open intervals.}$$

We use the standard symbols $\in$ and $\subset$ which are interpreted as follows:

$x \in A$ is read "x is a member (or element) of A,"

or "x belongs to A" or "x in A"

$A \subset B$ is read "A is a subset of B" or "A is contained in B."

To negate either of the symbols $\in$ or $\subset$ we write $\notin$ or $\not\subset$, respectively. Thus, for example,

$$5 \notin \{x : x \text{ an integer, } 0 < x^3 < 117\},$$

$$[-3, -2) \subset \{x \in \mathbf{R} : x^2 + 4x + 3 \le 0\},$$

$$\{3, 4\} \not\subset \{2, 4\}.$$

If X_i is a set for each $i = 1, 2, \ldots, n$, then the *product* of these sets,

$$X_1 \times X_2 \times \cdots \times X_n \quad \text{or} \quad \mathop{\times}_{i=1}^{n} X_i$$

is the set of all ordered *n*-tuples:

$$\{(x_1, \ldots, x_n): x_i \in X_i \quad \text{for} \quad i = 1, \ldots, n\}.$$

We shall sometimes denote an ordered *n*-tuple $(x_1, \ldots, x_n)$ simply by $\mathbf{x}$. Thus, for example, we have for the "open rectangle" in $\mathbf{R} \times \mathbf{R}$:

$$(\alpha, \beta) \times (\gamma, \delta) = \{\mathbf{x} = (x_1, x_2): \alpha < x_1 < \beta, \gamma < x_2 < \delta\}.$$

This last equation points up a defect in our notation. The same symbol (,) can mean an open interval or an ordered pair. Some authors avoid this difficulty by writing an open interval as $]\alpha, \beta[$ instead of (α, β), while others keep the parentheses for open intervals and write $\langle x_1, x_2 \rangle$ for an ordered pair. We shall stick with the more common parentheses notation for both open intervals and ordered pairs, and trust that the meaning will always be clear from the context.

In the special case when each $X_i = X$ we write simply X^n, instead of $X_1 \times X_2 \times \cdots \times X_n$ for the set of ordered *n*-tuples of members of X.

In $\mathbf{R}^n$, an *open ball* of radius $\delta > 0$ with center at a point $\mathbf{y} = (y_1, \ldots, y_n)$ is defined by

$$B_\delta(\mathbf{y}) = \left\{\mathbf{x} = (x_1, \ldots, x_n) \in \mathbf{R}^n: \sum_{i=1}^{n} |x_i - y_i| < \delta\right\}.$$

*DEFINITION

A set of $\mathbf{G} \subset \mathbf{R}^n$ is said to be an *open set* if for every $\mathbf{y} \in G$ there exists a number $\delta > 0$ such that $B_\delta(\mathbf{y}) \subset G$.

It is easy to verify that, if $-\infty < \alpha < \beta < \infty$, the "open interval" (α, β) is an open set in $\mathbf{R}$, but the intervals $[\alpha, \beta]$, $[\alpha, \beta)$, and $(\alpha, \beta]$ are not open. And in $\mathbf{R}^n$ it can be shown that an *open rectangle*

$$(\alpha_1, \beta_1) \times (\alpha_2, \beta_2) \times \cdots \times (\alpha_n, \beta_n)$$

is an open set, and that an open ball itself is an open set.

*DEFINITION

A set $A \subset \mathbf{R}^n$ is said to be a *closed set* if every convergent sequence of members of A has its limit in A, that is, if $\mathbf{x}_{(1)}, \mathbf{x}_{(2)}, \ldots \in A$ and $\lim_{k \to \infty} \mathbf{x}_{(k)} = \xi$ imply $\xi \in A$. [$\lim_{k \to \infty} \mathbf{x}_{(k)} = \xi$ means that each component $x_{(k)j}$ of $\mathbf{x}_{(k)}$ approaches the corresponding component ξ_j of ξ as $k \to \infty$.]

In the simplest case, when $n = 1$ and $-\infty < \alpha < \beta < \infty$, we find that the "closed interval" $[\alpha, \beta]$ is a closed set, but the intervals (α, β), $[\alpha, \beta)$, and

$(\alpha, \beta]$ are not closed. One can also show that, if $-\infty < \alpha_i < \beta_i < \infty$ for each i, the *closed rectangle*

$$[\alpha_1, \beta_1] \times [\alpha_2, \beta_2] \times \cdots \times [\alpha_n, \beta_n]$$

is a closed set in $\mathbf{R}^n$. An open ball $B_\delta(\mathbf{y})$ is not a closed set in $\mathbf{R}^n$.

Functions

If X and Y are two nonempty sets and if a rule f is specified which assigns to each x in X a unique element of Y, called $f(x)$, then f is called a *function* (or a *mapping* or a *transformation*). This will often be abbreviated

$$f: X \rightarrow Y.$$

We say that f maps X into Y and we refer to X as the *domain* of f.

Note that, strictly speaking, f *stands for the function*, while $f(x)$ *is an element of* Y. But in many books, including this one, the author occasionally refers to the function itself as $f(x)$.

Derivatives

For the derivative of a function $x: (\alpha, \beta) \rightarrow \mathbf{R}$, where (α, β) is an interval in $\mathbf{R}$, we shall generally use the notation x'. That is,

$$x'(t) = \lim_{h \to 0} \frac{x(t+h) - x(t)}{h} \quad \text{for} \quad \alpha < t < \beta.$$

On occasion it will be convenient to use Dx or dx/dt instead of x'.

The second derivative will be denoted by x'', or $D^2 x$, or d^2x/dt^2. For the kth derivative we will write $x^{(k)}$ or $D^k x$, or sometimes $d^k x/dt^k$.

If $D = (\alpha_1, \beta_1) \times \cdots \times (\alpha_n, \beta_n)$ is an open rectangle in $\mathbf{R}^n$, and v is a function mapping $D \rightarrow \mathbf{R}$, then we shall denote the first partial derivatives of v by $D_i v$ for $i = 1, \ldots, n$, where

$$D_i v(x_1, \ldots, x_n) = \lim_{h \to 0} \frac{v(x_1, \ldots, x_i + h, \ldots, x_n) - v(x_1, \ldots, x_i, \ldots, x_n)}{h},$$

that is, the partial derivative with respect to the ith independent variable. We shall sometimes write $\partial v/\partial x_i$ instead of $D_i v$.

Note that with the D_i notation we will have

$$D_2 v(t, x_1, x_2) = \frac{\partial v}{\partial x_1}(t, x_1, x_2).$$

In other words, D_i indicates differentiation with respect to the ith independent variable—and this need not be the variable called x_i.

2. SOME THEOREMS FROM CALCULUS

We shall state (mostly without proof) several important theorems from calculus. The proofs can be found in any book on advanced calculus, and in many introductory calculus texts. The references given below are to Landau [10] and Fulks [5].

THEOREM A (MEAN VALUE THEOREM)

Let $[\alpha, \beta]$ be a closed, bounded interval, that is, $-\infty < \alpha < \beta < \infty$. Let $f: [\alpha, \beta] \to \mathbf{R}$ be continuous and let f' exist on (α, β). Then there exists a number $\theta \in (\alpha, \beta)$ such that

$$f(\beta) - f(\alpha) = f'(\theta)(\beta - \alpha).$$

(Landau, Theorem 159; or Fulks, Theorem 3.10c.)

THEOREM B (TAYLOR'S THEOREM, OR THE GENERALIZED MEAN VALUE THEOREM)

Let $[\alpha, \beta]$ be a closed bounded interval. Let $f: [\alpha, \beta] \to \mathbf{R}$ be such that, for some integer $m \geq 1$, $f^{(m-1)}$ is continuous on $[\alpha, \beta]$ and $f^{(m)}$ exists on (α, β). Then for each t_0 and t in $[\alpha, \beta]$ there exists θ between t_0 and t such that

$$f(t) = \sum_{k=0}^{m-1} \frac{f^{(k)}(t_0)}{k!} (t - t_0)^k + \frac{f^{(m)}(\theta)}{m!} (t - t_0)^m. \tag{1}$$

(Landau, Theorems 177 and 178; or Fulks, Theorem 6.3c.)

[*Note:* In case $t_0 = \alpha$ or $t_0 = \beta$, we interpret $f'(t_0), \ldots, f^{(m-1)}(t_0)$ as appropriate one-sided derivatives. Observe that Theorem A is a special case of Theorem B.]

DEFINITION

Let $J = (\alpha, \beta) \subset \mathbf{R}$ and let $f: J \to \mathbf{R}$. We shall say f is *analytic* at $t_0 \in J$ if f has derivatives of all orders at t_0 and there exists $R > 0$ such that $(t_0 - R, t_0 + R) \subset J$ and

$$f(t) = \sum_{k=0}^{\infty} \frac{f^{(k)}(t_0)}{k!} (t - t_0)^k \quad \text{for all} \quad t \quad \text{in} \quad (t_0 - R, t_0 + R). \tag{2}$$

Comparing Eqs. (1) and (2) we see that f is analytic at t_0 if it has derivatives of all orders at t_0, *and* for some $R > 0$

$$\lim_{m \to \infty} \frac{f^{(m)}(\theta)}{m!} (t - t_0)^m = 0 \quad \text{for each} \quad t \quad \text{in} \quad (t_0 - R, t_0 + R),$$

where $\theta = \theta(t_0, t, m)$ is defined by Eq. (1). In this case, the infinite power series in (2) is called a *Taylor's series*.

Remark. It should be emphasized that the existence (and continuity) of all derivatives of f is *not* sufficient for analyticity. For example, if we define $f: \mathbf{R} \to \mathbf{R}$ by

$$f(t) = \begin{cases} 0 & \text{for } t \le 0 \\ e^{-1/t^2} & \text{for } t > 0, \end{cases}$$

then it can be shown that f has derivatives of all orders for all t and, in fact, $f^{(j)}(0) = 0$ for $j = 1, 2, \ldots$. Thus f cannot be analytic at $t_0 = 0$, since Eq. (2) would dictate $f(t) \equiv 0$ for $-R < t < R$ for some $R > 0$. (See Landau, Theorem 185; or Fulks, Section 15.3.) A function of similar form appears in the text in Example 31-5.

THEOREM C (CHAIN RULE FOR A FUNCTION OF n VARIABLES)

Let $D = (\alpha_1, \beta_1) \times \cdots \times (\alpha_n, \beta_n)$, an open rectangle in $\mathbf{R}^n$, and let f be a real-valued function on D having continuous first partial derivatives. Let J be an open interval in $\mathbf{R}$ and let $g_j: J \to (\alpha_j, \beta_j)$ be differentiable for each $j = 1, \ldots, n$. Then, for all t in J,

$$\frac{d}{dt} f(g_1(t), \ldots, g_n(t)) = \sum_{j=1}^{n} D_j f(g_1(t), \ldots, g_n(t)) g_j'(t). \tag{3}$$

(See Landau, Theorems 302 and 303 where $n = 2$; or Fulks, Theorems 9.2c and 9.4a.)

Remark. The chain rule theorem when f is a function of a single real variable (the case $n = 1$) has weaker hypotheses. Instead of requiring f to have a continuous derivative, it suffices in that case to merely require that the derivative of f exist at the point in question. (Landau, Theorem 101; or Fulks, Theorem 3.9b.) To see that this is *not sufficient* if $n > 1$, consider the following example with $n = 2$. Let f be defined by

$$f(\xi_1, \xi_2) = \begin{cases} \dfrac{\xi_1 \xi_2^2}{\xi_1^2 + \xi_2^2} & \text{for } (\xi_1, \xi_2) \ne (0, 0) \\ 0 & \text{for } (\xi_1, \xi_2) = (0, 0), \end{cases}$$

and let $g_1(t) = g_2(t) = t$. Then $f(t, t) = t/2$. Note that $D_1 f$ and $D_2 f$ exist everywhere, and yet

$$\frac{1}{2} = \frac{d}{dt} f(t, t) \bigg|_{t=0} \ne D_1 f(0, 0) g_1'(0) + D_2 f(0, 0) g_1'(0) = 0.$$

The function f is continuous at $(0, 0)$, although this is not quite obvious.

An even more bizarre example is defined by

$$f(\xi_1, \xi_2) = \begin{cases} \dfrac{\xi_1 \xi_2^2}{\xi_1^2 + \xi_2^4} & \text{for} \quad (\xi_1, \xi_2) \neq (0, 0) \\ \\ 0 & \text{for} \quad (\xi_1, \xi_2) = (0, 0). \end{cases}$$

This function has partial derivatives, $D_1 f$ and $D_2 f$ everywhere. And yet f is not even continuous since, for $t \neq 0, f(t^2, t) = \frac{1}{2} \nleftrightarrow f(0, 0)$ as $t \to 0$.

THEOREM D (MEAN VALUE THEOREM FOR A FUNCTION OF n VARIABLES)

Let $A = [\alpha_1, \beta_1] \times \cdots \times [\alpha_n, \beta_n]$ be a closed rectangle in $\mathbf{R}^n$ with $\alpha_j < \beta_j$ for each $j = 1, \ldots, n$. Let $f: A \to \mathbf{R}$ be continuous on A and have continuous first partial derivatives on the open rectangle $(\alpha_1, \beta_1) \times \cdots \times (\alpha_n, \beta_n)$. If $\xi = (\xi_1, \ldots, \xi_n)$ and $\tilde{\xi} = (\tilde{\xi}_1, \ldots, \tilde{\xi}_n)$ are any two points in A, then there exists a point $(\theta_1, \ldots, \theta_n)$ on the line segment connecting ξ and $\tilde{\xi}$ such that

$$f(\tilde{\xi}) - f(\xi) = \sum_{j=1}^{n} D_j f(\theta_1, \ldots, \theta_n)(\tilde{\xi}_j - \xi_j). \tag{4}$$

(This theorem is essentially in Landau, Theorem 304 for $n = 2$; and in Fulks, Theorem 9.5a. However, we shall give a proof here as an application of Theorems A and C. Note that, as in Theorem C, we require continuity of the first partial derivatives of f; not mere existence as in Theorem A.)

Proof. Let $h_j = \tilde{\xi}_j - \xi_j$ for each $j = 1, \ldots, n$, and let us define a new function $F: [0, 1] \to \mathbf{R}$ by setting

$$F(t) = f(\xi_1 + th_1, \ldots, \xi_n + th_n) \quad \text{for} \quad 0 \le t \le 1.$$

Then F, being a composition of continuous functions, is itself continuous on $[0, 1]$. The existence of F' on $(0, 1)$ follows from Theorem C, and (3) gives

$$F'(t) = \sum_{j=1}^{n} D_j f(\xi_1 + th_1, \ldots, \xi_n + th_n) \cdot h_j \quad \text{for} \quad 0 < t < 1.$$

From Theorem A we then conclude that there exists $\theta \in (0, 1)$ such that

$$F(1) - F(0) = F'(\theta).$$

But this is (4) with $(\theta_1, \ldots, \theta_n) = (\xi_1 + \theta h_1, \ldots, \xi_n + \theta h_n)$. ∎

THEOREM E (AN IMPLICIT FUNCTION THEOREM)

Let $D = (\alpha_1, \beta_1) \times (\alpha_2, \beta_2)$, and let $f: D \to \mathbf{R}$. Let f, $D_1 f$, and $D_2 f$ be continuous on D with $D_2 f(t, \xi) \neq 0$ for all $(t, \xi) \in D$. Let

$$f(t_0, x_0) = 0 \quad \text{for some} \quad (t_0, x_0) \in D.$$

Then it follows that for some $q > 0$ there exists a unique function $x: (t_0 - q, t_0 + q) \to \mathbf{R}$ such that

$$(t, x(t)) \in D \quad \text{and} \quad f(t, x(t)) = 0$$

for $t_0 - q < t < t_0 + q$. Moreover, x is continuously differentiable and

$$x'(t) = -\frac{(D_1 f)(t, x(t))}{(D_2 f)(t, x(t))}. \tag{5}$$

(This is a weak form of Landau's Theorems 314 and 315. Corresponding results in higher dimensions are found in Fulks, Theorem 10.6a and Section 9.4, Examples 1 and 2.)

If f is a function of two (or more) real variables we denote mixed second partial derivatives as follows:

$$D_2 D_1 f = D_2(D_1 f).$$

In general, one should *expect* that $D_2 D_1 f \neq D_1 D_2 f$. For example, if $f: \mathbf{R}^2 \to \mathbf{R}$ is defined by

$$f(\xi_1, \xi_2) = \begin{cases} \dfrac{\xi_1 \xi_2 (\xi_1^2 - \xi_2^2)}{\xi_1^2 + \xi_2^2} & \text{for} \quad (\xi_1, \xi_2) \neq (0, 0) \\ 0 & \text{for} \quad (\xi_1, \xi_2) = (0, 0), \end{cases}$$

then a straightforward computation shows that $D_1 D_2 f(0, 0) = -1$ while $D_2 D_1 f(0, 0) = 1$. (Landau, p. 211; or Fulks, p. 239.)

But under the mild conditions of the next theorem it turns out that the order of differentiation *is* interchangeable.

THEOREM F (INTERCHANGING ORDER OF PARTIAL DIFFERENTIATION)

Let f map some open set in $\mathbf{R}^2$ into $\mathbf{R}$. If f, $D_1 f$, $D_2 f$, and $D_2 D_1 f$ are continuous at (ξ_1, ξ_2), then $D_1 D_2 f(\xi_1, \xi_2)$ exists and

$$D_1 D_2 f(\xi_1, \xi_2) = D_2 D_1 f(\xi_1, \xi_2).$$

(See Landau, Theorem 299; or Fulks, Theorem 9.1a.)

THEOREM G (CONTINUOUS FUNCTION ON A CLOSED BOUNDED SET IN $\mathbf{R}^n$)

Let A be a closed bounded set in $\mathbf{R}^n$, and let $f: A \to \mathbf{R}$ be continuous. Then f is bounded on A. (The special case $n = 1$ is in Landau, Theorem 145. The general case, $n \geq 1$, is in Fulks, Theorem 8.6b.)

The study of power series in Section 34 assumes knowledge of the following properties of series of constants.

DEFINITION

A series of real or complex numbers $\sum_{k=0}^{\infty} a_k$ *converges* to S if $\lim_{m \to \infty} \sum_{k=0}^{m} a_k = S$. We write $S = \sum_{k=0}^{\infty} a_k$. (The quantity $S_m \equiv \sum_{k=0}^{m} a_k$ is called a *partial sum* of the series.)

This actually represents an abuse of notation. We are using the symbol $\sum_{k=0}^{\infty} a_k$ to stand for the series (convergent or not); and, in case of convergence, we use the same symbol to stand for S, the *limit* or *sum* of the series. This carelessness is common practice and it will not give us any trouble.

DEFINITION

If $\sum_{k=0}^{\infty} a_k$ converges (to some S) and $\sum_{k=0}^{\infty} |a_k|$ converges (to some T), then we say the series $\sum_{k=0}^{\infty} a_k$ *converges absolutely* or is *absolutely convergent*.

THEOREM H (SERIES OF CONSTANTS)

(i) (A Necessary Condition for Convergence). If $\sum_{k=0}^{\infty} a_k$ converges, then $\lim_{k \to \infty} a_k = 0$.

(ii) (The Comparison Test). If $\sum_{k=0}^{\infty} p_k$ is a convergent series of nonnegative terms and there exists K such that $|a_k| \leq p_k$ for all integers $k > K$, then $\sum_{k=0}^{\infty} a_k$ converges (absolutely). Moreover, $|\sum_{k=K}^{\infty} a_k| \leq \sum_{k=K}^{\infty} p_k$.

(iii) (Convergent Geometric Series). If $|r| < 1$ and a is a constant, then $\sum_{k=0}^{\infty} ar^k$ converges to $a/(1 - r)$.

(iv) (Ratio Test). Compute $r = \lim_{k \to \infty} |a_{k+1}|/|a_k|$ (if it exists). If $r < 1$, then $\sum_{k=0}^{\infty} a_k$ converges (absolutely). If $r > 1$ (or $r = \infty$), then $\sum_{k=0}^{\infty} a_k$ diverges (does not converge).

(v) (Addition of Series and Multiplication by Scalars). If $\sum_{k=0}^{\infty} a_k$ and $\sum_{k=0}^{\infty} b_k$ converge and if c_1 and c_2 are constants, then $\sum_{k=0}^{\infty} (c_1 a_k + c_2 b_k)$ converges and

$$\sum_{k=0}^{\infty} (c_1 a_k + c_2 b_k) = c_1 \sum_{k=0}^{\infty} a_k + c_2 \sum_{k=0}^{\infty} b_k.$$

(vi) (Multiplication of Series). If $\sum_{k=0}^{\infty} a_k = S$, $\sum_{k=0}^{\infty} b_k = T$, and at least one of these series converges absolutely, then $\sum_{k=0}^{\infty} c_k = ST$, where

$$c_k \equiv a_0 b_k + a_1 b_{k-1} + a_2 b_{k-2} + \cdots + a_k b_0.$$

(This last result is listed for completeness. We will not use it.)

It is useful to remember that the above definitions and results for series of constants remain valid when the constants are complex. Thus a_k, b_k, a, and r in the above series can be complex numbers, and so can the multipliers c_1 and c_2 in (v).

In Sections 28 and 35 we consider the limit of a sequence of functions, and then the notion of "uniform convergence" is important.

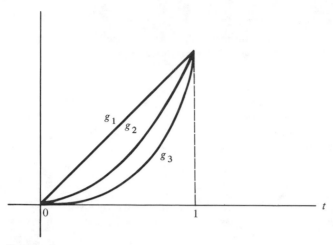

FIGURE 1

DEFINITION

Let S be any set and let g, g_1, g_2, ... be functions mapping $S \to \mathbf{R}$. Then we say $\{g_l\}$ *converges uniformly* to g on S if for every $\varepsilon > 0$ there exists K (independent of t) such that

$$l \geq K \quad \text{implies} \quad |g(t) - g_l(t)| < \varepsilon \quad \text{for all} \quad t \in S.$$

To appreciate the value of uniform convergence, let us first consider an example in which the convergence is *not uniform*: The sequence $\{g_l\}$ defined by $g_l(t) = t^l$ for $0 \leq t \leq 1$ ($l = 1, 2, \ldots$) converges to the function g defined by

$$g(t) = \begin{cases} 0 & \text{for} \quad 0 \leq t < 1 \\ 1 & \text{for} \quad t = 1. \end{cases}$$

(Figure 1.)

It is easy to see that each g_l is continuous while the limit g is discontinuous. With a little work one can show that the convergence of $\{g_l\}$ to g is not uniform.

The following important theorem shows that if the convergence had been uniform, the limit function could not have been discontinuous. Since this result is not presented in the usual first course on calculus, a complete proof is given below.

THEOREM I (UNIFORMLY CONVERGENT SEQUENCE OF CONTINUOUS FUNCTIONS)

Let J be an interval, and for each $l = 1, 2, \ldots$ let g_l be a continuous function mapping $J \to \mathbf{R}$. Let $\{g_l\}$ converge uniformly to another function $g : J \to \mathbf{R}$. Then

(i) g is continuous on J, and

(ii) if $J = [\alpha, \beta]$ is closed and bounded, then

$$\int_\alpha^\beta g(t)\, dt = \lim_{l \to \infty} \int_\alpha^\beta g_l(t)\, dt.$$

[Part (i) is essentially Landau's Theorem 231; or Fulks' Theorem 14.3a. Part (ii) is a very weak version of Landau's Theorem 412; or Fulks' Theorem 14.3c.]

Proof. (i) Let any t_0 in J and any $\varepsilon > 0$ be given. Choose l, in accordance with the definition of uniform convergence, sufficiently large so that

$$|g(t) - g_l(t)| < \frac{\varepsilon}{3} \quad \text{for all} \quad t \in J.$$

Then, using this value of l, choose $\delta > 0$ so small that $t \in J$ with $|t - t_0| < \delta$ implies $|g_l(t) - g_l(t_0)| < \varepsilon/3$. It follows that if $t \in J$ with $|t - t_0| < \delta$, then

$$|g(t) - g(t_0)| \le |g(t) - g_l(t)| + |g_l(t) - g_l(t_0)| + |g_l(t_0) - g(t)|$$

$$< \frac{\varepsilon}{3} + \frac{\varepsilon}{3} + \frac{\varepsilon}{3} = \varepsilon.$$

(ii) Now let $J = [\alpha, \beta]$ as indicated. The existence of all the integrals mentioned follows from the continuity of g, g_1, g_2, ... Let any $\varepsilon > 0$ be given. Choose K, in accordance with the definition of uniform convergence, sufficiently large so that

$$l \ge K \quad \text{implies} \quad |g(t) - g_l(t)| < \frac{\varepsilon}{\beta - \alpha} \quad \text{for all} \quad t \in [\alpha, \beta].$$

Then, for $l \ge K$,

$$\left| \int_\alpha^\beta g(t)\, dt - \int_\alpha^\beta g_l(t)\, dt \right| \le \int_\alpha^\beta |g(t) - g_l(t)|\, dt < \varepsilon. \quad \blacksquare$$

PROBLEMS

Find theorems in *your* calculus text which correspond to the following.

1. Theorem A.   2. Theorem B.   3. Theorem C.
4. Theorem E, or at least a method of calculation which leads to Eq. (5).
5. Theorem F.   6. Theorem G.   7. Each part of Theorem H.

REFERENCES

1. Bellman, R., and Cooke, K. L., *Differential-Difference Equations*. New York: Academic Press, Inc., 1963.
2. Coddington, E. A., *An Introduction to Ordinary Differential Equations*. Englewood Cliffs, N.J.: Prentice-Hall, Inc., 1961.
3. Coddington, E. A., and Levinson, N., *Theory of Ordinary Differential Equations*. New York: McGraw-Hill Book Company, Inc., 1955.
4. Driver, R. D., *Ordinary and Delay Differential Equations*. New York: Springer-Verlag, 1977.
5. Fulks, W., *Advanced Calculus: An Introduction to Analysis*, 2d ed., New York: John Wiley & Sons, Inc., 1969.
6. Halanay, A., *Differential Equations: Stability, Oscillations, Time Lags*. New York: Academic Press, Inc., 1966.
7. Hale, J., *Theory of Functional Differential Equations*, 2d ed., New York: Springer-Verlag, 1977.
8. Kamke, E., *Differentialgleichungen. Lösungsmethoden und Lösungen*. Band I. 3rd ed. Leipzig: Akademische Verlagsgesellschaft, 1944.
9. Krasovskiĭ, N. N., *Stability of Motion*. Stanford: Stanford University Press, 1963 (translated from Russian original, 1959).
10. Landau, E., *Differential and Integral Calculus*. New York: Chelsea Publishing Company, 1951 (translated from German original, 1934).
11. Lotka, A. J., *Elements of Mathematical Biology*. New York: Dover Publications, 1956 (originally published in 1925 as *Elements of Physical Biology*).
12. Minorsky, N., *Nonlinear Oscillations*. Princeton, N.J.: D. Van Nostrand Company, Inc., 1962. See Chapter 21.
13. Myškis, A. D., *Linear Differential Equations with Retarded Argument* (in Russian), 2d ed., Moscow: Izdatel'stvo "Nauka," 1972.
14. Ritger, P. D., and Rose, N. J., *Differential Equations with Applications*. New York: McGraw-Hill Book Company, Inc., 1968.
15. Volterra, V., *Leçons sur la Théorie Mathématique de la Lutte Pour la Vie*. Paris: Gauthier-Villars, 1931.
16. Wright, E. M., A non-linear difference-differential equation, *J. Reine Angew. Math.* **194**, 66–87 (1955).

ANSWERS

SECTION 2

3. $x(t) = 4e^{t-2} - 1$, $J = (-\infty, \infty)$. Unique.

4. $x(t) = (x_0 + 1)e^{-\cos t + \cos t_0} - 1$, $J = (-\infty, \infty)$. Unique.

5. $x(t) = t/2 - 1/4 + ce^{-2t}$, $J = (-\infty, \infty)$. c is arbitrary.

6. $i(t) = \begin{cases} (E/R)(1 - e^{-Rt/L}) + i_0 e^{-Rt/L} \text{ if } R \neq 0. \\ Et/L + i_0 \text{ if } R = 0. \end{cases}$

   In each case $J = (-\infty, \infty)$ and the solution is unique.

7. (a) $x(t) = t/2 + 1 - 3/2t$, $J = (0, \infty)$. Unique.

   (b) $x(t) = t/2 + 1 - 3/2t$, $J = (-\infty, 0)$. Unique.

   (c) The domains of definition are different.

8. $x(t) = 5e^{-t^2} + e^{-t^2} \int_0^t e^{s^2} \, ds$, $J = (-\infty, \infty)$. Unique.

9. $x(t) = t \ln|t| + ct$, $J = (-\infty, 0)$ or $(0, \infty)$. c arbitrary.

10. $x(t) = \frac{1}{2}(t - 1) + (t - 1)\ln\dfrac{1-t}{2} - 1$, $J = (-\infty, 1)$. Unique.

11. $x(t) = \sin t + \cos t$, $J = (-\pi/2, \pi/2)$. Unique.

12. $x(t) = \begin{cases} \sin t + \cos t & \text{for} \quad t \in (-\pi/2, \pi/2) \\ \sin t + c \cos t & \text{for} \quad t \in (\pi/2, 3\pi/2) \end{cases}$ for any c.

   No contradiction since $a(t) = \tan t$ is undefined at $t = \pi/2$.

13. (a) $x(t) = (\sin t - \cos t)/2 + ce^{-t}$.

   (b) The multiplier of c is e^{-t}, which is never zero.

14. $a = 0.0247$.   93 years.   15. $2 \ln 50 = 7.8$ min.

SECTION 3

1. (c) $x(t) = 1/(8 - t)$, $J = (-\infty, 8)$. Unique.

   (d) $x(t) = (7 + 2t - t^2)^{1/2}$, $J = (1 - 2\sqrt{2}, 1 + 2\sqrt{2})$. Unique.

   (e) $x(t) = \begin{cases} t^2/4 & \text{for} \quad t \geq 0 \\ -t^2/4 & \text{for} \quad t \leq 0. \end{cases}$ Not unique. Another solution is $x(t) \equiv 0$. Find still a third solution.

   (f) $x(t) = \pm(c^2 - t^2)^{1/2}$, $J = (-|c|, |c|)$ (two semicircles for each c). $c \neq 0$ arbitrary.

   (g) $x(t) = -1/t$, $J = (0, \infty)$. Unique.

   (h) $x(t) = \dfrac{c}{t} e^{-1/t}$, $J = (-\infty, 0)$ or $(0, \infty)$. c arbitrary.

SECTION 4

7. $te^x + \cos t = c$.

9. $t^2 e^x + x \sin t - 2t^3 - e^x = c$.

11. (a) Not exact.

    (b) Exact. $\cos t \sin x + \ln|\cos t| = c$.

    (c) Exact. $x \cos t + tx^2 + x + t = c$.

    (d) Exact. $tx^2 + t + 2x = c$.

    (e) Not exact.

    (f) Exact. $2tx^{3/2} + 4x^{3/2} - t^2 = c$.

    (g) Exact. $2tx + x^2 - t^2 = c$.

13. The equations $x' = 3x^{2/3}$ and $1 - \frac{1}{3}x^{-2/3}x' = 0$ are not "equivalent" since one has the solution $x = 0$ and the other does not. The equation $1 - \frac{1}{3}x^{-2/3}x' = 0$ is *not* exact on any rectangle including points $(t, 0)$, since $N(t, 0)$ is not even defined, much less continuously differentiable there.

14. $x(t) = \ln \dfrac{\pi - 1 - \cos t}{t}$ on $J = (0, \infty)$. Unique.

15. $x(t) = \left(\dfrac{t^2 + 1}{2t + 4} \right)^{2/3}$ on $J = (-2, \infty)$. Unique.

16. No solution.

17. $x(t) = \begin{cases} \dfrac{-1 + (1 + 4t - t^2)^{1/2}}{t} & \text{for } t \neq 0 \\ 2 & \text{for } t = 0. \end{cases}$

    $J = (2 - \sqrt{5}, 2 + \sqrt{5})$. Unique. (Use l'Hôpital's rule to prove continuity at $t = 0$).

18. The unique continuous solution of the implicit equation $(t^2 - 1)e^{x(t)} + x(t) \sin t - 2t^3 + 1 = 0$ with $x(0) = 0$, on some open interval J such that $0 \in J$ and $e^{x(t)}(t^2 - 1) + \sin t$ remains negative for $t \in J$. For example, we must have $J \subset (-\pi, 1)$. The solution is unique.

SECTION 5

2. $x(t) = \ln \dfrac{1 + ce^{-t}}{t}$ (where this is defined).

3. (a) No solution.   (b) A unique solution exists on $(-\infty, -\ln 2)$.

4. The implicit equation obtained in Example 2 applies only to solutions for which $tx(t) \neq 0$.

5. (a) $\mu(t, x) = t^{-2}x^{-1}$. $\ln\left|\dfrac{x}{t}\right| + tx = c$ for $tx \neq 0$.

    Another solution is $x(t) \equiv 0$. Did you miss this?

    (b) $\mu(t) = e^t$. $x(t) = 2 - \frac{1}{2}e^t + ce^{-t}$ on $J = (-\infty, \infty)$.

    (c) $\mu(x) = e^{-x^2/2}$. $txe^{-x^2/2} = c$.

    (d) $\mu(t, x) = t \cos(tx)$. $\sin(tx) = \dfrac{t^2}{2} + c$.

(e) $\mu(t, x) = (1 + tx^2)^{-1}.$ $\ln|1 + tx^2| - t = c$ for $tx^2 \neq -1.$ Also $x(t) = (-t)^{-1/2}$ for $t < 0.$

(f) $\mu(x) = e^{x^2}.$ $e^{x^2}(\sin t + 1) = c.$

(g) $\mu(x) = x^{-2}$ or $\mu(t, x) = \dfrac{1}{t(x + x^2)}.$ $x = \dfrac{-t}{t + c}.$

(h) $\mu(t, x) = (t^2 + x^2)^{-1}.$ $x(t) = t \tan\left(\dfrac{t^2}{2} + c\right).$

6. $x(t) = \dfrac{1}{t}\left(\pi - \text{Arcsin}\,\dfrac{t^2 - 2}{2}\right)$ for $0 < t < 2.$ Unique.

7. $2t + x^4 \cos t + t^2 x \cos t + (3x^2 + x^3 \sin t + t^2 \sin t)x' = 0$ is just one among infinitely many such equations.

8. *Hint*: This problem illustrates the fact that computations sometimes become easier if one generalizes the problem. Let $\mu(t, \xi)$ stand for $(t\xi)^{-2}$ and let $g(t, \xi)$ represent $-1/(t\xi) + \ln|\xi|$ so that $U(t, \xi) = v(g(t, \xi))\mu(t, \xi).$ Then use the fact that $D_2(\mu M) = D_1(\mu N)$ in proving that $D_2(UM) = D_1(UN).$

9. If $t_0 x_0 = -1,$ there is no solution, since $x'(t_0)$ is then undefined. If $t_0 x_0 \neq -1,$ then a unique solution exists on some interval and is determined by the implicit equation $-\dfrac{1}{tx(t)} + \ln|x(t)| = -\dfrac{1}{t_0 x_0} + \ln|x_0|.$ However such a solution is not valid at any point where $tx(t)$ becomes $-1.$ Note that this restriction, $tx(t) \neq -1,$ is not obvious from the implicit equation for $x.$

SECTION 6

1. $x^2 = t - \frac{1}{2} + ce^{-2t}.$

5. $x(t) = \dfrac{2\sqrt{2t}}{(1 - 4t^4)^{1/2}}$ valid and unique on $(0, 1/\sqrt{2}).$

6. $x(t) = t \sin(\ln|t| + c)$ for $t > 0$ or $t < 0.$

9. From $N' = kN - kN^2/P,$ one finds $N'' = k^2 N(1 - 2N/P)(1 - N/P).$ Thus $N'' > 0$ when $0 < N < P/2$ or $N > P,$ and $N'' < 0$ when $P/2 < N < P.$

10. (a) If $a_2 \neq 0,$ then $v'(t) = a_2 + b_2 f\left(\dfrac{a_1}{a_2} + \dfrac{a_2 c_1 - a_1 c_2}{a_2 v(t)}\right).$

If $b_2 \neq 0,$ then $v'(t) = a_2 + b_2 f\left(\dfrac{b_1}{b_2} + \dfrac{b_2 c_1 - b_1 c_2}{b_2 v(t)}\right).$

Similarly, if $a_1 \neq 0$ or $b_1 \neq 0.$ (The original equation is elementary if $a_1 = a_2 = b_1 = b_2 = 0.$)

(b) $z'(s) = f\left(\dfrac{a_1 s + b_1 z(s)}{a_2 s + b_2 z(s)}\right).$ Then introduce $v(s) = \dfrac{z(s)}{s}.$

11. (a) $s = t - \frac{3}{2}$ and $z(s) = x(s + \frac{3}{2}) + \frac{1}{2}$ leads to $z'(s) = -\left(\dfrac{s + 5z(s)}{s + 9z(s)}\right)^2.$

(b) $v(t) = t - 5x(t) + 1$ satisfies $v'(t) = \dfrac{9v(t) - 8}{4v(t) - 8}.$

12. $2e^x - x - t - 1 = 0.$

13. $x(t) = 1 - t + \tan t$ on $\left(-\dfrac{\pi}{2}, \dfrac{\pi}{2}\right)$.

14. $x(t) = (1 - ce^t)^2$ if $ce^t > 1$.

15. You may find $x(t) = (1 + e^t)^2$ and $x(t) = (1 - 3e^t)^2$, but only the latter is a solution. Why?

17. $x(t) = \dfrac{1}{2t} + \dfrac{1}{t \ln t + 2t}$ for $t > 0$. Unique.

SECTION 7

1. $x(t) = c_1 \ln|t| + c_2$ for $t > 0$ or $t < 0$.

2. $x(t) = c_1 e^{-t} + c_2 t + c_3$ for all t.

3. $x(t) = t + c_1 \displaystyle\int e^{-t^2/2}\, dt + c_2$ for all t.

5. $x(t) = \dfrac{2}{\omega} \sin \omega t$ $\left(\text{unique at least for } -\dfrac{\pi}{2\omega} < t < \dfrac{\pi}{2\omega}\right)$.

6. *Outline*: Note that $x'(t)$ is always increasing. Obtain an equation analogous to (5) relating x' and x. Now suppose (for contradiction) that $x'(t) \leq -a < 0$ for all $t \geq 0$.

7. $x'(t) \to c$ as $t \to \infty$ if $K > 0$, $x'(t) \to -c$ as $t \to \infty$ if $K < 0$. Does your proof show that $x'(t)$ cannot approach c or $-c$ as t approaches a *finite* value?

SECTION 8

4. One solution is defined by $x(t) = (1 + t/3)^3$ for all t.

  Another is defined by $x(t) = \begin{cases} 0 & \text{for } t < -3 \\ (1 + t/3)^3 & \text{for } t \geq -3. \end{cases}$

  There are infinitely many more. The solution is unique only on $(-3, \infty)$.

7. (a) $D = (0, \infty) \times \mathbf{R}$. (b) $D = \mathbf{R}^2$.

  (c) $D = \left(-\dfrac{\pi}{2}, \dfrac{\pi}{2}\right) \times \mathbf{R}$. (d) $D = \mathbf{R}^2$.

  (e) $D = \mathbf{R} \times (0, \infty)$. (f) $D = \mathbf{R} \times (-\infty, 0)$.

  (g) $D = \mathbf{R}^2$. (h) $D = \mathbf{R}^2$.

8. If $x(t) = 0$ for some t, then by uniqueness $x(t)$ must be zero for all t.

9. Yes, at least for $t > 0$. Consider $x' = f(t, x)$ where $f(t, \xi) = -\dfrac{\xi}{t + t^2 \xi}$ on

  $D = \left\{(t, \xi) \in \mathbf{R}^2 : t > 0,\ \xi > -\dfrac{1}{t}\right\}$.

10. The function f defined by $f(t, \xi) = \xi^{2/3}$ does not satisfy a Lipschitz condition on *any* rectangle containing $(t, 0)$. For if it did, with Lipschitz constant K, we would have $|f(t, \xi) - f(t, 0)| \leq K|\xi|$ or $|\xi^{2/3}| \leq K|\xi|$ for all sufficiently small $|\xi|$. This means $|\xi^{-1/3}| \leq K$ when $|\xi|$ is small, which is nonsense.

11. Prove that $f(t, \xi) = 1 + \xi^{2/3}$ is *not* Lipschitzian (as in Problem 10). But the solution is unique anyway, by the method of Section 3.

14. (a) $p/q \geq 1$.

  (b) If $p/q > 1$, the solution is valid on $\left(-\infty, \dfrac{q}{p-q} x_0^{(q-p)/q}\right)$. If $p/q \leq 1$, the solution is valid (but not necessarily unique) on $(-\infty, \infty)$.

16. If $f(t, \xi) = |\xi|$, then $D_2 f(t, 0)$ is undefined. But $|f(t, \xi) - f(t, \tilde{\xi})| = ||\xi| - |\tilde{\xi}||$ $\leq |\xi - \tilde{\xi}|$. Why?

17. If f were Lipschitzian on D we would have

$$|f(t, \xi) - f(t, \tilde{\xi})| \leq K|\xi - \tilde{\xi}|$$

  so that

$$\left|\frac{f(t, \xi) - f(t, \tilde{\xi})}{\xi - \tilde{\xi}}\right| \leq K \quad \text{whenever } \xi \neq \tilde{\xi}.$$

By the definition of $D_2 f$, it then follows that $|D_2 f(t, \xi)| \leq K$ whenever it exists.

SECTION 9

4. Given any $(t_0, \mathbf{x}_0) = (t_0, x_{01}, x_{02}) \in D$, there is at most one solution of $\mathbf{x}' = \mathbf{f}(t, \mathbf{x})$—on any subinterval of $(-1, 2)$—with $\mathbf{x}(t_0) = \mathbf{x}_0$.

5. (a) $D = \mathbf{R} \times (0, \infty) \times \mathbf{R}$.

  (b) $D = \mathbf{R} \times (-\infty, 0) \times \mathbf{R}$ or $D = \mathbf{R} \times \mathbf{R} \times (-\infty, 0)$.

  (c) $D = \mathbf{R} \times (-\sqrt{2}, \sqrt{2}) \times \mathbf{R} \times \mathbf{R}$.

  (d) $D = (-\infty, 0) \times \mathbf{R} \times \mathbf{R}$.

SECTION 10

2. $c_1 = 19/7$, $c_2 = 1/7$.

3. $x(t) = e^{-2t}$, unique on any interval containing $t = 0$.

5. If $x(t_0) = b_0$ and $x'(t_0) = b_1$ are given, there is at most one solution on:

  (a) the interval $(n\pi, n\pi + \pi)$ containing t_0 where $n = 0$, or ± 1, or ± 2, $\ldots$ (assuming t_0 is not a multiple of π).

  (b) any open interval containing t_0.

  (c) any open interval (containing t_0) on which $x(t) \neq 0$ (assuming $b_0 \neq 0$).

  (d) any open interval (containing t_0) on which $x'(t) \neq 1$ (assuming $b_1 \neq 1$).

  (e) any open interval (containing $t_0 = 2$) on which $x(t) > 0$.

11. (a) The soluton will not be unique.

  (b) No solution will exist (in general).

SECTION 11

1. $x(t) = x_0 \left(1 - \dfrac{t}{\beta}\right) + x_\beta \dfrac{t}{\beta} + \dfrac{1}{2} g(\beta t - t^2)$.

2. (a) $\omega\beta \neq (n + \tfrac{1}{2})\pi$ for each nonnegative integer n.

(b) $x(t) = x_0 \cos \omega t + \dfrac{v_\beta/\omega + x_0 \sin \omega\beta}{\cos \omega\beta} \sin \omega t.$

(c) Let $\omega\beta = (n + \tfrac{1}{2})\pi$ for some integer $n \geq 0$. Then there will be no solution if $v_\beta \neq (-1)^{n+1}\omega x_0$, but infinitely many solutions if $v_\beta = (-1)^{n+1}\omega x_0$ for some integer $n \geq 0$.

3. (a) $\sin \omega\beta - \omega \cos \omega\beta \neq 0.$

(b) $x(t) = \dfrac{-\sin \omega(t - \beta)}{\sin \omega\beta - \omega \cos \omega\beta}.$ \qquad (c) No solution.

4. (a) $k\beta > 0$ (which was already assumed).

(b) $x(t) = \dfrac{-x_0 \sinh k(t - \beta)}{\sinh k\beta} + \dfrac{x_\beta \sinh kt}{\sinh k\beta}.$

5. (a) $k > 0$ (which was already assumed).

(b) $x(t) = \dfrac{x_0 \cosh k(t - \beta)}{\cosh k\beta} + \dfrac{v_\beta \sinh kt}{k \cosh k\beta}.$

6. (a) $k \cosh k\beta - \sinh k\beta \neq 0.$

(b) $x(t) = \dfrac{\sinh k(t - \beta)}{k \cosh k\beta - \sinh k\beta}.$ \qquad (c) No solution.

7. Apply the comparison test for improper integrals, using the fact that

$$(1 - u^3)^{-1/2} < (1 - u)^{-1/2}, \quad \text{for} \quad 0 < u < 1.$$

SECTION 12

2. This problem illustrates a common mistake made by beginners. One *cannot differentiate an inequality.* For example, $\sin t \leq 1$ certainly does not imply $\cos t \leq 0.$

4. The proof of (iv) for $\|\cdot\|_\infty$ is as follows. For each i,

$$|\xi_i + \eta_i| \leq |\xi_i| + |\eta_i| \leq \|\xi\|_\infty + \|\eta\|_\infty.$$

Thus $\|\xi + \eta\|_\infty = \max_i |\xi_i + \eta_i| \leq \|\xi\|_\infty + \|\eta\|_\infty.$

5. For all ξ and η, $\|\xi\| = \|\xi - \eta + \eta\| \leq \|\xi - \eta\| + \|\eta\|.$
Thus $\|\xi\| - \|\eta\| \leq \|\xi - \eta\|.$ Similarly, $\|\eta\| - \|\xi\| \leq \|\xi - \eta\|.$ From these

$$|\|\xi\| - \|\eta\|| \leq \|\xi - \eta\|.$$

6. (b) $K = 8.$

7. $|x(t) - \tilde{x}(t)| = |x_0 - \tilde{x}_0| e^{-\int_{t_0}^t a(s)\, ds}.$ This is smaller, in general, than the right-hand side of inequality (10).

9. Theorem C gives

$$|x(t) - \tilde{x}(t)| \leq (|x(t_0) - \tilde{x}(t_0)| + |x'(t_0) - \tilde{x}'(t_0)| + |x''(t_0) - \tilde{x}''(t_0)|)e^{2|t - t_0|}.$$

(A sharper estimate can be obtained by direct solution of $x''' - x'' = 0.$)

10. (a) Use Theorem 8-A with $D = (-\infty, 1) \times \mathbf{R}.$

(b) If $x(0) = x_0 \neq 1$ (by any slight amount), the solution of the linear equation is found to be

$$x(t) = 1 + (x_0 - 1)/(1 - t) \quad \text{which} \quad \to \pm\infty \quad \text{as} \quad t \to 1-.$$

(c) No contradiction since f is not Lipschitzian on D.

11. (a) The solution $x(t) \equiv 1$ is unique as before.

(b) If $x(0) = x_0 \neq 1$ the solution is

$$x(t) = 1 + (x_0 - 1)e^{(3/2)[1 - (1 - t)^{2/3}]}.$$

Here $|x(t) - 1| \leq |x_0 - 1|e^{3/2}$ for all $t < 1$.

(c) Again f is not Lipschitzian on $D = (-\infty, 1) \times \mathbf{R}$. But this time the solution *does* "depend continuously" on x_0. While the inequality in (b) does not follow from Theorem C it could be proved directly from Lemma A with $k(s) = (1 - s)^{-1/3}$.

SECTION 13

1. (a) $21 - 8i$, (b) $-\dfrac{8}{13} - \dfrac{25}{13}i$, (c) $-2 - 8i$.

2. $2 - 11i$.

4. To finish proof of the triangle inequality, note that $\mathrm{Re}(z_1 \bar{z}_2) \leq |z_1 \bar{z}_2| = |z_1| \cdot |z_2|$.

5. $1 + i, 1 - i, -1, 2$.

8. $-1, \dfrac{1}{2} + i \dfrac{\sqrt{3}}{2}, \dfrac{1}{2} - i \dfrac{\sqrt{3}}{2}$.

SECTION 14

4. To show that the equations $x_0 = c_1 t_0 e^{\lambda t_0} + c_2 e^{\lambda t_0}$ and $v_0 = c_1(1 + \lambda t_0)e^{\lambda t_0} + c_2 \lambda e^{\lambda t_0}$ uniquely determine c_1 and c_2, one must show that

$$\begin{vmatrix} t_0 e^{\lambda t_0} & e^{\lambda t_0} \\ (1 + \lambda t_0)e^{\lambda t_0} & \lambda e^{\lambda t_0} \end{vmatrix} \neq 0.$$

5. (b) *Outline*: First evaluate c_1 and c_2 in Equation (12) to verify the asserted expression for $x(t)$ in terms of x_0 and v_0. Now observe that, if x is not identically zero, then in order that $x(t)$ *ever* be zero one must have $\lambda_2 x_0 - v_0 \neq 0$ and $\lambda_1 x_0 - v_0 \neq 0$. Why? Assuming these two inequalities to be satisfied, the equation $x(t) = 0$ is equivalent to

$$\frac{\lambda_2 x_0 - v_0}{\lambda_1 x_0 - v_0} = e^{(\lambda_2 - \lambda_1)(t - t_0)}. \tag{*}$$

Since the right-hand side of (*) is a strictly monotone decreasing function of t, there can be at most one value of $t > t_0$ which satisfies (*). There *will* be such a t if and only if the left-hand side of (*) is a number between 0 and 1. Show that this, in turn, holds if and only if v_0 and x_0 have opposite signs and $|v_0/x_0| > |\lambda_2|$.

(c) *Outline*: Evaluate c_1 and c_2 in Equation (13) to verify the expression for $x(t)$ in terms of x_0 and v_0. Now if x is not identically zero, then in order that

$x(t)$ ever be zero one must have $x_0 \neq 0$ and $v_0 - \lambda x_0 \neq 0$. Why? Assuming these two inequalities are satisfied, the equation $x(t) = 0$ is equivalent to

$$\left(\lambda - \frac{v_0}{x_0}\right)(t - t_0) = 1.$$

Show that the latter has no more than one solution $t > t_0$, and it *does* have such a solution if and only if $\lambda - v_0/x_0 > 0$.

6. $x(t) = e^{3t} + c_1 t e^{2t} + c_2 e^{2t}$ (on any interval).
7. (a) $(D + tI)(D + t^{-1}I) = D^2 + (t + t^{-1})D + (1 - t^{-2})I.$
   (b) $x(t) = t + c_1 t^{-1} e^{-t^2/2} + c_2 t^{-1}$ for $t \neq 0.$
   (c) $(D + t^{-1}I)(D + tI) = D^2 + (t + t^{-1})D + 2I.$

8. $x(t) = c_1 e^{-t^2/2} \int_0^t e^{s^2} \, ds + c_2 e^{-t^2/2}.$

10. $D^2 + \left(t + \frac{1}{t}\right)D + \left(1 - \frac{1}{t^2}\right)I = D\left[D + \left(t + \frac{1}{t}\right)I\right].$

SECTION 15

2. (a) $y(t) = c_1 e^{3t} + c_2 e^{-3t}.$
   (b) $y(t) = c_1 \cos 3t + c_2 \sin 3t.$
   (c) $y(t) = c_1 e^{3t} + c_2 t e^{3t}.$
   (d) $y(t) = c_1 e^{-2t} \cos \sqrt{5}\,t + c_2 e^{-2t} \sin \sqrt{5}\,t.$
   (e) $y(t) = c_1 e^{-t} + c_2 t e^{-t} + c_3 t^2 e^{-t}.$
   (f) $y(t) = c_1 e^t + c_2 t e^t + c_3 \cos t + c_4 \sin t.$
   (g) $y(t) = c_1 \cos t + c_2 \sin t + c_3 t \cos t + c_4 t \sin t.$
3. (a) $y(t) = e^{3t} - e^{-3t}.$
   (b) $y(t) = -\frac{1}{3} \cos 3t - 2 \sin 3t.$
   (c) $y(t) = -6e^{-3} e^{3t} + 5e^{-3} t e^{3t}.$

SECTION 16

1. (a), (c), and (d) are linearly independent sets. Sets (b) and (e) are linearly dependent.
2. If v_1 satisfied an equation $y'' + a_1(t)y' + a_0(t)y = 0$ with a_1 and a_0 continuous on $(-1, 1)$, then the fact that $v_1(0) = 0$ and $v_1'(0) = 0$ would imply $v_1(t) \equiv 0$ Why? Similarly for v_2.
4. $x(t) = t - 2 + c_1 e^{-t} + c_2 t e^{-t}.$
7. $y(t) = c_1 t^{\lambda_1} + c_2 t^{\lambda_2}$ where $\lambda_1, \lambda_2 = \dfrac{1 - b_1 \pm [(1 - b_1)^2 - 4b_0]^{1/2}}{2}.$
9. *Hint:* See the answer to Problem 2.
11. $W(y_1, y_2, y_3)(t) = (\lambda_2 - \lambda_1)(\lambda_3 - \lambda_1)(\lambda_3 - \lambda_2)e^{(\lambda_1 + \lambda_2 + \lambda_3)t}.$

SECTION 17

1. (a) $x(t) = c_1 e^{-2t} + c_2 t e^{-2t} + c_3 t^2 e^{-2t}$.
   (b) $x(t) = \frac{1}{2} t \sin t + c_1 \cos t + c_2 \sin t$.
   (c) $x(t) = -\frac{1}{2} \cos t + \frac{1}{2} \sin t + c e^{-t}$.
   (d) $x(t) = -\frac{1}{5} + \frac{1}{20} \cos 2t - \frac{3}{20} \sin 2t - \frac{1}{2} e^t + c_1 e^{2t} + c_2 e^{-t}$.
   (e) $x(t) = e^t + c_1 + c_2 e^{2t} + c_3 t e^{2t}$.
   (f) $x(t) = \frac{1}{4} t^2 e^{2t} + c_1 + c_2 e^{2t} + c_3 t e^{2t}$.
   (g) $x(t) = -\frac{1}{10} t^2 - \frac{1}{50} t - \frac{1}{500}$
       $\qquad + c_1 e^{2t} + c_2 e^{-t} \cos 2t + c_3 e^{-t} \sin 2t$.
   (h) $x(t) = -\frac{11}{500} t - \frac{1}{100} t^2 - \frac{1}{30} t^3$
       $\qquad + c_1 + c_2 e^{(-1+\sqrt{41})t/2} + c_3 e^{(-1-\sqrt{41})t/2}$.
   (i) $x(t) = -\frac{7}{9} t e^{-t} - \frac{1}{6} t^2 e^{-t} - \frac{1}{3} t e^{2t} + c_1 e^{-t} + c_2 e^{2t}$.
   (j) $x(t) = \frac{1}{2} t e^{-2t} \sin t + c_1 e^{-2t} \cos t + c_2 e^{-2t} \sin t$.
       $x(t) \to 0$ as $t \to \infty$.
   (k) $x(t) = -\frac{3}{40} \cos 2t - \frac{1}{40} \sin 2t + c_1 e^{-t} + c_2 e^{-2t}$.
   (l) $x(t) = \frac{1}{8} - \frac{1}{8} t \sin 2t + c_1 \cos 2t + c_2 \sin 2t$.
2. (a) $x(t) = 2e^{-2t} + 4t e^{-2t} - t^2 e^{-2t}$.

   (b) $x(t) = \frac{1}{2} t \sin t - \frac{1}{2} \cos t - \frac{\pi}{4} \sin t$.

   (c) $x(t) = -\frac{1}{4} \cos t + \frac{1}{2} \sin t + \frac{3}{2} e^{\pi-t}$.
3. (a) $\tilde{x}(t) = At \cos t + Bt \sin t + Ct^2 \cos t + Dt^2 \sin t$.
   (b) $\tilde{x}(t) = A + Bt + Ct^3 e^{-3t} + Dt^4 e^{-3t} + Et^5 e^{-3t}$.

SECTION 18

5. (a) For free oscillations $\omega = 3{,}999{,}998$ sec^{-1}.
   (b) $C = 160 \times 10^{-12}$ farads. Then in x_{ss} the ω_1 term has amplitude $a/563$ and the ω_2 term has amplitude $a/2$. For free oscillations $\omega = 4{,}999{,}998.4$ sec^{-1}.
6. $x(t) = \frac{1}{8} \cos 7t + \frac{1}{8} \sin 7t + c_1 e^{-14t} \cos 7t + c_2 e^{-14t} \sin 7t$.

7. $ka[(k - m\omega_1^2)^2 + b^2 \omega_1^2]^{-1/2}$.   Maximum when $\omega_1 = \left(\dfrac{k}{m} - \dfrac{b^2}{2m^2} \right)^{1/2}$.

8. The right-hand side of Eq. (3) contains a factor of ω_1 which is not present in the mechanical case.

SECTION 19

1. $W(y_1, y_2)(t) = t^{-3} \neq 0$.
2. (a) $W(y_1, y_2, y_3)(t) = t^{\lambda_1+\lambda_2+\lambda_3-3}(\lambda_2 - \lambda_1)(\lambda_3 - \lambda_1)(\lambda_3 - \lambda_2)$.
3. (a) $x(t) = c_1 |t|^{1/2} + c_2 |t|^{-1/2}$ on $(0, \infty)$ or on $(-\infty, 0)$.
   (b) $x(t) = c_1 |t|^{-1/2} + c_2 |t|^{-1/2} \ln|t|$ on $(0, \infty)$ or $(-\infty, 0)$.
   (c) $x(t) = c_1 + c_2 t^2 + c_3 t^2 \ln|t|$ on $(0, \infty)$ or $(-\infty, 0)$.
   (d) $x(t) = c_1 t + c_2 t \ln|t| + c_3 t (\ln|t|)^2$ on $(0, \infty)$ or $(-\infty, 0)$.

(f) $x(t) = c_1 \cos(t^2/2) + c_2 \sin(t^2/2)$ on $(0, \infty)$ or $(-\infty, 0)$. How do you conclude that this solution is valid beyond $\sqrt{2\pi}$?

4. Compute $x(t) = w(\ln|t|)$, $x'(t) = t^{-1}w'(\ln|t|)$,
   $x''(t) = t^{-2}[w''(\ln|t|) - w'(\ln|t|)]$, $x'''(t) = t^{-3}[w'''(\ln|t|) - 3w''(\ln|t|) + 2w'(\ln|t|)]$. Then you will find that w satisfies a linear homogeneous equation with constant coefficients.
   (a) Find $4w''(s) - w(s) = 0$. So $w(s) = c_1 e^{s/2} + c_2 e^{-s/2}$ or $x(t) = c_1|t|^{1/2} + c_2|t|^{-1/2}$ on $(0, \infty)$ or $(-\infty, 0)$.

5. (a) $x(t) = c_1 t^{-2} \cos(\ln|t|) + c_2 t^{-2} \sin(\ln|t|)$.
   (b) $x(t) = c_1 t^2 + c_2 t^{-1} \cos(\ln|t|^3) + c_3 t^{-1} \sin(\ln|t|^3)$.

6. (a) $x(t) = -1 + t + te^{-t} + c_1 e^{-t} + c_2 e^{-2t}$.
   (b) $x(t) = t - 1 + ce^{-t}$. Can you recognize your work as being equivalent to the method of Section 2?
   (c) $x(t) = -\dfrac{t}{2} \cos t + c_1 \cos t + c_2 \sin t$.
   (d) $x(t) = \frac{1}{2}e^t + c_1 \cos t + c_2 \sin t$.
   (e) $x(t) = -(\cos t) \ln|\sec t + \tan t| + c_1 \cos t + c_2 \sin t$.
   (f) $x(t) = (\cos t) \ln|\cos t| + t \sin t + c_1 \cos t + c_2 \sin t$.
   (g) $x(t) = -1 + (\sin t) \ln|\sec t + \tan t| + c_1 \cos t + c_2 \sin t$.
   (h) $x(t) = -te^{-t} + c_1 + c_2 e^{-t} + c_3 e^{-2t}$
   (i) $x(t) = te^t + c_1 t + c_2 t^2$.
   (j) $x(t) = (\ln|t|)^2/2t + c_1 t^{-1} + c_2 t^{-1} \ln|t|$.

7. $y(t) = c_1 t + c_2 t^2 + c_3\left[t^2 e^{2-1t-2} + t \displaystyle\int_1^t s^{-2} e^{2-1s-2} ds\right]$.

8. (a) $x(t) = \dfrac{1}{6} \displaystyle\int_{t_0}^t h(s)[e^{t-s} - e^{-5(t-s)}]\, ds + c_1 e^t + c_2 e^{-5t}$.
   (b) $x(t) = \dfrac{1}{8} \displaystyle\int_{t_0}^t h(s)[e^{-2(t-s)} - \cos 2(t-s) + \sin 2(t-s)]\, ds$
   $\qquad + c_1 e^{-2t} + c_2 \cos 2t + c_3 \sin 2t$.
   (c) $x(t) = \displaystyle\int_{t_0}^t h(s)\sin(t-s)\, ds + c_1 \cos t + c_2 \sin t$.
   (d) $x(t) = \displaystyle\int_{t_0}^t h(s)e^{-2(t-s)} \sin(t-s)\, ds + c_1 e^{-2t} \cos t + c_2 e^{-2t} \sin t$.

SECTION 20

1. The choice of $\theta(t)$ is arbitrary in the sense that we can modify any correct choice by the addition of an integral multiple of 2π. Choosing $\theta(t)$ in a natural "way" means that we should never introduce a jump discontinuity of 2π or more. (Then θ will actually be continuous).

2. If $A(t)$ is the area swept out during time t, show that $A'(t) = \frac{1}{2}r^2(t)\theta'(t)$.

4. Putting $r_0 = R$, the radius of the earth, condition (19) becomes $v_{or} \geq \sqrt{2kM/R}$. Since $kM = gR^2$, this is equivalent to the condition found in Example 7-2.

5. The assumption that the earth is a point mass. To correct this defect, simply declare solutions to be invalid whenever $x_1^2(t) + x_2^2(t) \leq R^2$, where $R > 0$ is the radius of the earth.

6. $M_s/M_e = 3.5 \times 10^5$. To compute the mass of Jupiter, you would want to know the period and the length of the semimajor axis of one of Jupiter's moons.
7. M is replaced by $M + m$ where m is the mass of the satellite. Thus the ellipse becomes smaller, the relative velocity required for escape becomes larger, and the period of an elliptical orbit becomes smaller.

SECTION 21

1. (a) Linearly independent.
   (b) Linearly independent.
   (c) Linearly dependent, $\xi_{(1)} + \xi_{(2)} + \xi_{(3)} - 2\xi_{(4)} = 0$.
   (d) Linearly dependent, $2\xi_{(1)} - \xi_{(2)} + \xi_{(3)} = 0$.
2. The ith component of $A(c_1\xi + c_2\eta)$ is

$$\sum_{j=1}^{n} a_{ij}(c_1\xi_j + c_2\eta_j) = c_1 \sum_{j=1}^{n} a_{ij}\xi_j + c_2 \sum_{j=1}^{n} a_{ij}\eta_j,$$

which is the ith component of $c_1 A\xi + c_2 A\eta$.

3. $AB = \begin{pmatrix} 0 & 9 & 3 \\ 5 & -4 & 3 \\ 8 & 6 & -5 \end{pmatrix}$, $\quad BA = \begin{pmatrix} -8 & 6 & 1 \\ 0 & 1 & 6 \\ 1 & 12 & -2 \end{pmatrix}$, $\quad B^T = \begin{pmatrix} -2 & 2 & 1 \\ 1 & 0 & 4 \\ 3 & 1 & 0 \end{pmatrix}$.

4. (a) $(A + B)C = \begin{pmatrix} -12 & 5 \\ 4 & 11 \end{pmatrix} = AC + BC$.

   (b) $(AB)C = \begin{pmatrix} -24 & 2 \\ 36 & -7 \end{pmatrix} = A(BC)$.

5. The ijth elements of $A(cB)$, $(cA)B$, and $c(AB)$ are

$$\sum_{k=1}^{n} a_{ik}(cb_{kj}), \quad \sum_{k=1}^{n} (ca_{ik})b_{kj}, \quad \text{and} \quad c\sum_{k=1}^{n} a_{ik}b_{kj}$$

respectively. But these three expressions are all equivalent.

7. $A^{-1} = \dfrac{1}{19}\begin{pmatrix} 7 & 2 & -6 \\ 2 & 6 & 1 \\ 6 & -1 & 3 \end{pmatrix}$, $\quad B^{-1} = \dfrac{1}{33}\begin{pmatrix} -4 & 12 & 1 \\ 1 & -3 & 8 \\ 8 & 9 & -2 \end{pmatrix}$.

SECTION 22

4. *Note*: There are infinitely many correct choices for P besides those listed below.

   (a) If $P = \begin{pmatrix} 1 & 1 \\ 1 & -2 \end{pmatrix}$, then $P^{-1} = \dfrac{1}{3}\begin{pmatrix} 2 & 1 \\ 1 & -1 \end{pmatrix}$ and $P^{-1}AP = \begin{pmatrix} 1 & 0 \\ 0 & 4 \end{pmatrix}$.

   (b) If $P = \begin{pmatrix} 1+i & 1-i \\ -1 & -1 \end{pmatrix}$, then $P^{-1} = \dfrac{1}{2}\begin{pmatrix} -i & -1-i \\ i & -1+i \end{pmatrix}$ and

   $P^{-1}AP = \begin{pmatrix} i & 0 \\ 0 & -i \end{pmatrix}$.

(c) If $P = \begin{pmatrix} 0 & 3 & 1 \\ 1 & 0 & -1 \\ 0 & 2 & 1 \end{pmatrix}$, then $P^{-1} = \begin{pmatrix} -2 & 1 & 3 \\ 1 & 0 & -1 \\ -2 & 0 & 3 \end{pmatrix}$ and

$$P^{-1}AP = \begin{pmatrix} -1 & 0 & 0 \\ 0 & -1 & 0 \\ 0 & 0 & -2 \end{pmatrix}.$$

(d) If $P = \begin{pmatrix} 1 & 1 & 1 \\ -1 & 1+i & 1-i \\ 0 & -1 & -1 \end{pmatrix}$, then $P^{-1} = \frac{1}{2}\begin{pmatrix} 2 & 0 & 2 \\ -i & -i & -1-2i \\ i & i & -1+2i \end{pmatrix}$

and $P^{-1}AP = \begin{pmatrix} -1 & 0 & 0 \\ 0 & -1+i & 0 \\ 0 & 0 & -1-i \end{pmatrix}.$

5. If $P^{-1}AP = -2I$, then we must have

$$A = P(-2I)P^{-1} = -2PP^{-1} = -2I, \quad \text{a contradiction.}$$

6. If P^{-1} exists, then $(\det P^{-1})(\det P) = \det P^{-1}P = \det I = 1$.

Thus if $P^{-1}AP = \begin{pmatrix} a & 0 \\ 0 & b \end{pmatrix}$, then

$$\det(A - \lambda I) = \det P^{-1} \det(A - \lambda I)\det P$$
$$= \det[P^{-1}(A - \lambda I)P]$$
$$= \det(P^{-1}AP - \lambda I)$$
$$= (a - \lambda)(b - \lambda).$$

Hence a and b must be the eigenvalues of A. The result now follows from Problem 5.

7. The matrix A of Problem 4(c) has $\lambda = -1$ as an eigenvalue, and corresponding to this eigenvalue it has eigenvectors $\xi = \text{col}(3c_2, c_1, 2c_2)$ for arbitrary constants c_1 and c_2 not both zero.

SECTION 23

4. If $y_{(1)}, \ldots, y_{(n)}$ are linearly independent solutions of (3), then (by Theorem D) any other solution can be written as a linear combination of $y_{(1)}, \ldots, y_{(n)}$. So there cannot be $n + 1$ linearly independent solutions.

5. (a) $y(t) = c_1 e^{5t}\begin{pmatrix} 1 \\ 1 \end{pmatrix} + c_2 e^{-2t}\begin{pmatrix} 4 \\ -3 \end{pmatrix}.$

(b) $x(t) = \frac{1}{5}\begin{pmatrix} 2 \\ -3 \end{pmatrix} + \frac{1}{65}\begin{pmatrix} -22\cos t - 6\sin t \\ 4\cos t + 7\sin t \end{pmatrix} + y(t),$

where $y(t)$ is as in part (a).

(c) $x(t) = \dfrac{1}{4}\begin{pmatrix} -2+4t \\ 5-2t \\ -2 \end{pmatrix} + c_1 e^{-t}\begin{pmatrix} 1 \\ -1 \\ 0 \end{pmatrix} + c_2 e^{-2t}\begin{pmatrix} 2 \\ -1 \\ -2 \end{pmatrix} + c_3 e^{-3t}\begin{pmatrix} 1 \\ -1 \\ -2 \end{pmatrix}.$

6. (a) Put $c_1 = -1/7$ and $c_2 = 2/7$ in solution of 5(a).
   (b) Put $c_1 = 9/65$ and $c_2 = 1/5$ in solution of 5(b).

SECTION 24

5. (a) $x(t) = \begin{pmatrix} te^{4t} + e^{4t} - 2t - 2 \\ -2te^{4t} + e^{4t} - 2t - 2 \end{pmatrix} + c_1 e^{t}\begin{pmatrix} 1 \\ 1 \end{pmatrix} + c_2 e^{4t}\begin{pmatrix} 1 \\ -2 \end{pmatrix}.$

(b) $y(t) = C_1 e^{it}\begin{pmatrix} 1+i \\ -1 \end{pmatrix} + C_2 e^{-it}\begin{pmatrix} 1-i \\ -1 \end{pmatrix}$

$= c_1\begin{pmatrix} \cos t - \sin t \\ -\cos t \end{pmatrix} + c_2\begin{pmatrix} \cos t + \sin t \\ -\sin t \end{pmatrix}$

where $c_1 = C_1 + C_2$ and $c_2 = iC_1 - iC_2$.

(c) $x(t) = \dfrac{1}{4}\begin{pmatrix} 3t-3 \\ t \end{pmatrix} + \begin{pmatrix} -5e^{-t} \\ -2e^{-t} \end{pmatrix} + c_1 e^{-2t}\begin{pmatrix} 1 \\ 1 \end{pmatrix} + c_2 e^{-2t}\begin{pmatrix} 1+t \\ t \end{pmatrix}.$

(d) $y(t) = c_1 e^{t}\begin{pmatrix} 0 \\ 1 \\ 0 \end{pmatrix} + c_2 e^{t}\begin{pmatrix} 3 \\ 0 \\ 2 \end{pmatrix} + c_3 e^{2t}\begin{pmatrix} 1 \\ 4 \\ 1 \end{pmatrix}.$

(e) $y(t) = c_1 e^{t}\begin{pmatrix} 0 \\ 1 \\ 0 \end{pmatrix} + c_2 e^{t}\begin{pmatrix} 3 \\ -2t \\ 2 \end{pmatrix} + c_3 e^{2t}\begin{pmatrix} 1 \\ 3 \\ 1 \end{pmatrix}.$

(f) $y(t) = c_1 e^{-2t}\begin{pmatrix} 1+t+t^2 \\ t-t^2 \\ 2+t-t^2 \end{pmatrix} + c_2 e^{-2t}\begin{pmatrix} 1+2t \\ 1-2t \\ 1-2t \end{pmatrix} + c_3 e^{-2t}\begin{pmatrix} 1 \\ -1 \\ -1 \end{pmatrix}.$

(g) $y(t) = c_1 e^{-2t}\begin{pmatrix} 1+t \\ -t \\ -t \end{pmatrix} + c_2 e^{-2t}\begin{pmatrix} 1 \\ 0 \\ 1 \end{pmatrix} + c_3 e^{-2t}\begin{pmatrix} 2 \\ -1 \\ 0 \end{pmatrix}.$

(h) $y(t) = c_1 e^{-t}\begin{pmatrix} 1 \\ -1 \\ 0 \end{pmatrix} + C_2 e^{-t+it}\begin{pmatrix} 1 \\ 1+i \\ -1 \end{pmatrix} + C_3 e^{-t-it}\begin{pmatrix} 1 \\ 1-i \\ -1 \end{pmatrix}$

$= c_1 e^{-t}\begin{pmatrix} 1 \\ -1 \\ 0 \end{pmatrix} + c_2 e^{-t}\begin{pmatrix} \cos t \\ \cos t - \sin t \\ -\cos t \end{pmatrix} + c_3 e^{-t}\begin{pmatrix} \sin t \\ \sin t + \cos t \\ -\sin t \end{pmatrix}.$

6. In the answers to Problem 5:

(a) Set $c_1 = 2$, $c_2 = 1$.   (b) Set $c_1 = 1$, $c_2 = 0$.
(c) Set $c_1 = 2$, $c_2 = 23/4$.   (d) Set $c_1 = -26$, $c_2 = -2$, $c_3 = 7$.
(e) Set $c_1 = -19$, $c_2 = -2$, $c_3 = 7$.

7. (a) $y(t) = c_1 e^{-t} \begin{pmatrix} 1 \\ -1 \end{pmatrix} + c_2 e^{-3t} \begin{pmatrix} 1 \\ -3 \end{pmatrix}$. $z(t) = y_1(t)$.

(b) $y(t) = c_1 e^{-2t} \begin{pmatrix} 1 \\ -2 \end{pmatrix} + c_2 e^{-2t} \begin{pmatrix} t \\ 1 - 2t \end{pmatrix}$. $z(t) = y_1(t)$.

SECTION 25

3. All roots of Equations (a), (d), and (e) have negative real parts. Equations (b) and (c) have roots with nonnegative real parts.

4. The characteristic equations are

(a) $\det (A - \lambda I) = -(\lambda^3 + 3\lambda^2 + 3\lambda + 2) = 0$.
(b) $\det (A - \lambda I) = \lambda^4 + 7\lambda^3 + 17\lambda^2 + 20\lambda + 10 = 0$.

10. (a) $\tilde{x}(t) = \text{col} (2 \cos t, \sin t, -\cos t - \sin t)$.
(b) Every solution $x(t)$ is close to $\tilde{x}(t)$ for all sufficiently large t.

SECTION 26

3. Since $T(t, s) = T(t - s, 0)$ in each case, we list only $T(t, 0)$.

(a) $T(t, 0) = \dfrac{1}{3} \begin{pmatrix} 2e^t + e^{4t} & e^t - e^{4t} \\ 2e^t - 2e^{4t} & e^t + 2e^{4t} \end{pmatrix}$

(b) $T(t, 0) = \begin{pmatrix} \cos t + \sin t & 2 \sin t \\ -\sin t & \cos t - \sin t \end{pmatrix}$

(d) $T(t, 0) = \begin{pmatrix} 3e^t - 2e^{2t} & 0 & -3e^t + 3e^{2t} \\ 8e^t - 8e^{2t} & e^t & -12e^t + 12e^{2t} \\ 2e^t - 2e^{2t} & 0 & -2e^t + 3e^{2t} \end{pmatrix}$.

(e) $T(t, 0) = \begin{pmatrix} 3e^t - 2e^{2t} & 0 & -3e^t + 3e^{2t} \\ 6e^t - 2te^t - 6e^{2t} & e^t & -9e^t + 2te^t + 9e^{2t} \\ 2e^t - 2e^{2t} & 0 & -2e^t + 3e^{2t} \end{pmatrix}$.

(f) $T(t, 0) = e^{-2t} \begin{pmatrix} 1 + t & t - \frac{1}{2}t^2 & \frac{1}{2}t^2 \\ -t & 1 - 2t + \frac{1}{2}t^2 & t - \frac{1}{2}t^2 \\ -t & -2t + \frac{1}{2}t^2 & 1 + t - \frac{1}{2}t^2 \end{pmatrix}$.

4. (a) $x(t) = \begin{pmatrix} x_{01} \cos t + (x_{01} + 2x_{02})\sin t \\ x_{02} \cos t - (x_{01} + x_{02})\sin t \end{pmatrix} + \begin{pmatrix} 3 \cos t + \sin t - 3 \\ -\cos t - 2 \sin t + 1 \end{pmatrix}$.

(b) $x(t) = \begin{pmatrix} x_{01} \cos t + (x_{01} + 2x_{02})\sin t \\ x_{02} \cos t - (x_{01} + x_{02})\sin t \end{pmatrix} + \begin{pmatrix} t \cos t + (1 + t)\sin t \\ -t \sin t \end{pmatrix}$.

(c) $\mathbf{x}(t) = \begin{pmatrix} x_{01}\cos t + (x_{01} + 2x_{02})\sin t \\ x_{02}\cos t - (x_{01} + x_{02})\sin t \end{pmatrix}$

$\qquad + \begin{pmatrix} 2\sin t - 2\cos t \ln(\sec t + \tan t) \\ -1 + \cos t - \sin t + (\sin t + \cos t)\ln(\sec t + \tan t) \end{pmatrix}.$

(d) $\mathbf{x}(t) = \begin{pmatrix} x_{01}e^{-2t} + (x_{01} - x_{02})te^{-2t} \\ x_{02}e^{-2t} + (x_{01} - x_{02})te^{-2t} \end{pmatrix} + \begin{pmatrix} -\frac{1}{2}t^2 e^{-2t} \\ (t - \frac{1}{2}t^2)e^{-2t} \end{pmatrix}.$

6. $\mathbf{x}(t) = \begin{pmatrix} x_{01}\cos\dfrac{t^2}{2} + x_{02}\sin\dfrac{t^2}{2} \\[2mm] -x_{01}\sin\dfrac{t^2}{2} + x_{02}\cos\dfrac{t^2}{2} \end{pmatrix} + \begin{pmatrix} 2 - 2\cos\dfrac{t^2}{2} \\[2mm] 2\sin\dfrac{t^2}{2} \end{pmatrix}.$

SECTION 27

1. $\|A\| = 4$, $\|B\| = 5$, $\|A + B\| = 8$, $\|AB\| = 14$, $\|BA\| = 20$.
6. First show that $\|A^k\| \le \|A\|^k$.
8. $\|I\|_s = n$, which is less desirable than $\|I\| = 1$.
9. (a) If $\mathbf{y}_{(j)}$ is the solution of $\mathbf{y}' = A(t)\mathbf{y}$ with $\mathbf{y}(s) = \mathbf{I}_{(j)}$, then by Theorem F,
$\|\mathbf{y}_{(j)}(t)\| \le e^{K|t - s|}$ on J. Thus $\|T(t, s)\| \le \max_j \|\mathbf{y}_{(j)}(t)\| \le e^{K|t - s|}$.

(b) In order to compute $\left| \displaystyle\int_{t_0}^{t} Me^{K|t - s|}\,ds \right|$, one must consider separately the

cases $t < t_0$ and $t > t_0$.

(c) By the mean value theorem

$$e^{K|t - t_0|} - 1 = K|t - t_0|e^{\theta}$$

for some θ such that $0 < \theta < K|t - t_0|$.
10. (b) $\|\mathbf{x}(t)\| \le H(M/\delta + \|\mathbf{x}_0\|)$.

SECTION 28

1. (a) $x_{(1)}(t) = 1 + ct - c\dfrac{t^2}{2}$,

$\qquad x_{(2)}(t) = 1 + ct + \dfrac{1}{2!}(ct)^2 - \dfrac{1}{3!}c^2 t^3$,

$\qquad x_{(3)}(t) = 1 + ct + \dfrac{1}{2!}(ct)^2 + \dfrac{1}{3!}(ct)^3 - \dfrac{1}{4!}c^3 t^4$,

$\qquad x(t) = e^{ct} = 1 + ct + \dfrac{1}{2!}(ct)^2 + \dfrac{1}{3!}(ct)^3 + \cdots$.

(b) $x_{(1)}(t) = 2 + 3(t - 1) + \frac{1}{2}(t - 1)^2$,

$\qquad x_{(2)}(t) = 2 + 3(t - 1) + 2(t - 1)^2 + \dfrac{1}{3!}(t - 1)^3$,

$\qquad x_{(3)}(t) = 2 + 3(t - 1) + 2(t - 1)^2 + \dfrac{4}{3!}(t - 1)^3 + \dfrac{1}{4!}(t - 1)^4$,

$\qquad x(t) = 4e^{t-1} - t - 1$

$\qquad\qquad = 2 + 3(t - 1) + 4\left[\dfrac{1}{2!}(t - 1)^2 + \dfrac{1}{3!}(t - 1)^3 + \cdots\right].$

(c) $\mathbf{x}_{(1)}(t) = \begin{pmatrix} 1 - t \\ t \end{pmatrix}$, $\mathbf{x}_{(2)}(t) = \begin{pmatrix} 1 - t - \dfrac{1}{2} t^2 \\ t - \dfrac{5}{2} t^2 \end{pmatrix}$

$$\mathbf{x}(t) = \begin{pmatrix} 2e^{-2t} - e^{-3t} \\ e^{-2t} - e^{-3t} \end{pmatrix} = \begin{pmatrix} 1 - t - \dfrac{1}{2} t^2 + \dfrac{11}{6} t^3 - \cdots \\ t - \dfrac{5}{2} t^2 + \dfrac{19}{6} t^3 - \cdots \end{pmatrix}.$$

4. $x(t) = \dfrac{1}{1 - t}$,   for $t < 1$.

5. On **R**, $|\xi^2 - \tilde{\xi}^2| = |\xi + \tilde{\xi}| \cdot |\xi - \tilde{\xi}|$, where $|\xi + \tilde{\xi}|$ is unbounded. However, on $D \subset (-b, b)$, where $b > 0$,

$$|\xi^2 - \tilde{\xi}^2| \le 2b|\xi - \tilde{\xi}|.$$

8. Inequality (31) holds with $M(t) = |\sin t|$ and $N(t) = 1$ [or $M(t) = 2$ and $N(t) = 0$]. (Whenever there is one appropriate choice for M and N there are infinitely many appropriate choices.)

9. Corollary E can be applied with $M(t) = 18$ and $N(t) = \max \{3 + |t|, 2e^t, 1 + t^2\}$.

10. Let $x_1 = z$ and $x_2 = z'$ to obtain a system. Then apply Corollary E.

13. Using suggestion (30), one sets $v(t) = \frac{1}{2}(z'(t))^2 + \frac{1}{2} z^2(t)$ and calculates $v' = \mu(1 - z^2)z'^2$. Now when $|z(t)| > 1$, $v'(t) \le 0$; and when $|z(t)| \le 1$, $v'(t) \le \mu z'^2$. So, in any case, $v'(t) \le 2\mu v(t)$. This gives $v(t) \le v(t_0)e^{2\mu(t - t_0)}$ for $t \ge t_0$. Conclude by adapting the arguments of Example 3.

SECTION 29

1. $z(0) = 0$, $z(0.05) = 0.0500$, $z(0.10) = 0.1001$, $z(0.15) = 0.1506$, $z(0.20) = 0.2018$.

4.

t_i	(a) Euler $z(t_i)$	(b) Improved Euler y_i	Exact $x(t_i) = 4e^{t_i - 1} - t_i - 2$
1.0	1.00	1.00	1.000
1.2	1.60	1.68	1.686
1.4	2.36	2.55	2.567
1.6	3.31	3.66	3.688
1.8	4.49	5.06	5.102
2.0	5.95	6.81	6.873

5.

t_i	$z_1(t_i)$	$z_2(t_i)$	Exact	
			$x_1(t_i) = -t_i e^{-2t_i}$	$x_2(t_i) = (1 - t_i)e^{-2t_i}$
0.0	0.000	1.000	0.000	1.000
0.1	-0.100	0.700	-0.082	0.737
0.2	-0.160	0.480	-0.134	0.536
0.3	-0.192	0.320	-0.165	0.384
0.4	-0.205	0.205	-0.180	0.270
0.5	-0.205	0.123	-0.184	0.184

6.

t_i	(a) Euler, $z(t_i)$	(b) Improved Euler, y_i
0.0	0	0
0.2	0	0.004
0.4	0.01	0.024
0.6	0.04	0.076
0.8	0.11	0.179
1.0	0.24	0.356

SECTION 30

1. (a) $cr < 2 - \sqrt{2}$.

   (b) "No" is probably the correct answer at this stage. It appears difficult to find such a condition by the method of steps. The sufficient condition $cr \leq 1/e$ will be obtained by another method in Example 32-4.

   (c) Let $y(t) = qa/c - x(t)$ or $y(t) = x(t) - qa/c$.

3. $x(t) = -\dfrac{b}{a^2} - \dfrac{bt}{a} + \left(1 + \dfrac{b}{a^2}\right)e^{at}$ on $[0, 1]$,

$$x(t) = \frac{2b^2}{a^3} - \frac{b^2}{a^2} + \frac{b^2 t}{a^2} + \left[1 + \frac{b}{a^2} - \left(b + \frac{b}{a} + \frac{b}{a^2} + \frac{b^2}{a^2}\right)e^{-a}\right]e^{at}$$
$$- \frac{2b^2}{a^3} e^{a(t-1)} + \left(b + \frac{b^2}{a^2}\right)te^{a(t-1)} \text{ on } [1, 2].$$

4. (a) $a = \dfrac{e}{e-1}$, $b = \dfrac{-e}{e-1}$.

   (b) $x(t) = x_0 + c(e^t - 1)$ for arbitrary constant c.

   (c) The solution is still not unique, but this is not obvious. A much stronger non-uniqueness result will be proved in Section 31.

5. A unique solution exists for all $t \geq 0$.

6. A unique solution exists for *all* $t \geq 0$.

8. (a) On $[0, r]$ first solve the ordinary linear first-order equation

$$y'(t) = -a_2 y(t) + b_2 \theta_1(t - r)\theta_2(t - r)$$

with $y(0) = \theta_2(0)$. Then, having found y on $[0, r]$, solve the ordinary Bernoulli equation

$$x'(t) + [b_1 y(t) - a_1]x(t) = -(a_1/P)x^2(t)$$

on $[0, r]$ with $x(0) = \theta_1(0)$. Proceed similarly on $[r, 2r]$, $[2r, 3r]$, ...

(b) No.

9. One answer is $q = 1$, $k = \pi^2$, $\omega = \pi$. Then we will have $x(t) = \cos \pi t$ for $t \geq 0$, for example, if $x(t) = \theta(t) = \cos \pi t$ for $-1 \leq t \leq 0$.

SECTION 32

1. Two solutions on $[0, \infty)$ are $x(t) = 0$ and $x(t) = \left(\dfrac{t}{12}\right)^3$.

2. $x(t) = \tan\left(t - 1 + \dfrac{\pi}{4}\right)$ on $\left(1, 1 + \dfrac{\pi}{4}\right)$. The solution cannot be continued to (or past) $1 + \pi/4$. Why?

3.

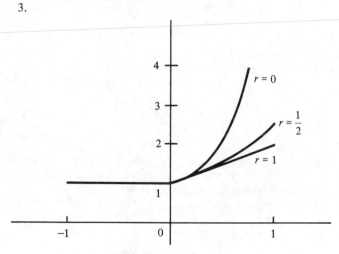

For small $r > 0$ the solution will be close to $x(t) = (1 - t)^{-1}$ on $[0, t_1]$, where $t_1 < 1$ gets larger as r gets smaller.

4. The solution exists (and is unique) on $[-3, \infty)$.

5. On $[0, r]$, $\mathbf{x}'(t) = \begin{pmatrix} 0 & 1 \\ -\dfrac{k}{m} & -\dfrac{b}{m} \end{pmatrix} \mathbf{x}(t) - \begin{pmatrix} 0 \\ \dfrac{q}{m}\theta_2(t-r) \end{pmatrix}$ with $\mathbf{x}(0) = \mathbf{\theta}(0)$. This linear system has a unique solution on $[0, r]$. The argument is analogous on $[r, 2r]$.

SECTION 33

1. Uniqueness follows from Theorem 9-A (or from Corollary 9-B or Theorem 12-B). Existence follows from Theorem 26-B combined with Theorem 24-B (or from Theorem 28-A).

3. In (4) one can take $c = \max\left\{2, \dfrac{l\pi^2}{2g}\right\}$.

5. If z is any solution of Equation (16), let $w(t) = e^{\delta t}z(t)$, where $\delta > 0$. Find the equation satisfied by w. Then Example 4 will show that w is bounded provided δ is so small that $b > 2\delta$, $k > b\delta - \delta^2$, $pe^{\delta} > -k + \delta(b + qe^{\delta r} + \delta)$ and $|q|(1 + \tau\delta) + \tau e^{\delta \tau}|p| \le b - 2\delta$. It follows that $z(t) = e^{-\delta t}w(t) \to 0$ as $t \to \infty$.

6. There are infinitely many correct answers for each part of this question. But the correct answers found most naturally are

   (a) $\tilde{x}(t) = -\dfrac{7 + 4\sqrt{2}}{34}\cos t + \dfrac{11 - \sqrt{2}}{34}\sin t$.

   (b) $\tilde{x}(t) = -\dfrac{2\pi}{\pi^2 + 4}\,t\cos t + \dfrac{4}{\pi^2 + 4}\,t\sin t$.

   (c) $\tilde{x}(t) = t - 2$.

   (d) $\tilde{x}(t) = t^2/4 + t/8$.

   (e) $\tilde{x}(t) = t^3/3 + t^2/3$.

7. If x is any solution,

$$x(t) = \frac{\sqrt{3}}{14}\cos t + \frac{5}{14}\sin t + z(t),$$

   where $z(t) \to 0$ as $t \to \infty$.

SECTION 34

1. (a) $|t| < \infty$.
   (b) $|t - 1| < 2$.
   (c) $|t + 2| < 3$.
   (d) $|t| < \infty$.
   (e) $|t| < 1$.

2. (a) $\displaystyle\sum_{k=0}^{\infty}(-1)^k t^k$, $\quad |t| < 1$.

   (b) $\displaystyle\sum_{k=0}^{\infty}\frac{(-1)^k}{2^{k+1}}(t - 1)^k$, $\quad |t - 1| < 2$.

   (c) $\displaystyle\sum_{j=0}^{\infty}(-1)^j t^{2j}$, $\quad |t| < 1$.

   (d) $\displaystyle\sum_{k=0}^{\infty}\left(2 - \frac{3}{2^{k+1}}\right)t^k$, $\quad |t| < 1$.

SECTION 35

1. $\operatorname{Arctan} t = \displaystyle\sum_{j=0}^{\infty}\frac{(-1)^j}{2j + 1}t^{2j+1}$ for $-1 < t < 1$.

2. $\dfrac{\pi}{6} = \operatorname{Arctan}\dfrac{1}{\sqrt{3}} = \dfrac{1}{\sqrt{3}} - \dfrac{1}{3}\left(\dfrac{1}{\sqrt{3}}\right)^3 + \dfrac{1}{5}\left(\dfrac{1}{\sqrt{3}}\right)^5 + \text{error} = 0.526 + \text{error}$,

where, using Theorem A2-H(ii),

$$|error| = \left| \sum_{j=3}^{\infty} \frac{(-1)^j}{2j+1} \left(\frac{1}{\sqrt{3}}\right)^{2j+1} \right| \le \sum_{j=3}^{\infty} \frac{1}{7\sqrt{3}} \left(\frac{1}{3}\right)^j = \frac{1}{9 \cdot 14\sqrt{3}} = 0.0046.$$

If one recalls the properties of "alternating series," one can easily find that our approximate value of 0.526 is really sharper than the above would indicate. In fact $-\dfrac{1}{7 \cdot 3^3\sqrt{3}} = -0.003 \le error \le 0$. Thus $0.523 \le \dfrac{\pi}{6} \le 0.526$.

3. $R = \infty$.

SECTION 36

3. Power-series solutions in powers of $t - 3$ exist for $|t - 3| < 2$. The easiest way to prove this is to write

$$\frac{1}{1 - t^2} = \frac{\frac{1}{2}}{1 - t} + \frac{\frac{1}{2}}{1 + t} = \frac{\frac{1}{2}}{-2 - (t - 3)} + \frac{\frac{1}{2}}{4 + (t - 3)}$$

$$= \frac{-\frac{1}{4}}{1 + \dfrac{t - 3}{2}} + \frac{\frac{1}{8}}{1 + \dfrac{t - 3}{4}},$$

and note that these have series expansions in powers of $t - 3$ which converge for $|t - 3| < 2$.

5. (a) $y(t) = \displaystyle\sum_{j=0}^{\infty} \frac{(-1)^j}{(2j)!} t^{2j} - 2 \sum_{j=0}^{\infty} \frac{(-1)^j}{(2j+1)!} t^{2j+1}$

   $= \cos t - 2 \sin t$,   valid for all t.

   (b) $y(t) = -1 - \frac{1}{4}t^2 + \displaystyle\sum_{j=2}^{\infty} (-1)^j \frac{1 \cdot 3 \cdots (2j - 3)}{(2j)!} t^{2j} + 3t$

   $= \displaystyle\sum_{j=0}^{\infty} \frac{(-1)^j}{2^j(2j - 1)j!} t^{2j} + 3t$,   valid for all t.

   (c) $x(t) = c_0 + (1 + c_0) \displaystyle\sum_{j=1}^{\infty} \frac{(-1)^{j+1}}{2^j(2j - 1)j!} t^{2j} + c_1 t$

   $+ \displaystyle\sum_{j=1}^{\infty} (-1)^{j+1} \frac{2^{j-1}(j - 1)!}{(2j + 1)!} t^{2j+1}$,   valid for all t.

   (d) $y(t) = 17\left(1 - \dfrac{1}{2!} t^2 - \dfrac{5}{4!} t^4 - \dfrac{5 \cdot 19}{6!} t^6 - \cdots\right)$

   $+ \pi\left(t + \dfrac{1}{3!} t^3 + \dfrac{11}{5!} t^5 + \dfrac{11 \cdot 29}{7!} t^7 + \cdots\right),$

   valid for $-1 < t < 1$. *Note:* Even though we have not obtained a general expression for either c_{2j} or c_{2j+1}, it is easy to test for convergence of the series by using the recursion relation itself

$$c_{k+2} = \frac{k^2 + k - 1}{(k + 2)(k + 1)} c_k, \qquad k = 0, 1, 2, \ldots.$$

In fact, for k even or odd, we find

$$\left|\frac{c_{k+2} t^{k+2}}{c_k t^k}\right| = \left|\frac{k^2 + k - 1}{(k+2)(k+1)}\right| t^2 \to t^2 \quad \text{as} \quad k \to \infty.$$

(e) $y(t) = \sum_{j=0}^{\infty} \frac{(-1)^j}{j! 2^j} t^{2j}$, valid for all t. $y(t) = e^{-t^2/2}$.

6. (a) $y_1(t) = 1 - 2t^2$, $y_2(t) = t - \frac{1}{2} t^3 - \sum_{j=2}^{\infty} \frac{1 \cdot 3 \cdot 5 \cdots (2j-3)}{j! 2^j} t^{2j+1}$,

    valid for $-1 < t < 1$.

(b) $y_1(t) = 1 + \sum_{j=1}^{\infty} (-1)^j \frac{1 \cdot 4 \cdot 7 \cdots (3j-2)}{(3j)!} t^{3j}$,

    $y_2(t) = t + \sum_{j=1}^{\infty} (-1)^j \frac{2 \cdot 5 \cdot 8 \cdots (3j-1)}{(3j+1)!} t^{3j+1}$,   valid for all t.

(c) $y_1(t) = t$, $y_2(t) = 1 - \sum_{j=1}^{\infty} \frac{1}{2j-1} t^{2j}$,   valid for $-1 < t < 1$.

(d) $y_1(t) = 1 - \frac{1}{2!} \left(\frac{3}{4}\right) t^2 - \frac{1}{4!} \left(\frac{3}{4}\right) \left(\frac{21}{4}\right) t^4 - \cdots$,

    $y_2(t) = t + \frac{1}{3!} \left(\frac{5}{4}\right) t^3 + \frac{1}{5!} \left(\frac{5}{4}\right) \left(\frac{45}{4}\right) t^5 + \cdots$,

    valid for $-1 < t < 1$. See the note to the solution of Problem 5(d).

(e) $y_1(t) = 1 + \sum_{j=1}^{\infty} (-1)^j \frac{2^j p(p-2) \cdots (p-2j+2)}{(2j)!} t^{2j}$,

    $y_2(t) = t + \sum_{j=1}^{\infty} (-1)^j \frac{2^j (p-1)(p-3) \cdots (p-2j+1)}{(2j+1)!} t^{2j+1}$,

    valid for all t. (Note that if p is a nonnegative integer, then one of these solutions is just a polynomial—a "Hermite polynomial.")

7. $y(t) = c_1 t + c_2 t \int_1^t \frac{1}{s^2} e^{-s^2/2} \, ds$,   valid for all $t > 0$. (With a little care one can show that this is actually valid for all t.)

8. (a) $y_1(t) = 1 - \sum_{k=2}^{\infty} \frac{(-1)^k}{k(k-1)} (t-1)^k$,

    $y_2(t) = (t-1) + \sum_{k=2}^{\infty} \frac{(-1)^k}{k(k-1)} (t-1)^k$,   valid for $0 < t < 2$.

(b) $y_1(t) = t - t \ln t$, $y_2(t) = t \ln t$.

10. (a) $y(t) = -1 + 3t - \frac{1}{2!} t^2 + \frac{1}{4!} t^4 - \frac{3}{6!} t^6 + \cdots$,   valid for all t.

(b) $y(t) = 1 + 2t - \frac{1}{3!} t^3 - \frac{4}{4!} t^4 + \frac{1}{5!} t^5 + \frac{12}{6!} t^6 + \cdots$,   valid for all t.

SECTION 37

2. You will find $c_0 = c_1 = \cdots = 0$, giving only the trivial solution. *Note*: This had to happen since the nontrivial solutions of Equation (4) have the form $y(t) = ce^{a/t}$ for $t \neq 0$, and it can be shown that $e^{a/t}$ is not equal to $t^\lambda \times$ an analytic function for any choice of λ.

4. (a) $q_1(t) = 1, q_0(t) = t^2 - p^2$.

   (b) $q_1(t) = \dfrac{r - (1 + p + q)t}{1 - t}, q_0(t) = -\dfrac{pqt}{1 - t}$.

5. (a) $y_1(t) = t^p + t^p \displaystyle\sum_{j=1}^{\infty} \frac{(-1)^j}{j!(1 + p)(2 + p) \cdots (j + p)} \left(\frac{t}{2}\right)^{2j}$ with $p = \frac{1}{3}$.

   $y_2(t)$ is like $y_1(t)$ but with p replaced by $-\frac{1}{3}$. Both are valid for all $t > 0$.

   (b) $y_1(t) = \dfrac{1}{t}$ for $t > 0$, $\quad y_2(t) = \displaystyle\sum_{k=0}^{\infty} \frac{1}{k + 1} t^k$ for $0 < t < 1$.

6. $y_2(t) = -\dfrac{1}{t} \ln(1 - t)$ for $0 < t < 1$.

SECTION 38

3. (b) $y(t) = cJ_{1/2}(t)$, for some constant c. $t^{1/2}y'(t) \to c/\sqrt{2\pi}$ as $t \to 0+$. So $y'(t) \to \pm\infty$.

5. (c) $y(t) = cJ_0(t)$ for some constant c. $y'(t) \to 0$ as $t \to 0+$.

6. (a) $R = \infty, \lambda_1 = p, \lambda_2 = -p$.

   (b) $R = \infty, \lambda_1 = \lambda_2 = 0$.

   (c) $R = 2, \lambda_1 = \lambda_2 = 0$.

   (d) $R = 1, \lambda_1 = 0, \lambda_2 = 1 - r$.

7. (a) $y_1(t) = t^p + t^p \displaystyle\sum_{j=1}^{\infty} \frac{(-1)^j}{j!(1 + p)(2 + p) \cdots (j + p)} \left(\frac{t}{2}\right)^{2j}$ with $p = \frac{3}{2}$.

   $y_2(t)$ is like $y_1(t)$ but with p replaced by $-\frac{3}{2}$.

   Both are valid for all $t > 0$. [Compare with Problem 37-5(a).]

   (b) $y_1(t) = 1$ for all t, $y_2(t) = \ln t + \displaystyle\sum_{k=1}^{\infty} \frac{1}{k \cdot k!} t^k$ for $t > 0$.

   (c) $y_1(t) = 1 - t$ for all t,

   $y_2(t) = (1 - t)\ln t + 3t - \displaystyle\sum_{k=2}^{\infty} \frac{1}{(k - 1)kk!} t^k$ for $t > 0$.

   (d) $y_1(t)$ as in part (a) with $p = 1$, valid for all t. Seek a second linearly independent solution in the form

   $$y_2(t) = y_1(t)\ln t + \sum_{k=0}^{\infty} d_k t^{k-1}.$$

   Here d_2 is arbitrary. Why? By convention we set $d_2 = -\frac{1}{2}$. Then one finds

   $$y_2(t) = y_1(t)\ln t - \frac{2}{t} - \frac{t}{2}$$

   $$+ \sum_{j=2}^{\infty} \frac{(-1)^j}{j!(j - 1)!} \left[\left(1 + \frac{1}{2} + \cdots + \frac{1}{j}\right) + \left(1 + \frac{1}{2} + \cdots + \frac{1}{j - 1}\right)\right] \left(\frac{t}{2}\right)^{2j-1}$$

   valid for all $t > 0$.

   (e) $y_1(t) = t$ for all t.

   $$y_2(t) = t \ln(t - 1) - 5\left(\frac{t - 1}{2}\right) + \sum_{k=2}^{\infty} \frac{(-1)^k(k + 2)}{2(k + 1)k} \left(\frac{t - 1}{2}\right)^k$$

   valid for $0 < t - 1 < 2$.

8. $y(t) = c_1 t + c_2 \left[1 + \frac{1}{2} t \ln \frac{t-1}{t+1} \right]$ for all $t > 1$.

[Could you use this information to find the sum of the series involved in $y_2(t)$ in the answer to Problem 7(e)?]

SECTION 39

5. (a) $G(s) = \dfrac{s}{s^2 + 9} + \dfrac{2}{(s-2)^2}$ (for Re $s > 2$).

(b) $G(s) = \dfrac{\mu}{s^2 - \mu^2}$ (for Re $s > |\mu|$).

(c) $G(s) = \dfrac{1}{s}(1 - e^{-as})$ (for Re $s > 0$).

(d) $G(s) = -\dfrac{a}{s} e^{-as} - \dfrac{1}{s^2} e^{-as} + \dfrac{1}{s^2}$ (for Re $s > 0$).

(c) $G(s)$ does not exist.

(f) $G(s) = \dfrac{(s-\mu)^2 - \omega^2}{[(s-\mu)^2 + \omega^2]^2}$ (for $s > \mu$).

(g) $G(s) = \dfrac{2\omega(s-\mu)}{[(s-\mu)^2 + \omega^2]^2}$ (for $s > \mu$).

(h) $G(s) = \dfrac{2s^3 - 6\omega^2 s}{(s^2 + \omega^2)^3}$ (for $s > 0$).

6. (a) $x(t) = 4e^{-2t} - 3te^{-2t} + t^2 e^{-2t} - 4e^{-3t}$.
(b) $x(t) = 3e^{-t} \cos 2t - \frac{3}{2} e^{-t} \sin 2t$.
(c) $x(t)$ does not exist.
(d) $x(t) = \frac{2}{5} e^t + \frac{3}{5} e^{-t} \cos t - \frac{4}{5} e^{-t} \sin t$.

7. (i) We have not proved that a solution of exponential order exists, and (ii) once having found $X(s)$ we have not shown that there cannot be more than one function x such that $\mathscr{L}[x] = X$.

9. $G(s) = \operatorname{Arctan} s - \dfrac{\pi}{2}$ for $s > 0$.

SECTION 40

3. (a) $x(t) = \cos 2t - \sin 2t$.
(b) $x(t) = \cos 2t - \frac{7}{8} \sin 2t - \frac{1}{4} t \cos 2t$.
(c) $x(t) = -\frac{1}{3} + 2e^{2t} - 4te^{2t} + \frac{1}{3} e^{3t}$.
(d) $x(t) = \frac{5}{3} e^t - \frac{2}{3} e^{-2t} - te^{-2t} - \frac{1}{2} t^2 e^{-2t}$.
(e) $x(t) = 2e^{-t} \cos 2t + \frac{1}{2} e^{-t} \sin 2t$.

SECTION 41

4. (a) $x(t) = \dfrac{t}{a} - 2u(t-a)\dfrac{t-a}{a} + u(t-2a)\dfrac{t-2a}{a}$

$$= \begin{cases} t/a & \text{for } 0 \le t < a \\ 2 - t/a & \text{for } a \le t < 2a \\ 0 & \text{for } t \ge 2a. \end{cases}$$

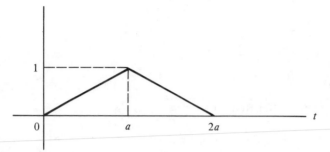

(b) $x(t) = \dfrac{t}{a} - 2u(t-a)\dfrac{t-a}{a} + 2u(t-2a)\dfrac{t-2a}{a} - \cdots.$

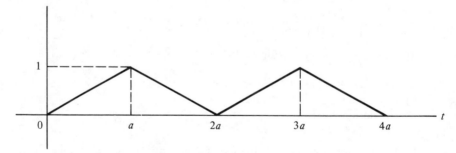

(c) $x(t) = h(t) - u(t-a)h(t-a) + u(t-2a)h(t-2a) - \cdots,$

where $h(t) = -\dfrac{1}{\lambda}(1 - e^{\lambda t}).$

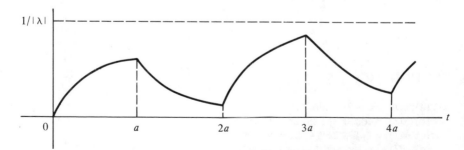

(d) $x(t)$ is not defined.

5. (a) $x(t) = \dfrac{1}{R} e^{-Rt/L} + \dfrac{t - L/R}{R} [1 - u(t - \pi)]$

$\qquad + \dfrac{L}{R^2} e^{-Rt/L} + \dfrac{\pi - L/R}{R} e^{-R(t-\pi)/L} u(t - \pi).$

(b) $x(t) = h(t) - u(t - T/2)h(t - T/2) + u(t - T)h(t - T) - \cdots,$

$\qquad$ where $h(t) = \dfrac{E}{R}(1 - e^{-Rt/L})$. See the answer to Problem 4(c).

SECTION 42

2. (a) Same as Problem 26-4(d).
   (b) Same as Problem 26-4(a).
   (c) Same as Problem 26-4(b).

INDEX

78   79   80   81   9   8   7   6   5   4   3   2   1